Cowpea Research, Production and Utilization

Edited by
S. R. Singh
Director, Grain Legume Improvement Program,
International Institute of Tropical Agriculture,
Ibadan, Nigeria

and

K. O. Rachie
Consultant, World Cowpea Research Conference,
International Institute of Tropical Agriculture,
Ibadan, Nigeria

A Wiley-Interscience Publication

International Institute of Tropical Agriculture

JOHN WILEY & SONS
Chichester · New York · Brisbane · Toronto · Singapore

Invited Papers from The World Cowpea Conference, IITA, Ibadan

Cosponsored by:
International Institute of Tropical Agriculture
and
Bean/Cowpea Collaborative Research Support Program

Library of Congress Cataloguing in Publication Data:
Main entry under title:

Cowpea research, production, and utilization.

'A Wiley–Interscience publication
Bibliography: p.
Includes index.
1. Cowpea—Congresses. I. Singh, S. R.
II. Rachie, Kenneth O.
SB205.C8C74 1985   633.3'3   85–9368

ISBN 0 471 90802 9 (U.S.)

British Library Cataloguing in Publication Data:

Singh, S. R.
   Cowpea research, production and utilization.
   1. Cow-peas
   I. Title   II. Rachie, Kenneth O.
   583'.322   SB205.C8

ISBN 0 471 90802 9

Printed and bound in Great Britain.

Cowpea Research,
Production and Utilization

Dedicated to our wives Manju and Mary

Contents

List of Contributors

V. D. Aggarwal *IITA-SAFGRAD, Ouagadougou, Burkina Faso*

I. O. Akobundu *International Institute of Tropical Agriculture, Ibadan, Nigeria*

M. Ashraf *International Institute of Tropical Agriculture, Ibadan, Nigeria*

E. A. Baker *University of Georgia, Experiment, Georgia, USA*

J. P. Baudoin *Faculté des sciences agronomiques de l'état, 5800 Gembloux, Belgium*

R. Bressani *INCAP, Guatemala. Bean/Cowpea CRSP*

F. E. Caveness *International Institute of Tropical Agriculture, Ibadan, Nigeria*

R. B. Chalfant *Coastal Plain Experiment Station, Tifton, Georgia, USA. Bean/Cowpea CRSP*

M. S. Chhinnan *University of Georgia, Experiment, Georgia, USA. Bean/ Cowpea CRSP*

R. A. Daoust *CNPAF, Goiânia, Goiás, Brazil, and Boyce Thompson Institute for Plant Research, Cornell University, Ithaca, USA. Bean/Cowpea CRSP*

B. P. Das Neves *CNPAF, Goiânia, Goiás, Brazil*

J. P. P. de Araújo *EMBRAPA/CNPAF, Goiânia, Goiás, Brazil, Bean/ Cowpea CRSP*

F. A. Elazegui *International Rice Research Institute, College, Laguna, Philippines*

A. M. Emechebe *Faculty of Agriculture, Institute for Agricultural Research, Ahmadu Bello University, Zaria, Nigeria*

H. C. Ezumah *International Institute of Tropical Agriculture, Ibadan, Nigeria*

G. C. J. Fernandez   *Department of Horticultural Sciences, Texas A&M University, College Station, Texas, USA*

R. L. Fery   *United States Department of Agriculture, Agricultural Research Service, US Vegetable Laboratory, Charleston, South Carolina, USA*

A. E. Hall   *University of California, Riverside, USA. Bean/Cowpea CRSP*

L. E. N. Jackai   *International Institute of Tropical Agriculture, Ibadan, Nigeria*

S. J. B. A. Jayasekara   *Agricultural Research Station, Maha Illupallama, Sri Lanka*

M. B. Kennedy   *University of Georgia, Experiment, Georgia, USA*

E. A. Kueneman   *International Institute of Tropical Agriculture, Ibadan, Nigeria*

M. T. Lin   *University of Brasilia, Brasilia*

K. H. McWatters   *University of Georgia, Experiment, Georgia, USA. Bean/ Cowpea CRSP*

R. Maréchal   *Faculté des sciences agronomiques de l'état, 5800 Gembloux, Belgium*

T. W. Mew   *International Rice Research Institute, College, Laguna, Philippines*

J. C. Miller Jr   *Department of Horticultural Sciences, Texas A&M University, College Station, Texas, USA*

S. N. Mishra   *G. B. Pant University of Agriculture and Technology, Nainital, India*

N. Muleba   *IITA-SAFGRAD, Ouagadougou, Burkina Faso*

K. Mulongoy   *International Institute of Tropical Agriculture, Ibadan, Nigeria*

N. Q. Ng   *Genetic Resources Unit, International Institute of Tropical Agriculture, Ibadan, Nigeria*

A. T. Ngarm   *Faculty of Agriculture, Khon Kaen University, Thailand*

B. R. Ntare   *International Institute of Tropical Agriculture, Ibadan, Nigeria*

A. O. Ogunfowora   *Institute of Agricultural Research and Training, Ibadan, Nigeria*

R. K. Pandey   *IITA/IRRI, Manila, Philippines*

J. S. Pate   *University of Western Australia, Nedlands, Australia*

P. N. Patel   *University of California, Riverside, USA. Bean/Cowpea CRSP*

R. D. Phillips   *University of Georgia, Experiment, Georgia, USA. Bean/ Cowpea CRSP*

J. A. Poku   *International Institute of Tropical Agriculture, Ibadan, Nigeria*

A. K. Raheja   *Institute for Agricultural Research, Ahmadu Bello University, Zaria, Nigeria*

V. N. M. Rao   *University of Georgia, Experiment, Georgia, USA*

Y. P. S. Rathi   *G. B. Pant University, Pantnagar, India*

G. P. Rios   *EMBRAPA/CNPAF, Goiânia, Goiás, Brazil*

D. W. Roberts   *Boyce Thompson Institute for Plant Research, Cornell University, Ithaca, USA. Bean/Cowpea CRSP*

E. H. Roberts   *Department of Agriculture, University of Reading, Reading, UK*

H. W. Rossel   *International Institute of Tropical Agriculture, Ibadan, Nigeria*

S. A. Shoyinka   *CEC/IITA Project, Institute for Agricultural Research, Ahmadu Bello University, Zaria, Nigeria*

B. B. Singh   *International Institute of Tropical Agriculture, Ibadan, Nigeria*

S. R. Singh   *International Institute of Tropical Agriculture, Ibadan, Nigeria*

R. J. Summerfield   *Department of Agriculture, University of Reading, Reading, UK.*

G. Thottappilly   *International Institute of Tropical Agriculture, Ibadan, Nigeria*

J. S. Verma   *G. B. Pant University of Agriculture and Technology, Nainital, India*

E. E. Watt   *International Institute of Tropical Agriculture, Ibadan, Nigeria*

F. Wiedijk   *International Institute of Tropical Agriculture, Ibadan, Nigeria*

H. C. Wien   *New York State College of Agriculture and Life Sciences, Cornell University, Ithaca, USA*

Foreword

The International Institute of Tropical Agriculture (IITA) welcomed the opportunity to be co-sponsor and host for the first World Cowpea Research Conference from 5 to 9 November 1984 at its headquarters in Ibadan, Nigeria. Proof of the worldwide interest in the conference was in the numbers of scientists, extension personnel, government officials and others who attended—203 persons representing 34 countries in Africa, Asia, North and South America and Europe.

It was a good decision to organize this meeting at IITA in Nigeria. Within the Consultative Group on International Agricultural Research, IITA has worldwide responsibility for research on cowpeas, and in carrying out its mandate it maintains close links with national programmes. Farmers throughout the world grow the crop (a total of more than 8 million hectares), but Africa produces more cowpeas than any other continent.

The Bean/Cowpea CRSP cosponsor is the major effort on the part of the United States of America (US) to make the professional resources of the US Land-Grant University system and the US government's agricultural research community available to address important constraints to cowpea production and utilization. As such, it works through support from the US Agency for International Development (AID) to build collaborative research and training relationships between Host Country scientists and scientists in US institutions.

Individual and coordinated efforts of the national programs, the Bean/Cowpea CRSP and IITA have resulted in major contributions to increased availability of cowpeas. Sharing and discussing these accomplishments among the international cowpea research community was all the more critical as the timing of the conference paralleled the severe droughts and the growing food crisis in Africa. Further development and progress in cowpea improvement could help add to food supplies and provide more quality protein for millions of people who depend on the crop as a major source of this essential ingredient. Some of the new IITA high-yielding varieties released for use in many countries have protein content as high as 23 per cent, and improved lines now being tested in trials around the world rate as high as 24.4 per cent.

Participants in the conference pointed out that if progress in cowpea improvement can continue at its present rate, world production could triple by the turn of the century. The scientists participating in the research programs represented at this meeting have a deep commitment to helping sustain this progress. The research reports and recommendations presented at the conference and published in this book merit close attention and appropriate action by all concerned with improving food supplies, especially in developing countries where food production is inadequate to feed a growing population.

We are grateful to the many donors who generously supported this conference. Their contributions made possible a stimulating scientific meeting, having considerable potential for increasing the availability of this important crop in food deficit areas of the world.

ERMOND HARTMANS
Director General, IITA

Preface

Cowpea, also called southernpea and blackeyed pea, has been cultivated for many centuries in the developing world. The crop is well adapted to the stressful growing conditions of the tropics and has excellent nutritional qualities; its drawback has been its low yields.

Recent advances in genetic improvement, however, have made cowpea into practically a new crop, one that is attractive to smallholders throughout the world. The new elite strains have resistance to many diseases and insects; they mature early (within 60 days) and produce well. Optimal cultural practices for a broad range of growing conditions and uses have also been developed. The whole package of technology promises to expand greatly the role of cowpea in subsistence agriculture, permitting more intensive cultivation of the crop in complex cropping systems with a minimum of inputs.

In recognition of burgeoning global interest in cowpea, and of the need to take stock of developments in research on the crop, the first world cowpea conference was held at the International Institute of Tropical Agriculture (IITA), Ibadan, Nigeria, 5–9 November 1984. The purpose of the conference was to improve communication among cowpea researchers and promote greater collaboration in the planning of research strategies. The meeting also provided the opportunity to record current production practices and assess the state of the art in plant improvement. Both those tasks have been ably accomplished by the authors of the invited papers presented in this volume (Volume I). Twenty-nine additional voluntary papers were read at the conference and will be published in a separate proceedings (Volume II).

It was particularly appropriate that the world's cowpea experts should come together in Africa. Much of the recent progress in research has been made on this continent, and this is where most of the world's cowpeas are produced and where the crop has the greatest potential for alleviating food shortage. Even so, cowpea is gaining popularity in many other parts of the tropics.

This book is the first comprehensive review of cowpea-production constraints and of research being conducted around the world to overcome them. It explains the origin, taxonomy, genetics and physiology of the plant and describes current efforts to breed cultivars that are disease- and pest-resistant, drought-tolerant, efficient in nitrogen fixation, high-yielding, and well suited to the climate and soils of Africa, Asia, Latin America and the United States. Papers on the major insects and nematodes that attack cowpea cover their distribution, life cycles, economic importance in terms of grain-yield losses, and the prospects for their control. The chapter on pathology gives the most up-to-date information on viral, fungal and bacterial diseases prevalent in the major production areas. There are also thorough treatments of agronomic practices and of innovations in processing designed to produce convenient, nutritious foods that are acceptable to consumers in different parts of the world.

The need for a book such as this has been evident for some time, but it simply was not possible to undertake the project until an international effort had been mounted to improve and develop cowpea. This has now been brought about through close cooperation among (1) IITA, which has a world mandate for cowpea research within the Consultative Group on International Agricultural Research, (2) the Bean/Cowpea Collaborative Research Support Program (CRSP) which opens an avenue to the research establishment of the US Land-Grant University system and the US Government Agricultural Research Community and (3) cowpea scientists and development specialists in 55 countries throughout the world. In a relatively short time these researchers have established a solid record of achievement and formed a rapidly growing body of information about the crop.

We hope that this book, in addition to being a valuable source of information about cowpea improvement, will lead to other reference works on particular aspects of the crop.

This first volume could not have been written without the enthusiastic and dedicated efforts of the conference participants and the generous support of several donors. The information presented here represents the joint contribution of many scientists and organizations, who along with IITA, included numerous regional programmes in Africa, Asia and Latin America; the University of Gembloux in Belgium; the University of Reading and Imperial Chemical Industries in England; Rhone-Poulenc Agrochimie, France; and in the United States, the US Department of Agriculture, the University of California at Riverside, the Boyce Thompson Institute at Cornell University, the University of Georgia, Michigan State University and Texas A&M University. Many of the papers from the United States, and several of those from national programmes, reported research wholly or partially supported by the Bean/Cowpea Cooperative Research Support Project (CRSP). We are particularly indebted to Dr P. W. Barnes-McConnell, Director, Bean/Cowpea CRSP, and the members of the CRSP Technical Committee for co-sponsoring this conference with IITA.

We also owe special thanks to the entire IITA Grain Legume Improvement Program staff for their assistance, to Dr L. E. N. Jackai, entomologist, and Bernadette Bakare, conference officer, both of IITA, for their unflagging support before, during and after the conference; to Amy Chouinard for her careful editing of the conference papers; to Gerry Cambier, Eloy Molinero and Godfrey Spencer for simultaneous French–English interpretation during the conference; and, finally, to Dr E. H. Hartmans, director general of IITA, without whose continuing support of cowpea research the meeting would not have been possible.

We sincerely hope that this publication will be of value to agricultural researchers, administrators and funding agencies around the world.

S. R. SINGH and K. O. RACHIE

Introduction

K. O. RACHIE
Consultant, International Institute of Tropical Agriculture, Ibadan, Nigeria

In some areas of the semihumid tropics the cowpea, *Vigna unguiculata* (L.) Walp., provides more than half the plant protein in human diets. In fact it is a key staple for the poorest sector of many developing countries of the torrid zone—a pulse crop high in protein as well as other essential nutrients. As a dietary staple it greatly improves an otherwise bland, unbalanced diet. As a food it is eaten in the form of dry seeds, green pods, green seeds and tender green leaves. It is also utilized for fodder and as a quick-growing cover crop under a wide range of conditions. On account of its ability to fix nitrogen efficiently—up to 240 kg N/ha—the cowpea provides a high proportion of its own nitrogen requirements, besides leaving a fixed-N deposit in the soil of up to 60–70 kg/ha for the succeeding crop. The cowpea is also highly compatible as a companion with a wide range of food and fibre crops. Moreover, it is early—some new varieties mature in 60–65 days—and drought-tolerant as well as being tolerant of most soil stresses.

PRODUCTION

Worldwide cowpea production in 1981 was estimated at 2.27 million tonnes from 7.7 million hectares. Cowpeas are grown extensively in 16 African countries, with this continent producing two-thirds of the total. Two countries—Nigeria and Niger—produce 850,000 t and 271,000 t annually or 49.3 per cent of the world crop. The second-highest producing country is Brazil where 600,000 t of dry seeds or 26.4 per cent of the worldwide total was produced in 1981. Other major producers in Africa include Burkina Faso (95,000 t), Ghana (57,000 t), Kenya (48,000 t), Uganda (42,000 t), and Malawi (42,000 t). Tanzania, Senegal and Togo each produce annually from 20,000 t to 22,000 t.

Current estimates of production vary widely according to source, but the statistics are probably conservative. For example, Asian production, including that of long beans as a vegetable, may be underestimated by a factor of 10 or at a level of about 1 million hectares, concentrated in India, Sri Lanka, Burma, Bangladesh, Philippines, Indonesia, Thailand, Pakistan, Nepal, China and Malaysia. India alone is estimated to cultivate more than half or 500,000 ha in all forms—dry seed, fodder, green pod, green manure and cover crops. Similarly, production estimates may be low for Africa and the Western Hemisphere where cowpeas are traditionally included as associated crops in peasant-farming systems. Thus, realistic production levels may approach or exceed 2.5 million tonnes of dry seeds on about 9 million hectares. The only developed country producing large amounts of cowpea is the USA (60,000 t).

Low yields are a significant attribute of production estimates, particularly in Africa and Asia where 240–300 kg/ha are typical. The reasons include heavy biotic pressures—particularly from insects and other pests, which often affect the plant throughout its life cycle and the seeds in storage. Literally dozens of insects attack cowpea, but the chief ones in the major growing region are: bud–flower thrips, pod borers, aphids, leafhoppers, curculio, pod suckers and bruchids. In addition, several diseases can be devastating.

Other factors contributing to low yields are problem soils—especially low-pH, high-Al, fertility-depleted or high-pH, saline soils—excessively high temperatures, and drought or excessive moisture, and inadequate management and plant protection. Suboptimal planting dates, low plant populations, poor soil physical properties, low-fertility soils, poor weed control and mixed cropping all reduce yields.

This long list of problems indicates the urgency for research and national production programmes geared to help farmers obtain inputs and adopt improved technologies. Unfortunately there are comparatively few researchers working on this crop, and most projects were established or strengthened within the past 2 or 3 decades.

Regional utilization

In tropical Africa cowpeas are primarily used in the form of dry seed cooked as a pulse in a large variety of dishes. Preference is for brown, white or cream seeds with a small eye and wrinkled or rough seed coat. In many areas of both West and East Africa the tender green leaves are cooked like spinach or as a relish. Generally the grower also likes to obtain some dry-seed yields from the plucked crop. Green beans or cut green pods used as a vegetable are of secondary importance. Cowpeas are also grown for fodder, groundcover or green manure but to a much lesser extent than for pulse.

Cowpea is practically never grown as a sole crop in Africa; it is a component of cropping systems, frequently being grown in association or relays with sorghum, millet and maize. However, some workers discourage the growing of cowpeas with roots and tubers, as the system appears to favour the build-up of nematodes. Short-season cowpeas (60–65 days) are sometimes grown in a monoculture system or as a late relay crop in sorghum or maize to use residual moisture in the soil.

In Asia the pulse uses of cowpeas are important primarily in the drier regions such as India where increasing amounts of the crop are used in dhal. In the more humid areas of Southeast Asia and southern China the long bean is the predominant form and is used as cut green beans. Seed-quality preferences are somewhat broader than in West Africa where brown and white seeds are preferred. The dry-seed types may be grown both in mixed-cropping systems and in monoculture, but long beans are mainly grown with supports—trellises or standing-crop residues like maize stalks.

In South America cowpeas are grown mainly as pulse with a wide range of seed colours, blackeye-whites, reds, yellows, browns, black and mottled. Although the major producing areas are the drier lowlands like northeastern Brazil, there is increasing interest in cowpeas for the humid interiors like the Amazon Basin where cowpeas are better adapted to low-pH, high-Al soils than *Phaseolus* beans and other crops. Cowpeas are also utilized in cropping systems similar to those in Africa and Asia. Sometimes they are used as forage, hay and cover; and the seeds may replace soybean or other protein sources used in animal feed.

In the USA dry seed is grown mainly in California and western Texas, whereas green peas for canning, freezing or fresh peas are mainly produced from eastern Texas to Florida and the Carolinas. Sometimes cowpeas are also grown for forage and as a cover crop, particularly in the southeastern USA.

RESEARCH NEEDS

Rapid progress in cowpea improvement has been made during the 2 decades up to the mid-1980s. Most spectacular contributions are the multiple disease- and pest-resistant, high-yielding and short-duration strains together with more efficient production practices designed for a broad range of growing conditions. These advances are combining both to increase production in traditional growing regions and to allow expansion into new areas. For example, the cowpea appears to have a particularly bright future in rice-based systems of Southeast Asia. However, there remains much to be done on both current problems and in new areas.

Continuing challenges

The cowpea is remarkably attractive to pests—especially in southern Nigeria. There are at least 12–14 important diseases, an equal number of harmful insects, and two or three nematode species afflicting the crop in this one area.

Throughout the world, production and improvement are currently being constrained by:

○ Insect pests, which will continue to be the most serious biological problem in the tropics. Excellent progress has been made in identifying sources of genetic resistance, and the breeding approach is clear—enhance or pyramid resistance genes and combine with resistance to as many insects and diseases as possible. However, it would be risky to short-circuit other methods of pest management as progress is made on host-plant resistance. A valuable lesson came with the several biotypes of the rice brown leafhopper in Southeast Asia: when resistance breaks down a national disaster occurs.

○ Diseases, which are shifty enemies—new ones emerging just when a major problem is solved. Both pathologists and plant breeders must be vigilant even as they pyramid several resistances in advanced lines of commercial varieties. As with insects, diseases will continue to require study—not only their epidemiology but also alternative approaches to their control. In addition, root-knot and lesion nematodes and the devastating plant parasite *Striga* merit study in endemic regions.

○ Problem soils, which characterize not only lands subjected to continuous cropping but also the frontiers of the future—the vast tracts of humid forests being rapidly settled from Southeast Asia to the Amazon. These are often highly leached, low-pH, high-Al soils called red–yellow podsols in Asia and oxisols or ultisols in the Amazon region. These new areas may have the greatest potential for expanding cowpea production, but the problems are acute: high Al sometimes combines with toxic levels of Fe and Mn; low availability of soil nutrients, especially N, P and Ca, and, at times, K; and often excessive moisture but sometimes drought.

○ The symbiont, which is the bacterium providing legumes with the capacity to fix nitrogen. Not only must the plant have the capacity to attract and maximize the association but also the N-fixers must receive attention. Although good progress has been made in developing high-temperature strains of *Rhizobium*, to date there have been no strains efficient in problem soils or in competition with indigenous soil rhizobia.

○ The yield barrier, which derives from the uncertainty about the definition of high-yielding idiotypes. Should plants be erect, large, small, spreading, profusely branching and tolerant of high populations? Or are numbers of branches, peduncles, pods, seeds/pod and seed size paramount—and in what order?

○ Cultural practices, which depend ultimately on the prevalent growing conditions in a producing area. A large amount of information and experience on optimal management of cowpeas has been accumulated for a wide range of growing conditions. However, there always seem to be new permutations of environmental conditions, fertilizer levels to test, uses to consider and varieties to grow. On these are superimposed the varied needs of different cropping systems: associations, relays and sequences. Studies must include the economic parameters and social factors for different production areas.

○ Product quality, which must be tailored to consumer preferences in particular areas. The dry-seed preferences of West African consumers are well defined; however, elsewhere, consumer tastes are less clearcut, and in many areas mottled, buff, yellow, red or black seeds are preferred for certain purposes. Maintaining these 'off-colours' could speed the adoption of cowpeas as a 'new' crop in such areas. Besides the seed size and shape, colour and pattern, and testa texture, the qualities for cowpea as a fresh vegetable cannot be ignored. Moreover, the yard-long bean is widely grown in Eastern Asia—a favourite vegetable with much wider consumption than the dry-seed crop.

These are problem areas, but each one has several components when fragmented genetically or otherwise. In any event, there is much left to be done.

Special needs

Besides broad problems and objectives, some special needs and uses may deserve attention in the years ahead as a focus for secondary objectives:

○ Climbing cowpeas, which are not as strong climbers as some other grain legumes. Most climbing legumes are substantially higher-yielding than the bush forms, and they better fit some cropping associations, may be less prone to certain pests, and tend to have better-quality seeds under humid conditions. For these reasons the role for climbing cowpeas will probably continue and may increase in intensive, peasant-cropping systems.

○ Long beans, which are vegetable types with greater acceptance in some areas, such as Southeast Asia, than is generally recognized. In parts of Indonesia, Thailand, Philippines, Taiwan and China the long bean is among the most important vegetables in season. It is normally grown intensively and ubiquitously on trellises and is unimproved in terms of pest resistance, cooking quality, increased yields, and longer-bearing duration.

○ Rustic creepers, which are prostrate and can be planted or can emerge voluntarily in off-seasons, waste places, or where other crops are not cultivated—a kind of easy-care, catch crop. These forms might be developed from recombination with wild or semiwild species by researchers focusing on improving seed, synchronizing fruiting and enhancing yield while retaining a high degree of rusticity.

○ Enhanced stress tolerance, which is already a focus of some special breeding programmes but may be expanded to cover low-pH, high-Al, depleted soils; high-pH, alkaline, and saline conditions; drought conditions; and waterlogged soils.

○ Optimized cropping systems, which involve cowpeas for different purposes. Studies should include sequences and associations with warm-weather cereals, such as cowpeas following irrigated rice; associations with upland rice, maize and cassava; climbers with annual crops and trellises; climbers with perennials as living trellises; and climbers and bush types with root crops. In addition, attention must be given to the special needs of bush–fallow rotations, semiarid cropping, intercropping with long-term perennials like oil palm, rubber and coffee. Superimposed on these systems' studies will be the all-pervasive problems of weed control, soil and water conservation, pest management and postharvest handling.

○ Pest management, which may involve methods integrating biological, cultural and chemical control as well as the use of resistant varieties. Models need to be developed for controlling or managing pests in new areas or new pests in defined areas, and these should focus on widely applicable cultural practices and on the natural enemies for biological control.

○ Food preparations, which incorporate dry-seed types in areas where the crop is new and where methods of preparation are neither familiar nor understood. For example, consumers of other grain legumes may tend to overcook cowpeas. Some excellent cowpea dishes like Akara balls and Moi-moi are virtually unknown outside West Africa. Such preparations could be introduced through demonstrations in training courses or at international conferences and meetings. Recipe books are also likely to improve popularity.

THE FUTURE

The excellent progress on improving cowpeas in recent years, and the rapidly increasing worldwide interest in this crop, bode well for the future. The stage is now set for a dramatic increase in production both through increased yields in existing producing regions and the spread of cowpeas to new areas and niches. Specifically, virtually unexplored is the potential for cowpeas in Latin America and the Caribbean, the Middle East and Southeast Asia (from India to Indonesia and mainland China).

Considerable effort on cowpea improvement will be needed over the next 15–20 years to sustain the apparently bright future for this crop. Cowpea researchers need to collaborate more closely at the national level and with the joint programme pooling the resources of the Bean/Cowpea CRSP (Cooperative Research Support Program), advanced laboratories and IITA (the International Institute of Tropical Agriculture).

The joint programme should aim at developing a group of superior new varieties together with agronomic and pest-management practices for optimal plant growth. The composite package might be described broadly as involving:

○ The plant, which must be stronger and more robust, quicker-growing and tolerant of stress from drought, waterlogging, pH extremes and unfavourable soil conditions. It should fruit profusely and carry large, well-packed pods (crowders). Superior types will be needed in at least four different plant types: an elite-branching, population-responsive bush yielding up to 5.0 t/ha or more under favourable conditions; an intermediate spreading bush (weakly climbing) yielding 3–4 t/ha; prostrate/creeping types, rustic with long peduncles with a yield potential of 1.0–2.5 t/ha; and vigorously climbing forms yielding 6.0 t/ha or more under good growing conditions.

○ Insect- and disease-resistance, which combine genetic resistance and tolerance for all the important growing regions of the world. Breeders and pathologists should strive for effective levels of resistance to all major diseases of each region in addition to maximizing resistance to as many insects and nematode species as possible. Certainly the painfully sought and developed resistance and tolerance to *Maruca*, pod suckers, thrips, aphids and other pests should be routinely transferred to most improved types. Even tolerance to stored-grain pests may be included if the tolerance factor does not adversely affect the nutritional and organoleptic properties of grain.

○ The seeds, which should be medium-large, e.g. 18–20 g/100 seeds, mostly light or creamy-white (but other colours if needed) with an easily removed rough or wrinkled seed coat, and possibly a small eye; the flatulence factor and metabolic inhibitor should be 'bred out'; and the peas should be quick to cook, e.g. 20–25 minutes; taste and cooking qualities must be preserved and efforts made to enhance the nutritional qualities of the seeds, mainly through increased sulphur amino acids. Vegetable types should be developed with larger, string-free pods and with more attractive cooking and taste qualities. Similarly, special varieties for leaf vegetables that produce reasonable seed yields need to be developed.

○ Pest management, which is likely to involve more than just breeding for resistant varieties and the use of 'convenient' cultural practices. The predators and biological agents should be studied to find ways of reducing the rapid build-up of specific pests, and the screening of new, more effective chemical pesticides should be continued, especially for formulations with low mammalian toxicities and residual activity.

○ The varied permutations of optimal management practices for different growing conditions and uses. For instance, researchers must deal with particular tillage problems involving surface residues, water- and nutrient-holding capacity of the soil, and plant roots. Soil microbiologists will have an important role here. They must also address weed control—particularly in low-tillage systems—with breakthroughs sought in the form of inexpensive, new and more effective herbicides; in appropriate mechanization; and in more judicious use of associated species. These cultural practices need to be combined with special biological agents such as free-living nitrogen-fixers, more efficient *Rhizobium*, and soil inoculation with Mycorrhizae to improve the total growing environment. These are some of the obvious goals to be considered for the future; they demand continued all-out efforts and cooperation between social and biological scientists as well as food technologists. However, if recent progress is an indication, cowpeas will become the golden crop of the twenty-first century. If the momentum of present advances were sustained total worldwide production would be increased by 25 per cent by the end of this decade and would more than triple—to about 2 million tonnes—by the turn of the century.

PART 1

Origin, Taxonomy and Morphology

1

Genetic diversity in Vigna

J. P. BAUDOIN and R. MARÉCHAL
Faculté des sciences agronomiques de l'état, 5800 Gembloux, Belgium

In the developing nations, breeding better plants is the focal point around which all programmes to increase crop yields develop. The achievement of this objective depends mainly on the genetic diversity available for the crop, but some useful characters may be found only outside the limits of the species. In contrast with cereals, the possibility of using related species in plant improvement is restricted in food legumes because of the presence of strong and early isolating mechanisms between species. *Vigna unguiculata* (L.) Walpers, one of the major pulses in the subhumid and humid tropics, is even more constrained in this respect than other species of the subtribe. This paper is intended to define the taxonomic status of cowpea within the Phaseoleae tribe and to assess the actual range of genetic diversity available in the different gene pools of the species.

TAXONOMY—PHYLOGENY WITHIN THE TRIBE PHASEOLEAE

The international legume conference held at Kew, UK, in 1978 markedly improved classification of the Leguminosae at generic and tribal levels. The traditional classification based on simple morphological criteria provided a stable, practical system for plant identification. The phyletic concept expanded by the 1978 conference imposed consideration of new information deriving from technical advances in biochemical analysis, cytology, palynology, blastogeny, anatomy, etc. From this information, long-standing uncertainties are being resolved and new outlines are emerging; for example, in one of the largest and most advanced tropical tribes the Phaseoleae. Although the overall pattern of the tribe remains difficult to comprehend, the study of Lackey (1977, 1978, 1981) resulted in a much better understanding.

The system presented by Baudet (1978), based on ecologic–geographic concepts and the style morphology, departs further from the traditional classification and impacts on the thinking about the evolution of Phaseoleae. Within the tribe, Baudet recognized two subtribes, Phaseolinae and Cajaninae, with internal conformity and a high level of specialization resulting from the extension in the savanna conditions and the expanded subtribe, Glycininae, mainly confined to the forest regions.

The subtribe Phaseolinae, which contains the largest number of economic legumes, has been the subject of much recent taxonomic study. The past confusion about the delimitation of some genera in this subtribe was such that a brief summary of the taxonomic history of *Phaseolus* and *Vigna* is useful.

Bentham and Hooker (1865) first distinguished the two genera on the basis of keel morphology, symmetric in *Vigna* and coiled or curved in *Phaseolus*, but the distinction soon proved to be unsatisfactory. In 1954 Wilczek, followed by Hepper (1958), gave a new circumscription of *Vigna* based on two characters: produced stipules and the style prolonged beyond the stigma by a beak. This concept allowed a better natural grouping of the Old World *Vigna* with the consequence of transferring into the genus all the Asiatic *Phaseolus* section *Ceratotropis*. Yet the system was not always reliable, and later, in 1970, Verdcourt brought a substantial improvement to the generic limits. His conclusion, from the observation of a wide range of species and many different characters (morphologic, chemical, cytologic and palynologic) resulted in a restricted concept of *Phaseolus* and in a much enlarged concept of *Vigna*.

Inside the Phaseolinae subtribe, Baudet and Maréchal (1976) distinguished two groups based on style morphology: the Phaseolastrae containing all those genera with the style bearded on the inner part below the stigma, and the Dolichastrae, a rather more heterogeneous entity grouping all other types of style hairiness. The review of all Phaseolastrae by Maréchal and co-workers (1978a) has not only vindicated the systematic reorganization proposed by Verdcourt, but has also provided a much clearer understanding of this economically important group. Their numerous and new data, observed both on living plants and herbarium specimens, were carefully computed. The different characters, evaluated objectively, provided a neat boundary between *Phaseolus* and *Vigna*.

The restricted genus *Phaseolus*, of New World origin, appears relatively well isolated, while the large genus *Vigna*, of pantropical distribution, seems *a priori* fairly heterogeneous. Yet the different intrageneric groups show a certain cohesion within this genus. The relations of similarity help to give a picture of the most probable working hypothesis of the phyletic relationships, and have been graphically represented in the publication by Maréchal *et al.* (1978a). A neotropical initiation of the generic differentiation is quite plausible with the large and rather primitive subgenus *Sigmoidotropis*, including most of the American *Vigna*. Still confined to the neotropical regions, the subgenus *Lasiospron* accumulates most of the *Vigna* characteristics. All the characters are present in the subgenus *Plectotropis*, which seems to have originated in the Old World. Two evolutionary tendencies appeared:

○ In Asia, a differentiation toward a specialized floral morphology led to the homogenous subgenus *Ceratotropis*; and

○ In Africa, a relative simplification of the floral morphology restored bilateral symmetry in the species-rich subgenus *Vigna*. Cowpea belongs to the latter.

More or less closely related with this African group, the subgenus *Haydonia* probably represents a relatively recent evolutionary trend expressed by the loss of some typical *Vigna* characteristics and acquisition of some new ones. The last subgenus of *Vigna*, *Macrorhyncha*, appears remote from the others and is maintained in the genus for convenience.

Smartt (1980) was reluctant to accept such a broad sense for one genus and suggested a dismembering of it. However, Maréchal (1982) strongly supported the large genus as having biological merit because of continuity inside the group as well as close phyletic affinities. A strong argument for keeping the genus as it is, is the palynologic one (the granular nature of the infratectum layer of the exine). Smartt (1980) argued that the genus *Vigna* included too many effectively isolated gene pools and should be split. We will demonstrate, however, that this argument cannot be applied in the taxonomy of the food legume, taking, for example, the gene pool of *V. unguiculata* and some other pulses.

GENE POOL

On the basis of reproductive affinity, the classification of Harlan and de Wet (1971) assigns related races or species to primary, secondary or tertiary gene pools. The primary gene pool corresponds with the species of the taxonomic hierarchy and includes both the cultigen and the wild forms, all of them hybridizing freely. The secondary gene pool comprises other species that are relatives of the crop and are suitable for interspecific hybridization. They are isolated from the species partly by barriers such as chromosomal and genic sterility. The tertiary gene pool involves still greater barriers to hybridization, embracing taxa that display either unviable or sterile hybrids with the cultivated plant and do not permit gene flow by normal introgression. The range of total gene pool varies with each crop, and unfortunately the secondary and tertiary gene pools in food legumes are rather restricted, coinciding with the genera, except perhaps for *Cajanus atylosia* (Smartt, 1980).

Vigna unguiculata

We have applied Harlan and de Wet's classification to cowpea. According to Maréchal *et al.* (1978a), this species belongs to the section Catiang of the subgenus *Vigna*. So far this section involves two species: *V. unguiculata* and *V. nervosa* Markötter. The taxonomy of *V. unguiculata* has been refined, with the four subspecies constituting the primary gene pool: *unguiculata*, the cultigen, being roughly divided into four cultigroups (cv-gr.):

— Unguiculata: the cowpea, the most important group in Africa;

— Biflora: the catjang, a fodder and seed type, mainly grown in Southeast Asia;

— Sesquipedalis: the yard-long bean or asparagus bean, a green pod vegetable, grown in India, Southeast Asia and China; and

— Textilis: a very small group found in West Africa (Niger, north Nigeria) and grown for the textile fibres of the long peduncles.

The wild forms, spread throughout subsaharan Africa, are subdivided as:

O Subspecies *dekindtiana* (Harms) Verdc.:

—Var. *dekindtiana* (Harms) Verdc., the most common wild form in Guinean and Zambezian savanna regions.

—Var. *mensensis* (Schweinf.) M.M.&S., a variant with large flowers and long calyx lobes, probably found mostly in East Africa;

—Var. *pubescens* (R. Wilczek) M.M.&S., with a short pubescence of leaves and stalks, limited to East Africa;

—Var. *protracta* (E. Mey.) Verdc., with small leaflets and long pubescence, confined to Natal and Transvaal;

O Subspecies *tenuis* (E. Mey.) M.M.&S., with small glabrescent lobed leaflets confined to Natal; and

O Subspecies *stenophylla* (Harvey) M.M.&S., with narrow hastate leaflets, common in Transvaal.

This intraspecific classification is still provisional. Further biotaxonomic research is likely to reveal other groupings and to bring about changes at the subspecies level.

Southern Africa and East Africa as far north as Ethiopia are the regions of primary diversity for the wild forms. The species has probably been domesticated primarily in these regions, and samples should be collected as a source of new diversity. The last working group meeting of the International Board for Plant Genetic Resources on *Vigna*, held in New Delhi (IBPGR, 1981), recommended, as a priority, collection of both wild and cultivated forms in South Africa, Zimbabwe, Transvaal and Natal.

West Africa, from Cameroon to Senegal, is a secondary centre of diversity: it contains the largest number of primitive cultivars and weedy types.

Another centre of diversity lies in Southeast Asia where people have intensively selected African domesticated varieties, prompting development of specialized forms such as *V. biflora* and *V. sesquipedalis*.

No secondary and tertiary gene pools have been identified for *V. unguicula-ta*. Only one species, *V. nervosa* from Southern Africa (Zimbabwe to Natal), is likely to belong to the secondary gene pool and, unfortunately, does not exist in any living collections. No attempts have yet been made to hybridize it with cowpea, and efforts to cross cowpeas with other species of *Vigna* have been unsuccessful (Sen and Bhowal, 1960a; Faris, 1965; Evans, 1976).

Other important food legumes

In other important food legumes of *Phaseolus* and *Vigna*, as in cowpea, each cultigen has a wild counterpart with which it hybridizes easily (Maréchal *et al.*, 1978a; Smartt, 1984). However, in contrast to cowpea, these species have secondary or tertiary gene pools and sometimes both (Table 1). Relationships and possibilities for interspecific crosses have been reviewed or investigated by Ahn and Hartmann (1978), Jain and Mehra (1980), Chen *et al.* (1983) and Smartt (1984) for the Asiatic *Vigna* and by Baudoin (1981a) and Smartt (1984) for *Phaseolus*.

The gene pool of the Asiatic *Vigna* species appears to comprise three rather isolated complexes: *V. radiata–mungo*, *V. umbellata–angularis*, and *V. aconi-tifolia–trilobata* (Lukoki, 1980). In *Phaseolus* the *P. vulgaris–coccineus* complex forms one gene pool well separated from *P. lunatus* and its wild relatives. *Phaseolus acutifolius* and *P. filiformis* are intermediate, showing a closer affinity with *P. vulgaris*. The nature or extent of the gene pool, however, is probably not completely known. Results of crosses depend on the different genotypes involved, and future crosses, particularly with the wild species, may reveal other affinities and additional genetic resources.

WIDE CROSSING WITH COWPEAS

Wild forms and species of the genus *Vigna* have great potential for the improvement of *V. unguiculata*, and priority should be given to collections and crosses of them. The primary gene pool for a species like cowpea encompasses a huge range of morphologic and agronomic attributes, exceed-ing the variability present in primary gene pools of many other food legumes, and so far free gene exchange has proved successful between all wild and cultivated forms (Evans, 1976; Rawal, 1975b; Maréchal *et al.*, 1978a). Although some regions of genetic diversity have been well explored in search of useful characteristics, other areas, including southeastern Africa, have been neglected. We are convinced that the lack of any serious collecting activities in these areas limits not only knowledge of the crop evolution but also the possibilities of new and outstanding genetic improvement. We believe that much greater attention should be devoted to collection efforts before investigation of crosses with other *Vigna* species.

Table 1.   Secondary and tertiary gene pools in the cultivated *Ceratotropis Vigna* and *Phaseolus*

Species	Secondary pool	Tertiary pool
Vigna radiata (L.) R. Wilczek	*V. mungo*	*V. umbellata, V. angularis, V. glabrescens, V. trilobata*
V. mungo (L.) Hepper	*V. radiata*	*V. umbellata, V. angularis, V. glabrescens, V. trilobata*
V. umbellata (Thunb.) Ohwi & Ohashi	*V. angularis*	*V. radiata, V. mungo*
V. angularis (Willd.) Ohwi & Ohashi	*V. umbellata*	*V. radiata, V. mungo*
V. aconitifolia (Jacq.) Maréchal	*V. trilobata*	—
Phaseolus vulgaris L.	*P. coccineus*	*P. acutifolius, P. filiformis* Bentham, *P. ritensis* Jones
P. coccineus L.	*P. vulgaris*	*P. acutifolius*
P. acutifolius A. Gray	—	*P. vulgaris, P. coccineus*
P. lunatus L.	—	*P. ritensis, P. metcalfei* Woot & Standley, *P. polystachyus*

So far, no interspecific crosses have been successful with *V. unguiculata*, but, provided living specimens are collected, *V. nervosa*, from the same section Catiang, could be included in the secondary gene pool. The only other section that shows a relatively high similarity coefficient with Catiang is the section Liebrechtsia, with *V. frutescens* A. Richard (Maréchal *et al.*, 1978a). Unfortunately the different chromosome numbers between the two sections are likely to restrict the gene flow in combinations.

The presence of strong isolating barriers blocks exploitation of the tertiary gene pool and is a general characteristic for Phaseolinae: they appeared early in the evolution of the species and the major incompatibility mechanism sets up rapidly in the postzygotic life cycle of the hybrids (Le Marchand *et al.*, 1976; Rabakoarihanta *et al.*, 1979; Maréchal, 1982; Chen *et al.*, 1983). No doubt intensive use of some sophisticated methods such as cell, tissue and embryo culture; protoplast fusion; and chromosome engineering will reap progress in interspecific crosses. Then other constraints will emerge, such as hybrid sterility and expression of the transferred genes in novel genotypic backgrounds. Such constraints have been well demonstrated in the study of segregating populations in interspecific *Phaseolus* hybrids (Wall, 1970; Nassar, 1978; Baudoin, 1981b; Pratt, 1983).

Incompatibility barriers present in many Phaseolinae species should be kept constantly in mind by researchers reassessing the value of taxonomic systems. Some authors believe that a taxonomic classification should reflect the phyletic distances between the taxa of a well-defined group. This would greatly help breeders to organize attempts at interspecific hybridization. First, however, a good taxonomic classification should aim at showing the logical evolution of a group of species, and at least in Phaseolinae, the neat evolutionary trends demonstrated in the genera do not necessarily correlate with the degree of incompatibility between species—another reason that one should not divide the large genus *Vigna* into smaller units. The study of Maréchal *et al.* (1978a), based on an objective evaluation of discriminant characters, has, in fact, enlarged the genus with a geocarpic species of importance in dry, savanna Africa: the Bambarra groundnut (formerly *Voandzeia subterranea* Thouars). There is no doubt that this crop belongs to the *Vigna* genus, its most closely related wild species being the amphicarpic *V. hosei* (Craib.) Backer of the section *Vigna*. This wild species may belong to the secondary, if not the tertiary, gene pool of *Vigna subterranea*.

Cowpea Research, Production and Utilization
Edited by S. R. Singh and K. O. Rachie
© 1985 John Wiley & Sons Ltd

2

Cowpea taxonomy, origin and germ plasm

N. Q. Nɢ* and R. Maréchal†
* Genetic Resources Unit, International Institute of Tropical Agriculture, Ibadan, Nigeria; and †Faculté des sciences agronomiques de l'état, 5800 Gembloux, Belgium.

A sound foundation of taxonomy, origin, diversity and interrelationships of a crop and its wild progenitors and relatives is essential if the diversity necessary for plant breeders is to be explored, assembled and utilized. Knowledge of the origin and the centre of diversity of the crop and its related species will help researchers to determine the areas and strategies for plant exploration and collection.

This paper reviews the recent concepts and opinions about the taxonomy and origin of cowpeas and the achievements in cowpea germ-plasm exploration and conservation at IITA.

TAXONOMY

Because of the number of distinctive forms, synonyms and common names of cowpea and its affinities with *Phaseolus* and *Dolichos*, there has been confusion in classification and nomenclature. Most nomenclatural problems in the cultivated cowpeas and related wild species have recently been satisfactorily resolved (Verdcourt, 1970; Summerfield *et al.*, 1974; Westphal, 1974; Maréchal *et al.*, 1978a,b). Experts agree that cowpeas belong to the botanical species *Vigna unguiculata* (L.) Walp. but still debate the classification and nomenclature of taxa at the intraspecific level. There are more than 20 synonyms for *V. unguiculata* (Table 1) and many common names.

Vigna is a large and immensely variable genus subdivided by Verdcourt (1970) in eight subgenera: *Vigna, Sigmoidotropis, Cochliasanthus, Plectotropis, Ceratotropis, Dolichovigna, Macrorhynchus* and *Haydonia*. This classification was recently modified by Maréchal and his colleagues, who reduced

Table 1.   Synonyms of cultivated *Vigna unguiculata*

cv-gr. Unguiculata	cv-gr. Biflora	cv-gr. Sesquipedalis
V. unguiculata subsp. *unguiculata* (L.) (Walpe.) Verdc. *V. sinensis* (L.) Hassk. subsp. *sinensis* *Dolichos unguiculata* L. *D. biflorus* L. *D. sinensis* L. *Phaseolus unguiculata* (L.) Piper *P. sphaerospermus* L.	*V. unguiculata* subsp. *cylindrica* (L.) *Van Eseltine*, *V. catjang* (Burm. f.) Walp., *V. sinensis* (L.) Hassk. var. *catjang* (Burm. f.) Chiov., *V. cylindrica* (L.) Skeels., *V. unguiculata* (L.) Walp. subsp. *catjang* (Burm. f.) Chiov., *V. unguiculata* var. *cylindrica* (L.) Ohashi., *D. catjang* Burm. f., *D. tranquebaricus* Jacq., *D. monachalis* Brot., *D. biflorus* L., *P. cylindrica* L.	*V. unguiculata* subsp. *sesquipedalis* (L.) Verdc., *V. sinensis* (L.) Hassk. var. *sesquipedalis* (L.) Ascherson and Schweinf., *V. sinensis* (L.) Hassk. subsp. *sesquipedalis* (L.) Van Eseltine, *V. sesquipedalis* (L.) Fruhw., *V. unguiculata* var. *sesquipedalis* (L.) Ohashi, *Dolichos sesquipedalis* L.

the genus to seven subgenera: *Vigna, Sigmoidotropis, Plectotropis, Macrorhyncha, Ceratotropis, Haydonia* and *Lasiocarpa* (Maréchal *et al.*, 1978a). The African subgenus *Vigna*, which was divided into nine sections by Verdcourt, has been reduced to six sections: Vigna (20 spp.), Comosae (2 spp.), Macrodontae (2 spp.), Reticulatae (9 spp.), Liebrechtsia (1 spp.), and Catiang (2 spp.). Cowpea belongs to section Catiang, which is relatively simple and is typified by its prominent spur stipules. It includes only two distinct species, *V. unguiculata* and *V. nervosa* Markötter.

Verdcourt subdivided *V. unguiculata* into five subspecies, of which three are cultivated and two are closely related wild species. These cultivated subspecies are *unguiculata, cylindrica* and *sesquipedalis*, while the wild subspecies are *dekindtiana* and *mensensis*. Maréchal and his colleagues (1978a) do not consider these three cultivated subspecies as distinct; they adopted and expanded the idea of Westphal (1974) to lump them under one subspecies *V. unguiculata* subsp. *unguiculata* and to differentiate them by the intraspecific category 'cultigroup'. Thus, the subspecies *unguiculata, cylindrica* and *sesquipedalis* were renamed by Maréchal and his colleagues cultigroups Unguiculata, Biflora and Sesquipedalis, respectively. In addition, they included cultigroup Textilis (synonym *V. sinensis* var. *textilis* A. Cheval.) to describe a cultivar grown in northern Nigeria, for its strong fibre, which is obtained from the erect peduncles (Dalziel, 1937, quoted by Westphal, 1974). They also lumped the two wild subspecies *dekindtiana* and *mensensis* recognized by Verdcourt under a single subspecies *dekindtiana* and used the varietal category to distinguish them. They expanded this wild subspecies *dekindtiana* to include the species formerly known as *V. pubescens* Wilczek and *V. unguiculata* var. *protracta* (E. Mey.) Verde. and categorically classified these two taxa as varieties *pubescens* and *protracta* of the subspecies *dekindtiana* respectively. Two other wild taxa previously known as *V. tenuis* (E. Mey.) Dietr. and *V. angustifoliolata* Verdc., that appeared to be very similar and closely related to the cultivated species, were reclassified by Maréchal and his colleagues as subspecies *tenuis* and *stenophylla* of *V. unguiculata* respectively.

The two different schemes of classification and nomenclature at the intraspecific level of section Catiang by Verdcourt and Maréchal and his colleagues are quite similar (Table 2), but the one by Maréchal and colleagues is clearer, simplifying visualization of the interrelationships among the different groups of taxa within the section.

All four cultigroups of the subspecies *unguiculata* and the varieties of the subspecies *dekindtiana* are interfertile, although hybridization between var. *protracta* and the cultivated subspecies has not been attempted (Evans, 1976; Maréchal, 1976; Rachie and Rawal, 1976). No living collection of this variety is available. Similarly whether the two subspecies *tenuis* and *stenophylla* and the species *V. nervosa* are interfertile with cultivated subspecies *unguiculata* is

Table 2. Classification and nomenclature of the taxa within section Catiang of the subgenus *Vigna* (Savi) Verdc.

Maréchal et al. (1978a)	Verdcourt (1970)	Status
V. unguiculata	*V. unguiculata*	
subsp. *unguiculata*		
cv-gr. Unguiculata E. Westphal	subsp. *unguiculata* (L.) (Walp) Verdc.	Cultivated
cv-gr. Biflora E. Westphal	subsp. *cylindrica* (L.) Van Eseltine	Cultivated
cv-gr. Sesquipedalis E. Westphal	subsp. *sesquipedalis* (L.) Verdc.	Cultivated
cv-gr. Textilis E. Westphal		Cultivated
subsp. *dekindtiana*		
var. *dekindtiana*	subsp. *dekindtiana* (Harms) Verdc.	Wild
var. *mensensis* (Schweinf.) Maréchal, Mascherpa & Stainer	subsp. *mensensis* (Schweinf.) Verdc.	Wild
	V. unguiculata (L.) Walp.	
var. *protracta* (Wilczek) Maréchal, Mascherpa & Stainer	var. *protracta* (E. Mey.) Verdc.	Wild
var. *pubescens* (Wilczek) Maréchal, Mascherpa & Stainer	*V. pubescens** Wilczek	Wild
subsp. *stenophylla* (Harv.) Maréchal, Mascherpa & Stainer	*V. angustifoliolata* Verdc.	Wild
subsp. *tenuis* (E. Mey.) Maréchal, Mascherpa & Stainer	*V. tenuis** (E. Mey.) Dietr.	Wild
V. nervosa Markötter	*V. nervosa* Markötter	Wild

* Verdcourt noted that this species might be a variant or variety of *V. unguiculata*.

uncertain. Further plant exploration and biosystematic research would clarify the taxonomic status and interrelationships of these taxa.

Chromosome number of *V. unguiculata* is 2n = 2x = 22 (Faris, 1964; Frahm-Leliveld, 1965; Maréchal, 1970).

DIVERSITY AND ORIGIN OF CULTIVATED COWPEAS

Cultivated species are usually variable because of artificial selection under diverse environments, and cowpea is no exception (Table 3).

Cultigroup Unguiculata is the most diverse of the cultivated subspecies *unguiculata* and has the widest distribution. It is commonly called cowpea and is widely grown in Africa, India and Brazil. Varieties are prostrate, semierect, erect or climbing, and pods are coiled, round, crescent or linear.

Peduncles are from less than 5 cm to more than 50 cm long. Six different patterns of pigmentation of flower and pod, and 62 eye colours and 42 eye patterns on the seeds, have been described (Porter *et al.*, 1974). The number of days from sowing to pod maturity of different cultivars varies from 53 days to more than 120 days when grown at Ibadan during the second growing season (sown at the end of August).

Cowpea was known in India before Christ. It has a Sanskrit name and is mentioned in the Mahabhashya of Patanjali, an early treatise (*ca.* 150 BC) (cited in Steele and Mehra, 1980). The wide distribution and early cultivation of cowpea in Asia, where probable wild progenitors are absent, suggests an ancient origin and basic status for it among cultigroups of subspecies *unguiculata*. It is also believed that cowpea reached southwestern Asia about 2300 BC and came to southern Europe early enough for the Greeks and Romans to grow it under the names of *Phaseolos, Phaseolus* and *Phaselus* (Burkhill, 1953; Purseglove, 1976). Probably, it reached India more than 2000 years ago from East Africa, through the Sabanean lane along with other crops such as sorghum, finger millet and bulrush millet. It was introduced to the New World in the late seventeenth century by the Spanish, and more cultivars were transported there from West Africa with the slave trade (Steele, 1976). It probably reached the southern United States in the early nineteenth century (Wight, 1907) (Figure 1).

The suggestion that cowpea originated in Asia could not be supported because of the absence of progenitors there. All evidence points to its originating in Africa, although where the crop was first domesticated is uncertain. Ethiopia (Vavilov, 1951; Sauer, 1952; Steele, 1976), Central Africa (Piper, 1913), Central and South Africa (Zhukovskii, 1962), and West Africa (Murdock, 1959; Faris, 1965; Rawal, 1975b; Maréchal *et al.*, 1978a; Steele and Mehra, 1980; Lush and Evans, 1981) have all been considered probable centres of domestication.

Steele (1976) believed that cowpea was domesticated with sorghum and

Table 3. Characteristics of various cv-gr. of subspecies *unguiculata* and varieties of subspecies *dekindtiana*

Character	cv-gr. Unguiculata	cv-gr. Biflora	cv-gr. Sesquipedalis	cv-gr. Textilis	var. mensensis	var. dekindtiana	var. pubescens
Flower colour	White, purple	White, purple	White, purple	White, purple	Purple	Purple	Purple
Standard petal Width (mm)	24-30	24-28	28-35	24-27	33-42	23-42	28-32
Standard petal Length (mm)	18-23	18-24	21-24	19-21	—	17-34	19-21
Pod Length (cm)	6.5-25	7-13	15-90	7-14	—	6.0-11.6	—
Pod Width (mm)	3-12	4-6	5-11	—	—	2.8-7	—
Pod Orientation	Mostly pedant, some vertical	Mostly vertical	All pendant	Vertical	Vertical	Vertical	Vertical
Texture	Fibrous, hard, firm: not inflated when young	Fibrous, hard, firm: not inflated when young	Succulent: inflated toward maturity, shrinking after maturity	Fibrous, hard, firm: not inflated when young	Fibrous, hard, firm: not inflated when young	Fibrous, hard, firm: not inflated when young	Fibrous, hard, firm: not inflated when young
Dehiscence	Nil	Nil to moderate	Nil	Nil to moderate	Shatters	Shatters	Shatters
Locules/pod	7-23	12-16	15.8-23	—	16-19*	14-17	
Calyx lobe (mm)	<5	<5	<5	<5	>5	<5	<5
Seed Length (mm)	6-11	5-7	7-11	5.1-7.6	—	3-6	—
Seed Width (mm)	4-9	3-5	5-8	4-5.6	—	2-4	—
Seed Orientation in pod	Crowded	Crowded	Far apart	Crowded	Crowded	Crowded	Crowded
Breeding system	Inbreeder	Inbreeder	Inbreeder	Inbreeder	Outbreeder*	Inbreeder	Inbreeder
Growth habit	Erect, prostrate, climbing	Prostrate, climbing	Semierect, climbing	Prostrate	Prostrate, climbing	Semierect, prostrate, climbing	Semierect, prostrate
Shoots	Glabrous	Glabrous	Glabrous	Glabrous	Glabrous	Glabrous	Pubescence
Inverted V-shaped pigmentation on leaves	Nil	Nil	Nil	Nil	Some	Some	All

* Information quoted from Lush and Evans, 1981.

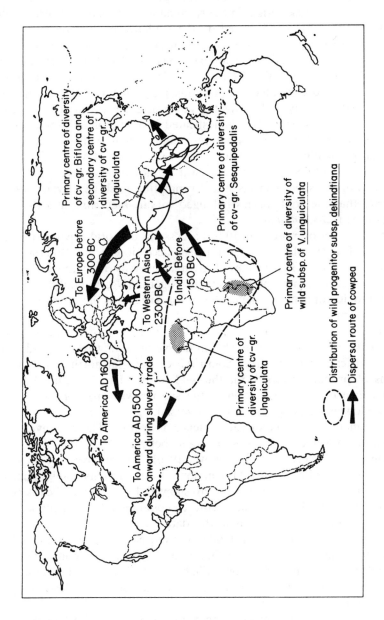

Figure 1. Centre of diversity and dispersal of cowpeas

pearl millet in Ethiopia because of the close association with these crops in early African farming. He also believed that it was equally probable that cowpeas had a 'diffuse' origin in the savanna after dispersal of cereals. Faris (1965), on the other hand, based on the evidence of the presence of wild progenitors of cowpeas in West and Central Africa, postulated that West Africa was the centre of domestication of cowpeas. This belief is strengthened by the studies of Rawal (1975b), who demonstrated that introgression between wild and cultivated cowpea took place in West Africa, and many forms of wild, weedy and cultivated cowpeas grew in this region. Rawal suggested that cowpeas were domesticated in the subhumid and semiarid regions of West Africa, from a natural colonizer that evolved from the wild perennial progenitor. This was supported by Steele and Mehra (1980), and Maréchal et al. (1978a). However, Rawal did not clearly specify the taxonomic status of the species in his study, prompting Lush and Evans (1981) to speculate that the wild species belonged to var. *mensensis* and the weedy species to var. *dekindtiana*.

Recent studies of more than 10,000 accessions of world cowpea collections held at IITA revealed that germ-plasm accessions from Nigeria, Niger, Burkina Faso and Ghana showed greater diversity than those from East Africa, with the centre of diversity in Nigeria (Ng, unpublished). These studies provided further evidence that West Africa was the primary centre of domestication.

Based on the present knowledge of the distribution of wild subspecies *unguiculata*, Baudoin and Maréchal in this volume, however, consider areas in East Africa, stretching from South Africa to as far north as Ethiopia, as regions of primary diversity for the wild forms. They concluded that the primary centre of domestication of the species was probably in this same region. However this opinion is controversial.

Unfortunately, archaeological evidence is scanty. There is evidence of prehistoric movement of crops or trade between East and West Africa (Harlan and Stemler, 1976; Purseglove, 1976). The discovery of carbonized pericarps of oil palm, a West African domesticate at Shabeinahb in northern Sudan from about 3300 BC (Shaw, 1976) showed the existence of early contact between West and East Africa, so cowpea could have been introduced from one region to the other as early as this date.

The carbon dating (*ca.* 1450–1400 BC) of cowpea remains discovered from Kintampo rockshelter in central Ghana is, as far as we know, the oldest archaeological evidence of cowpea (Flight, 1970, 1976), and the remains (*ca.* AD 100) reported from Nkope in Malawi are the second oldest (Robinson, 1970, cited by Shaw, 1976).

Biflora is commonly called catjang bean and is used as both dry seeds and fodder. It is frequently prostrate and sometimes climbing. Its pods are smaller than average Unguiculata, and they are held upright on the raceme axis. The seeds are frequently cylindrical and small. The pods and seeds are quite similar to those of var. *dekindtiana*. Biflora is cultivated mainly in India, Sri Lanka and, to some degree, in Southeast Asia.

Sesquipedalis, known as yard-long or asparagus bean, is used for its succulent young pods and sometimes for its leaves as a vegetable. Widely grown in India, Southeast Asia and China, it is mostly climbing, and its pods become deflated and flabby when mature. The pods are usually more than 20 cm long, and the seeds are widely spaced in locules. The flowers are normally larger than those in the other cultigroups.

In India, great variability is found in all three cultigroups. Unguiculata is less diverse than in West Africa and Sesquipedalis is less diverse than in Southeast Asia. Intermediate types between cultigroups Unguiculata and Biflora, and between cultigroups Unguiculata and Sesquipedalis, are predominantly found in north India, while those intermediate between Unguiculata and Sesquipedalis occur in the south (Steele and Mehra, 1980). These variations lead us to believe that the two cultigroups Biflora and Sesquipedalis were evolved from Unguiculata in India and Southeast Asia, respectively, under intensive human selection.

Cultigroup Textilis is an old, primitive cultivar and is rarely found in farmers' fields. It was reported to have been grown in northern Nigeria about half a century ago.

Within the more than 2000 samples of cowpeas from Nigeria collected or assembled by IITA's scientists, none resembled this cultigroup. One accession in IITA's collection looked similar to the description of Textilis; it had been obtained from a farmer's store in the northern part of Benin. It has a prostrate growth habit with strong, semierect peduncles, which are more than 60 cm long. The pods are small and upright, with smooth, small, black seeds.

This cultigroup was probably selected from cultigroup Unguiculata in West Africa for its long peduncles for fibre. Its morphological variation falls within Unguiculata.

GERM PLASM

Plant breeding is a creation of evolution at the will of humankind and is dependent on genetic variation (Frankel and Bennett, 1970). It requires a continuous supply of genes, and all germ plasm within genetic reach will be exploited in breeding programmes, including land races, wild and weedy species as well as the species included in secondary or tertiary gene pools defined by Harlan and de Wet (1971). The gene pool that can currently be exploited by cowpea breeders comprises the cultigroups and their land races as well as all the wild subspecies of *V. unguiculata*.

The genetic diversity is gradually diminishing because of humans changing or destroying the natural habitats to which wild species are adapted. Land races are being replaced by improved cultivars. In a very extreme case a crop may even be replaced in its home lands by an introduced crop. The West African rice (*Oryza glaberrima*) has been replaced by the Asian rice (*O. sativa*), and the Bambarra groundnut (*Vigna subterranea*) is gradually being replaced by the peanut (*Arachis hypogaea*) in Africa. Cowpea is no exception.

Fortunately, reasonably successful attempts have been made by IITA and other institutes to explore or assemble cowpea germ plasm. Due to excellent cooperation of many national programmes around the world, IITA has assembled more than 5000 accessions, through exchange or donation (Rachie and Rawal, 1976; Ng, 1982). In addition, IITA's scientists have mounted more than 40 plant exploration missions in 19 African countries, in collaboration with national scientists or agricultural authorities of the host countries and with IBPGR (International Board for Plant Genetic Resources); 39 of these exploration missions were carried out after 1975, when the genetic resources unit was established in IITA. As a result of this systematic germ-plasm exploration in Africa IITA's scientists have collected about 5000 samples of land races of cowpeas in this region. IITA has so far registered 13,270 cowpea germ-plasm accessions, but after elimination of duplicates and losses (Smithson *et al.*, 1980) it now holds 11,800 accessions of cowpea and 200 accessions of wild *Vigna* (Table 4).

Table 4.   Sources of IITA's world cowpea collection*

Country	Accessions	Country	Accessions
Benin	127	Somalia	3
Botswana	120	South Africa	108
Burkina Faso	197	Sudan	46
Cameroon	469	Tanzania	370
Egypt	345	Togo	75
Ethiopia	6	Uganda	70
Ghana	278	Zaire	15
Guinea Conakry	45	Zambia	550
Ivory Coast	129	Zimbabwe	7
Kenya	52	Brazil	32
Liberia	9	India	1935
Malawi	322	USA*	799
Mali	177	Yemen	20
Niger	700	Other	331
Nigeria	2492	Unknown	1085
Senegal	17		

* Original sources of the germ-plasm collections gathered from USA are yet to be clarified; 719 accessions are results of mixtures.

A small quantity of seeds of all the available germ-plasm collections at IITA are freely distributed to scientists around the world on request. During 1978–83 the genetic resources unit of IITA distributed 10,156 samples of its cowpea germ plasm to scientists in more than 50 countries around the world. This figure does not include the distribution of IITA's breeding lines for yield trials. IITA will continue operating on the principle of free availability of its germ-plasm materials and scientists who request the seeds should provide feedback on the uses of the germ plasm.

Not many accessions (or none at all) of IITA's world cowpea collections are from Burundi, Central African Republic, Congo, Ethiopia, Gabon, Madagascar, Mozambique, Zaire, or Zimbabwe. The germ plasm from Southeast Asia and Brazil is also not well represented. Further plant exploration or acquisition from existing collections in these areas is necessary.

The collections of the weedy and closely related wild species of cowpea are limited. The value of exploration and collection of these wild species, to enlarge the existing diversity of cowpea for plant breeding and research, cannot be overemphasized. Collaboration between national, regional and international institutes is needed for this worthy cause.

PART 2

Genetics

Cowpea Research, Production and Utilization
Edited by S. R. Singh and K. O. Rachie

3

The Genetics of Cowpeas: A Review of the World Literature

R. L. FERY
United States Department of Agriculture, Agricultural Research Service, US Vegetable Laboratory, Charleston, South Carolina, USA

Cowpea has been the subject of genetics research since the beginning of the 1900s. Although most of the early work was conducted in the United States, much of the current research effort is in the African and Asian countries. A total 158 specific genes have been identified in the cowpea. Additionally, many quantitatively inherited traits have been studied in the past two decades, and more than 225 heritability estimates published, with a large portion of the studied traits being of economic importance. A review of the published work will provide the plant breeder with much of the genetics information needed to devise or select efficient breeding procedures.

A multiplicity of scientific names has been applied to the cowpea, and the result has been considerable confusion among cowpea researchers. Incorrect scientific names are still frequently cited. The US Department of Agriculture has adopted Verdcourt's (1970) classification scheme (Gunn, 1973). Verdcourt (1970) stated that cowpea (formerly *V. sinensis* (L.) Savi ex Hassk.), catjang (fomerly *V. cylindrica* (L.) Skeels), and asparagus bean (formerly *V. sesquipedalis* (L.) Fruw. and *V. sinensis* (L.) Savi ex Hassk. subsp *sesquipedalis* (L.) van Eseltine) are not distinct species since they readily cross to form fully fertile hybrids.

The cowpea is almost ideal for plant breeding and genetics research. It is a diploid with a relatively short life cycle. The large flowers and untwisted keels of the cowpea make it one of the easiest legumes to emasculate and pollinate. Although it has been reported to undergo limited outcrossing (Harland, 1919a; Mackie and Smith, 1935; Purseglove, 1968; Rachie and Silvestre, 1977; Williams and Chambliss, 1980), it exhibits self-pollination under most environmental conditions.

Yarnell (1965), Rachie and Roberts (1974), and I (Fery, 1980) have reviewed genetics research on cowpea, with my review covering all the pertinent literature on cytologic, qualitative, and quantitative genetics. Additionally, it included an updated gene list that resolved the numerous discrepancies in the literature with respect to gene symbols.

GENE NOMENCLATURE

The literature has many examples of the same symbol being assigned to different genes, the same gene being assigned different symbols, or no symbol being assigned to a gene. Robinson *et al.* (1976) followed the recommendations of both the International Committee on Genetic Symbols and Nomenclature (Tanaka *et al.*, 1957) and the Tomato Genetics Cooperative (Barton *et al.*, 1955; Clayberg *et al.*, 1960, 1966, 1970) to construct a set of rules for the gene nomenclature of Cucurbitaceae. I adapted those rules and applied them to the genus *Vigna* (Fery, 1980: 315–316):

(1) The name of a gene should be descriptive of the characteristic phenotype conditioned by the gene. The name should be short and in either English or Latin.

(2) Genes are symbolized by italicized Roman letters, the first letter being the same as that for the name. A minimum number of additional letters is used to distinguish the symbol from other symbols already assigned to the species.

(3) The first letter of the symbol and name is capitalized if the gene is dominant to the normal allele; otherwise, all letters are in lower case. The normal allele is represented by the symbol '+' or if needed for clarity, the gene symbol followed by the superscript '+'. Except for instances in which a symbol has already been established for the normal allele, the primitive form of each species shall represent the + allele for each gene.

(4) A symbol shall not be assigned to a gene that is not firmly established by statistically valid segregation data.

(5) Mimics, i.e. different genes that condition similar phenotypes, may be assigned distinctive names and symbols or the symbol of the original gene followed by a hyphen and distinguishing Arabic numeral or Roman letter. The suffix '1' is used, or understood, for the original gene in a mimic series. Allelism tests should be made before a new gene symbol is assigned to a mimic.

(6) Multiple alleles are assigned the same symbol, followed by a Roman letter or Arabic number superscript. The allelism test must be made to establish multiple alleles.

(7) Indistinguishable alleles, i.e. alleles at the same locus that condition identical phenotypes, preferably should be given the same symbol. However, an allele that is an apparent recurrence of the same mutation can be given a distinctive symbol by adding to the symbol of the original allele

superscript Roman letters or Arabic numbers that are enclosed in parentheses. The superscript '(1)' is understood and not used for the original
allele.

(8) A modifying gene may be designated using a symbol for an
appropriate name, e.g. intensifier, suppressor, or inhibitor, followed by a
hyphen and the symbol of the allele affected. Alternatively, a modifying
gene may be given a distinctive name and symbol without reference to the
gene modified.

(9) In instances where the same symbol has been assigned to different
genes, or the same gene has been assigned different symbols, priority in
publication will be the primary criterion for establishing the preferred
symbol. Incorrectly assigned symbols will be enclosed in parentheses on
gene lists.

CYTOGENETICS AND INTERSPECIFIC HYBRIDIZATION

Several researchers have counted chromosomes in *V. unguiculata*, and the
preponderance of evidence indicates that $2n = 2x = 22$ (Faris, 1964;
Frahm-Leliveld, 1965; Yarnell, 1965). Mukherjee (1968) studied the
pachytene chromosomes and observed that the 11 bivalent complement
consisted of 1 short (19 μm), 7 medium (26–36 μm), and 3 long (41–45 μm)
chromosomes. He noted that the chromomeres were not distributed uniform-
ly along the arms.

Induced tetraploid forms of *V. unguiculata* have been studied by Sen and
Hari (1956) and Sen and Bhowal (1960b), who reported marked cultivar
differences in morphologic traits, pollen sterility, fruit setting, and meiotic
irregularities including the formation of quadrivalents, trivalents and univa-
lents and the presence of laggards and unequal distributions in the anaphase
stage. They concluded that pollen sterility was the factor most limiting to the
development of commercially useful tetraploid cultivars.

The cowpea has not been successfully hybridized with any other *Vigna* or
Phaseolus species. Singh *et al.* (1964) were unable to cross *V. unguiculata* with
V. umbellata (Thumb.) Ohwi & Ohashi. Evans (1976) failed in all efforts to
cross *V. unguiculata* with *V. vexillata* (L.) A. Rich., *V. mungo* (L.) Hepper
(black gram), *V. radiata* (L.) Wilczek (mung bean), *V. umbellata* (rice bean),
V. aconitifolia (Jacq.) Maréchal (moth bean), and *V. angularis* (Willd.) Ohwi
& Ohashi (adzuki bean). Attempts by Ballon and York (1959) to obtain
intergeneric crosses between *V. unguiculata* and *P. coccineus* L. (scarlet
runner bean) and *P. vulgaris* L. (common bean) were also unsuccessful.

HETEROSIS

A number of studies have shown that cowpea hybrids can exhibit considerable heterosis for seed yield (Hawthorne, 1943; Brittingham, 1950; Capinpin
and Irabagon, 1950; Agble, 1972; Singh and Jain, 1972; Ojomo, 1974;

Kheradnam *et al.*, 1975; Mak and Yap, 1977; Ojomo *et al.*, 1977; Bhaskaraiah *et al.*, 1980; Zaveri *et al.*, 1983). Ojomo (1974) speculated that the heterotic response for yield was due to the effect of dominance or epistasis. Kheradnam *et al.* (1975) concluded that the cowpea exhibits sufficient heterosis to justify commercial hybrid production if efficient emasculation and crossing procedures can be developed.

The cowpea also exhibits heterosis for a number of traits other than yield. Hofmann (1926) observed heterosis for plant height and stem diameter, Premsekar and Raman (1972) for branch length and both leaf width and length, and Bhaskaraiah *et al.* (1980) for pod length. Both Tikka *et al.* (1976) and Erskine and Khan (1978) reported heterosis for earliness.

FLOWERING

Photoperiod

The reproductive ontogeny of many cowpea cultivars is greatly influenced by photoperiod. Although photoperiod response is of profound interest to plant breeders developing cultivars suited for specific environments, this trait has not received much attention by geneticists. Sene (1967) concluded that photoperiod response was probably conditioned by a pair of major genes. He showed that the short-day response was dominant over the photoperiod-insensitive response. Plant breeders have been quite successful in the development of photoperiod-insensitive cultivars (Singh *et al.*, 1974).

Time of flowering

Early maturity is a relatively important agronomic characteristic. Typically, the trait is measured by such criteria as days to flowering or days to maturity. Some reports indicate that early maturity is dominant or partially dominant over late maturity (Brittingham, 1950; Mak and Yap, 1980; Ojomo, 1971; Roy and Richharia, 1948). Other reports, however, indicate that late maturity is dominant over early maturity (Mackie, 1939; Capinpin and Irabagon, 1950). At least two genes condition early flowering. Roy and Richharia (1948) concluded that two complementary genes probably condition the trait, and Ojomo (1971) thought that duplicate dominant epistasis between two major genes, symbolized *Ef-1* and *Ef-2*, in the presence of some minor modifying genes is responsible (Table 1). Brittingham (1950) speculated that maturity was inherited quantitatively.

A number of quantitative studies of the genetics of earliness parameters have been published (Table 2); the broad-sense heritability estimates averaged 48.3 per cent for days to flowering and 47.8 per cent for days to pod maturity. Several authors concluded that additive gene action is responsible

Table 1. Cowpea gene index

Preferred symbol	Synonym	Character	Reference
A		Alfalfa-like pod shape	Spillman and Sando, 1930
ax		Axillary buds; active buds in axils of cotyledons	Krishnaswamy et al., 1945
B		Blue seed coat	Spillman, 1912
$Bc\text{-}1$		Bacterial canker resistance-1	Singh and Patel, 1977; symbol by Fery, 1980
$bc\text{-}2$		Bacterial canker resistance-2	Singh and Patel, 1977; symbol by Fery, 1980
bcm		Blackeye cowpea mosaic virus resistance; likely a synonym of blc	Taiwo et al., 1981
Bcy		Brown calyx colour; dominant to green	Kolhe, 1970
Bf	(N)	Buff seed coat	Harland, 1919a,b, 1920; symbol by Fery, 1980
Bg		Brown grain; dominant to white	Kolhe, 1970
Bk	(B)	Black pod; dominant to white pod colour	Capinpin, 1935; symbol by Fery, 1980
Bl	$(B, B^b), P^b$	Black seed coat; also conditions anthocyanin production in the pod tip, calyx, and peduncle; heterozygote produces mottled seeds	Harland, 1919a,b, 1920; Franckowiak and Barker, 1974; symbol by Fery, 1980
Bl^e	(B^e)	New Era seed-coat pattern	Franckowiak and Barker, 1974; symbol by Fery, 1980
Bl^f	B^f	Blue seed coat (fine speckling pattern)	Franckowiak and Barker, 1974; symbol by Fery, 1980
Bl^g	(B^g)	Grey on black seed coat	Franckowiak and Barker, 1974; symbol by Fery, 1980
Bl^l	(B^l)	Light spots on black seed coat	Franckowiak and Barker, 1974; symbol by Fery, 1980
Bl^P	(B^P)	Purple seed coat	Franckowiak and Barker, 1974; symbol by Fery, 1980
Bl^s	(B^s)	Black spots on seed coat	Franckowiak and Barker, 1974; symbol by Fery, 1980
Bl^t	(B^t)	Taylor seed-coat pattern	Franckowiak and Barker, 1974; symbol by Fery, 1980

continued

Table 1 cont.

Preferred symbol	Synonym	Character	Reference
bl^x	(b^x)	Modifier of Bl^f	Franckowiak and Barker, 1974; symbol by Fery, 1980
blc		Blackeye cowpea mosaic virus resistance; likely a synonym of bcm	Walker and Chambliss, 1981
Bp	(Y)	Brown pod; dominant to straw colour	Saunders, 1960b; symbol by Fery, 1980
bpd*	(bp)	Branching peduncle	Fawole and Afolabi, 1983
Bpl-1*	(Bp-1)	Bacterial pustule resistance-1	Patel, 1982a
Bpl-2*	(Bp-2)	Bacterial pustule resistance-2	Patel, 1982a
bpl-3*	(bp-3)	Bacterial pustule resistance-3	Patel, 1982a
bpl-4*	(bp-4)	Bacterial pustule resistance-4	Patel, 1982a
bpl-5*	(bp-5)	Bacterial pustule resistance-5	Patel, 1982a
Br	(B)	Brown seed coat	Spillman, 1912; Spillman and Sando, 1930
Bs	(B)	Basic pigmentation gene for seed coat	Capinpin, 1935; symbol by Fery, 1980
Bu		Buff seed coat	Brittingham, 1950
by		Bean yellow mosaic virus resistance	Reeder *et al.*, 1972; symbol by Fery, 1980
C	(R)	Colour factor (general)	Spillman, 1912; Harland, 1919a,b, 1920; Spillman and Sando, 1930
Cbr		Cocoa brown pod (dry)	Krishnaswamy *et al.*, 1945
cc		Cowpea chlorotic mottle virus resistance; the recessive allele at the *Mvi* locus is likely a synonym	Rogers *et al.*, 1973; symbol by Fery, 1980
Ci	(C)	Compound inflorescence	Sen and Bhowal, 1961; symbol by Fery, 1980
Ce	(C)	Cerise (reddish) pod	Saunders, 1960a; symbol by Fery, 1980
Cls-1	(Cls_1)	Cercospora leaf spot resistance-1	Fery *et al.*, 1976; Fery and Dukes, 1977a; symbol by Fery, 1980
cls-2	(cls_2)	Cercospora leaf spot resistance-2	Fery *et al.*, 1976; Fery and Dukes, 1977a; symbol by Fery, 1980

continued

Preferred symbol	Synonym	Character	Reference
Cm	(*C*)	Cucumber mosaic virus resistance	Sinclair and Walker, 1955; symbol by Fery, 1980
Co		Thick seed coat	Ojomo, 1972
cp		Constricted petal	Rachie *et al.*, 1975
Cr	(*C*)	Colour-modifying gene for seed coat	Capinpin, 1935; symbol by Fery, 1980
crpt		Crumpled petal	Apparao and Reddy, 1976
Cy		Cylindrical-length pod	Krishnaswamy *et al.*, 1945
D	*w*	Dark flower colour	Harland, 1919a
Db	(*V*)	Dark brown seed coat	Franckowiak and Barker, 1974; symbol by Fery, 1980
De	(*D*)	Dense speckling; characteristic of New Era	Spillman and Sando, 1930; symbol by Fery, 1980
df		Dwarf (slow growth, dark green leaves, short internodes)	Jones, 1965
Dgd-a	*Dgda*	Dense grain deposition-a	Kolhe, 1970; symbol by Fery, 1980
Dgd-b	*Dgdb*	Dense grain deposition-b	Kolhe, 1970; symbol by Fery, 1980
Dt	(*G*)	Dotting; converts Holstein spots into numerous small ones	Spillman and Sando, 1930; symbol by Fery, 1980
E		New Era seed-coat pattern; also conditions anthocyanin production in the pod tip, calyx, and peduncle	Harland, 1920
Ef-1	*Ef₁*	Early flowering-1	Ojomo, 1971; symbol by Fery, 1980
Ef-2	*Ef₂*	Early flowering-2	Ojomo, 1971; symbol by Fery, 1980
Er	(*E*)	Erect pod attachment; dominant to drooping pod attachment	Singh and Jindla, 1971; symbol by Fery, 1980
F		Fine and dense speckling; gives rise to blue seed coat	Spillman and Sando, 1930
G		Tinged flower	Harland, 1920
Gn	(*G*)	Green pod; dominant to gn^y and recessive to Gn^d; conditions similar colour response in leaf, calyx and dorsal surface of standard	Sen and Bhowal, 1961; symbol by Fery, 1980

continued

Table 1 cont.

Preferred symbol	Synonym	Character	Reference
Gn^d	(G^D)	Dark green pod; dominant to Gn and gn^y; conditions similar colour in leaf, calyx and dorsal surface of standard	Sen and Bhowal, 1961; symbol by Fery, 1980
gn^y	(g)	Yellow green pod; recessive to Gn and Gn^d; conditions similar colour in leaf, calyx and dorsal surface of standard	Sen and Bhowal, 1961; symbol by Fery, 1980
Gnp	(G)	Green pod; dominant to white	Singh and Jindla, 1971; symbol by Fery, 1980
Gp		Green pod; dominant to cream pod	Kolhe, 1970
Gr		Green bud; dominant to white bud	Singh and Jindla, 1971
gt		Green testa; recessive to white seed coat	Chambliss, 1974
$h\text{-}1$	(H)	Holstein seed-coat pattern-1	Spillman, 1911; Sen and Bhowal, 1961
$h\text{-}2$	(H_2)	Holstein seed-coat pattern-2	Harland, 1919a; symbol by Fery, 1980
Ha		Hastate leaf	Ojomo, 1977
hs		Hollow seed	Krishnaswamy et al., 1945
$In\text{-}T$	(X)	Inhibitor of Taylor seed-coat pattern	Spillman and Sando, 1930; symbol by Fery, 1980
k		Drab pod colour	Mortensen and Brittingham, 1952
Kh		Khaki seed coat	Krishnaswamy et al., 1945
L		Pale flower	Harland, 1919a
La	(L_1)	Lanceolate leaf	Krishnaswamy et al., 1945; symbol by Fery, 1980
$Le\text{-}1$	(L_1)	Lethal-1; complementary to $Le\text{-}2$	Saunders, 1952, 1960c; symbol by Fery, 1980
$Le\text{-}2$	(L_2)	Lethal-2; complementary to $Le\text{-}1$	Saunders, 1952, 1960c; symbol by Fery, 1980
$Le\text{-}3$	(L_3)	Lethal-3; complementary to $Le\text{-}4$	Saunders, 1952, 1960c; symbol by Fery, 1980
$Le\text{-}4$	(L_4)	Lethal-4; complementary to $Le\text{-}3$	Saunders, 1952, 1960c; symbol by Fery, 1980
Lf	(L)	Longitudinal furrowing on seed coat	Spillman and Sando, 1930; symbol by Fery, 1980

continued

Preferred symbol	Synonym	Character	Reference
lg		Light-green pod	Krishnaswamy *et al.*, 1945
Lgf-1	(*Lga*)	Light-green foliage-1; complementary to *Lgf-2*	Kolhe, 1970; symbol by Fery, 1980
Lgf-2	(*Lgb*)	Light-green foliage-2; complementary to *Lgf-1*	Kolhe, 1970; symbol by Fery, 1980
Llf		Long leaf	Kolhe, 1970
ls		Leaf size; small leaf recessive to large leaf	Krishnaswamy *et al.*, 1945
lt		Loose testa	Krishnaswamy *et al.*, 1945
M		Maroon seed coat	Harland, 1919a,b, 1920
ms-1	*ms*	Male sterile-1	Sen and Bhowal, 1962
ms-2	ms_2	Male sterile-2	Rachie *et al.*, 1975b; symbol by Fery, 1980
*ms-3**	ms_3	Male sterile-3	Ladeinde *et al.*, 1980
*ms-4**	ms_4	Male sterile-4	Ladeinde *et al.*, 1980
Mv		Mottled seed-coat pattern	Franckowiak and Barker, 1974
*Mvi**	(*Mv*)	Movement of cowpea chlorotic mottle virus within the plant; recessive allele at this locus is likely a synonym of *cc*	Kuhn *et al.*, 1981
Mvn^1*	*MVN-1*	Necrotic reactions to cowpea mosaic virus; epistatic to *Mvs* and hypostatic to *mvs*; dominant over Mvn^2 and Mvn^3	Patel, 1982b
Mvn^2*	*MVN-2*	Necrotic reactions to cowpea mosaic virus; epistatic to *Mvs* and hypostatic to *mvs*; recessive to Mvn^1; probably recessive to Mvn^3	Patel, 1982b
Mvn^3*	*MVN-3*	Necrotic reactions to cowpea mosaic virus; epistatic to *Mvs* and hypostatic to *mvs*; recessive to Mvn^1; probably dominant to Mvn^2	Patel, 1982b
*Mvs**	*MVS*	Susceptible reaction to cowpea mosaic virus; the *mvs* allele controls the resistant reaction; epistatic/hypostatic relationship with the *Mvn* locus	Patel, 1982b
N		Anthocyanin pigment factor	Spillman, 1912; Spillman and Sando, 1930

continued

Table 1 cont.

Preferred symbol	Synonym	Character	Reference
Na	(E)	Narrow eye seed-coat pattern	Spillman and Sando, 1930; symbol by Fery, 1980
Nlf		Narrow leaf; dominant to broad leaf	Kolhe, 1970
Nv		Necrotic synergistic reaction associated with cowpea stunt	Pio-Ribeiro *et al.*, 1980
o		Hilum ring seed-coat pattern	Saunders, 1960a
P		Purple pod; dominant to green; also causes anthocyanin production in the calyx and peduncle	Harland, 1920
P^b	Bl	Red-tip pod	Mortensen and Brittingham, 1952
p^g		Purple pod with green sutures, partly coloured stem, and petioles	Sen and Bhowal, 1961
p^p	P	Purple pod	Mortensen and Brittingham, 1952
p^t		Purple-tip pod and stems with purple spots at the internodes	Sen and Bhowal, 1961
p^v		Green pod with purple ventral suture; also conditions the spread of purple on the calyx	Sen and Bhowal, 1961
$pa\text{-}1$	pa_1	Pod appearance-1; wrinkled dry pod recessive to smooth appearance	Krishnaswamy *et al.*, 1945; symbol by Fery, 1980
$pa\text{-}2$	pa_2	Pod appearance-2; wrinkled dry pod recessive to smooth appearance	Krishnaswamy *et al.*, 1945; symbol by Fery, 1980
Pb		Purple petiole base	Sen and Bhowal, 1961
Pbr		Purple branch base	Sen and Bhowal, 1961
Pc^*		Perpendicular macroscisereid arrangement in the smooth seed coat; dominant to parallel arrangement (rough seed coat)	Rajendra *et al.*, 1979
$Pcot$		Purple cotyledon; dominant to white cotyledon	Krishnaswamy *et al.*, 1945
Pf		Purple flower	Kolhe, 1970
pg	(g)	Pale-green plant	Saunders, 1960a; symbol by Fery, 1980

continued

Preferred symbol	Synonym	Character	Reference
Pn		Peduncle length; long peduncles dominant to short	Krishnaswamy *et al.*, 1945
Pp-1	*(P$_1$)*	Purple plant pigmentation-1	Venugopal and Goud, 1977; symbol by Fery, 1980
Pp-2	*(P$_2$)*	Purple plant pigmentation-2	Venugopal and Goud, 1977; symbol by Fery, 1980
Pr	*(P)*	Purple seed coat	Spillman and Sando, 1930; symbol by Fery, 1980
Ps		Purple sutures on green pod; pod tip is purple and small purple patches scattered over pod surface	Sen and Bhowal, 1961
Pu		Purple pod; stem and petiole are completely purple	Sen and Bhowal, 1961
R		Red seed coat	Spillman, 1912
rh	*(Rh)*	Beetle resistance	Kolhe, 1970; symbol by Fery, 1980
Rk		Root-knot resistance; allelic to *rk^i*	Fery and Dukes, 1980, 1982
rk^i		Root-knot resistance — intermediate; allelic to *Rk*	Fery and Dukes, 1982
Rs	*(Lg)*	Reniform-shape seed	Kolhe, 1970; symbol by Fery, 1980
S		Spotting pattern; patches of black pigment on certain types of seed coat	Spillman and Sando, 1930
Sbm		Southern bean mosaic virus resistance	Brantley and Kuhn, 1970; symbol by Fery, 1980
sh	*(s)*	Spindly growth habit; marked elongation of the main stem and few side branches	Saunders, 1960b; symbol by Fery, 1980
shp		Shrunken pericarp	Premsekar and Raman, 1972; symbol by Fery, 1980
Sk	*(S)*	Speckled pod	Saunders, 1960a; symbol by Fery, 1980
Sm		Smoky seed-coat pattern	Krishnaswamy *et al.*, 1945
Sp		*sesquipedalis*-length pod	Krishnaswamy *et al.*, 1945
Spk	*(S)*	Speckled seed coat	Saunders, 1959; symbol by Fery, 1980

continued

Table 1 cont.

Preferred symbol	Synonym	Character	Reference
Sr		Stem-rot resistance	Purss, 1958; symbol by Fery, 1980
St		Standard petal exhibits full expression of colour	Krishnaswamy *et al.*, 1945
Stp	*(E)*	Stippling seed-coat pattern; characteristic of New Era	Saunders, 1959; symbol by Fery, 1980
stx		*sesquipedalis*-like texture of pod (soft)	Premsekar and Raman, 1972; symbol by Fery, 1980
Sw		Swollen stem base	Sen and Bhowal, 1961; Ojomo, 1977
Sy-1	*(I_1)*	Straw-yellow pod (dry)-1	Krishnaswamy *et al.*, 1945; symbol by Fery, 1980
Sy-2	*(I_2)*	Straw-yellow pod (dry)-2	Krishnaswamy *et al.*, 1945; symbol by Fery, 1980
T		Taylor seed-coat pattern; thinly scattered speckling of bluish-purple dots	Spillman and Sando, 1930
Th		Thick seed coat	Ojomo, 1972
Tr	*(T)*	Tobacco ringspot virus resistance	DeZeeuw and Ballard, 1959; DeZeeuw and Crum, 1963; symbol by Fery, 1980
U		Buff seed coat	Spillman, 1912; Spillman and Sando, 1930
un		Unifoliate leaf; petiole, all stipellae, and the two lateral leaflets with their petioles are missing	Rawal *et al.*, 1976
V		Seed-coat mottling; characteristic of Brabham	Saunders, 1959
Vi-1	*(T, V)*	Vining-1	Brittingham, 1950; Norton, 1961; Kolhe, 1970; Singh and Jindla, 1971; symbol by Fery, 1980
Vi-2	*(V, V_2, T_2)*	Vining-2	Norton, 1961; Kolhe, 1970; Singh and Jindla, 1971; symbol by Fery, 1980
Vi-3	*(T_3)*	Vining-3	Singh and Jindla, 1971; symbol by Fery, 1980
Vw		Verticillium wilt resistance	Moore, 1974; symbol by Fery, 1980

continued

Preferred symbol	Synonym	Character	Reference
w	*I, W, D*	Watson seed-coat pattern; eye with indefinite margin	Spillman, 1911; Harland, 1919a; Spillman and Sando, 1930; Sen and Bhowal, 1961
Wh	(*W*)	Whippoorwill-type of seed-coat pattern	Spillman and Sando, 1930; symbol by Fery, 1980
Wp-a	*Wpa*	Wrinkled pod-a	Kolhe, 1970; symbol by Fery, 1980
Wp-b	*Wpb*	Wrinkled pod-b	Kolhe, 1970; symbol by Fery, 1980
X		Anthocyanin coloration in the vegetative parts	Harland, 1919b, 1920
xn		Xantha seedling (chlorophyll-deficient); homozygous lethal	Krishnaswamy *et al.*, 1945
y		Very small eye seed-coat pattern	Harland, 1922
Ymr		Cowpea yellow mosaic virus resistance; conditions resistance to cowpea mosaic virus	Bliss and Robertson, 1971
Ystp		Yellow strip on petals	Kolhe, 1970

* Proposed new symbol.

for much of the genetic variation for earliness (Kheradnam and Niknejad, 1971; Lal *et al.*, 1976; Tikka *et al.*, 1976; Mak and Yap, 1977; Zaveri *et al.*, 1980). Other reports, however, indicate that action by non-additive genes and interactions between genotype and environment are important in some instances. Ojomo (1974) reported that specific combining ability and not general combining ability is significant, and he speculated that much of the genetic variation for days to flowering is due to dominance or epistasis. Tikka *et al.* (1977) and Trehan *et al.* (1970) observed that the high heritability estimates for earliness are associated with low estimates for genetic advance. This negative association, they pointed out, suggests that earliness is conditioned by effects other than additive gene action.

Compound inflorescence

Sen and Bhowal (1961) reported that the compound inflorescence is monogenically recessive to the normal simple inflorescence. They proposed

Table 2.   Estimates of broad-sense heritability (%) for earliness (days to flowering and days to pod maturity)

Reference	Heritability estimate (%)	
	Days to flowering	Days to pod maturity
Bapna and Joshi, 1973	94.3	89.5
	84.8	50.3
Bordia et al., 1973	60.5	56.3
Erskine and Khan, 1978	3.8	23.0
	14.9	13.6
	0.0	0.0
	17.0	0.0
	0.0	41.9
Kohli et al., 1971	66.8	—
	34.4	—
	26.5	—
	31.6	—
	25.6	—
Kumar and Mishra, 1981	88.7	—
Singh and Mehndiratta, 1969	88.8	78.3
Sohoo et al., 1971	91.9	87.3
	59.4	73.3
	91.4	89.2
Tikka et al., 1976	38.8	—
	47.1	—
Tikka et al., 1977	95.1	—
Trehan et al., 1970	4.4	19.6
Zaveri et al., 1980	44.9	46.8

the symbol C for this gene, but as C is the symbol for the general colour factor, I have proposed Ci for the symbol (Fery, 1980).

OUTCROSSING

The development of population-improvement schemes for self-pollinated crops and the possibility of hybrid seed production have stimulated the search for male sterility and other outcrossing mechanisms to facilitate hybridization procedures in cowpea. Four recessive genes causing genetic male sterility have been reported. The first, $ms-1$, was described by Sen and Bhowal (1962), the second, *ms-2*, by Rachie *et al.* (1975b), and the third and fourth, *ms-3* and *ms-4*, by Ladeinde *et al.* (1980). Ladeinde *et al.* (1980) tested the *ms-2, ms-3* and *ms-4* genes for allelism and concluded that the genes are located at separate unlinked loci. Rachie *et al.* (1975b) described a type of mechanical male sterility involving abnormal petals that restrict stamen development yet allow the stigma and style to emerge at an early, prereceptive stage of development. They found that the trait is conditioned by a single recessive gene, designated *constricted petal* and symbolized *cp*. They considered the *cp* gene to be less promising than the *ms-2* gene because of extremely poor fruit set. Apparao and Reddy (1976) reported a second type of mechanical male sterility, *crumpled petal*, conditioned by the recessive gene *crpt*. Self-fertilization is restricted in crumpled-petal plants because the anthers are enclosed in the petals.

PIGMENTATION

In 1914 Mann showed that two sap-soluble pigments, anthocyanin and a melanin-like substance, are responsible for colour in the cowpea. The anthocyanins produce shades of purple and rose in acidic cell sap and shades of blue and black in alkaline sap. Anthocyanins are responsible for all the colour in the flower petals, seed pods, peduncles, petioles, stems and leaves. The melanin-like pigment is found only in the seed coat and is responsible for a pale yellow to deep copper-red basal colour. The pigment is always present in the third cell layer and often in the palisade (outer cell) layer of all coloured seed coats. However, the amount and the location of the pigment vary. Palisade cells of some coloured seed coats also contain anthocyanins, and the colours of various plant parts are correlated. A general colour factor must be present before any of several pigment genes can be expressed, and the interactions of the genes are complex.

Seed coat

Spillman (1912) made the first attempt to explain the inheritance of seed-coat colour. He assumed that a general colour factor, designated *C*, is needed for colour expression and, in combination with the genes *R, U, Br, Br* and *N*, and *N* and *B*, conditioned red, buff, brown, black, and blue seed coats, respectively. Later, Harland (1919a,b, 1920) offered a similar model, with *R* functioning as a general colour factor and conditioning red in the seed coat. The genes *B, N, M* and *N* and *M* in the presence of *R* condition black, buff, maroon and brown, respectively. In a comprehensive review published in 1930, Spillman and Sando redesignated Spillman's (1912) general colour factor as *R*, described *N* as an anthocyanin pigment factor, and used the symbols *B, F, P* and *U* for brown, fine and dense speckling, purple and buff, respectively. They presented a revised model to show how the *R, N, B, F, P* and *U* genes interact to produce 10 different seed colours.

However, as Spillman (1912) had used *R* for red seed coat, the original designation, *C*, seems more appropriate (Fery, 1980) for the general colour factor. Also, Harland's (1919a,b, 1920) symbols *B* and *N* had been used previously by Spillman (1912) and could be redesignated *Bl* and *Bf*, respectively (Fery, 1980). Likewise, Spillman's (1912) original designation for brown seed coat, *Br*, is more appropriate than the *B* in the later model, and the gene conditioning purple seed coat could be redesignated *Pr* (Fery, 1980).

To me (Fery, 1980), Harland's (1919a,b, 1920) model is compatible with the more comprehensive model of Spillman and Sando (1930) (Table 3), Harland's (1919a,b, 1920) *Bf* and *M* genes probably corresponding to Spillman and Sando's (1930) *U* and *Br* genes, respectively. Harland's (1919a,b, 1920) *Bl* gene functions much as Spillman and Sando's (1930) *N* gene. However, Harland's (1919a) finding that the *Bl* gene does not condition anthocyanin production in the flowers, stems and leaves suggests that *Bl* is not the anthocyanin pigment factor *N*.

Later, several other studies on the genetics of seed colour were published, but I believe (Fery, 1980) that many of the genes identified in these studies are the same as the genes in the Harland (1920) or Spillman and Sando (1930)

Table 3. Interaction of Spillman and Sando's (1930) pigment genes for the production of colour in cowpea seed coats

Colour	Pigment gene*		
	Necessary	Permissible	Not permissible
Anthocyanin			
Purple	*C, N, Pr*	*Br, U*	
Black	*C, N, Br, U*		*Pr*
Dull black	*C, N, Br*		*Pr, U*
Blue	*C, N, F, U*		*Pr*
Red	*C, N*		*Pr, Br, U*
Melanin-like			
Coffee	*C, Br, U*	*Pr*	*N*
Maroon	*C, Br*	*Pr*	*N, U*
Buff or clay	*C, U*	*N* or *Pr*	*N* and *Pr, Br*
Pink	*C*	*Pr*	*N, Br, U*
White or cream		*N, Br, Pr, U*	*C*

* See cowpea gene index (Table 1).

model. Capinpin (1935) described the gene B as a basic pigmentation gene and the gene C as a colour-modifying gene. Because both these symbols had been assigned previously, I proposed (Fery, 1980) that Capinpin's symbols be redesignated Bs and Cr, respectively. Krishnaswamy *et al*. (1945) proposed the symbol Kh for khaki colour, Brittingham (1950) proposed Bu for buff, and Kolhe (1970) proposed the Bg for brown grain. Chambliss (1974) identified a recessive gene that conditions a green seed coat, and he proposed that the gene be symbolized gt. Louis and Sundaram (1975) induced a single gene mutation that inhibited the production of anthocyanin pigment in the seed coat of a blackeye cultivar. Saunders (1959) successfully used Harland's (1919a, 1920) model to explain an extensive set of data on seed-coat colour.

Franckowiak and Barker (1974) completed an extensive inheritance study of seed-coat colour but published their results in abstract form only. They assigned the gene symbols B^b, B^g, B^p, V, N, and M for black, black with grey, purple, dark brown, buff, and maroon, respectively. The genes B, N, and M are probably those of Harland (1919a,b, 1920) and, thus, the B and N genes could be redesignated Bl and Bf, respectively (Fery, 1980), with Bl rather than Bl^b being used for black. The genes Bl, Bl^g and Bl^p are allelic. Since the symbol V had been used previously, I (Fery, 1980) felt that the dark brown gene would be better symbolized Db. The gene Bf is dominant and inhibits red pigmentation; the gene M conditions maroon only in $Bf^+ BF^+$ genotypes.

Although the genetics of seed-coat pigmentation appear to be generally worked out, the complete mechanism is not yet fully understood. Capinpin and Irabagon (1950), for example, reported that the drab, pale purple is a simple, dominant over dark Corinthian purple. Although either of Spillman and Sando's (1930) permissible factors for purple, Br or U, may account for variation in the intensity of the colour, the exact identification of the gene described by Capinpin and Irabagon (1950) cannot be made by use of either the Harland (1920) or the Spillman and Sando (1930) model.

Flower

Anthocyanin is responsible for the purple or violet flowers that are characteristic of many cowpea cultivars. There are four principal flower colours: dark, pale, tinged, and white. Dark flowers contain a high concentration of anthocyanin in all of the principal flower parts, pale flowers contain small amounts in the wings, tinged flowers have a faint narrow band of pigmentation along the outer edge of the standard, and white flowers are completely devoid of anthocyanin. Flower colour is associated with seed-coat colour and pattern (Spillman, 1913; Harland, 1919a; Saunders, 1960a): dark flowers are characteristic of cultivars with self-coloured or Watson-eye seeds; pale flowers with all eye and Holstein seed-coat patterns; tinged flowers with the hilum ring seed-coat pattern; and white flowers with white or cream seed coats.

As originally observed by Spillman (1912, 1913), the presence of pigment in the flower is dependent upon the general colour factor C and the anthocyanin factor N. Harland (1919a) noted that the presence of anthocyanin in flower tissue is the result of an interaction of the two dominant factors L and D. The L factor conditions the pale colour; the D factor increases the intensity of the colour but has no effect except in the presence of L. Later, Harland (1920) reported that the tinged colour is governed by a single dominant gene: G. Interestingly, he noted that G produces tinged flowers in white-seeded plants. Harland (1919a) speculated that the dark flowers and the Watson seed-coat pattern are controlled by the same factor. Yarnell (1965) concluded that Harland's (1919a) L factor is probably the general colour factor C. Sen and Bhowal (1961) have published evidence that the tinged colour is a secondary characteristic of the h factor for the Holstein seed-coat pattern. As pointed out by Saunders (1960a), white flowers may result when either the general colour factor C or the complementary genes are absent. Additionally, Saunders (1960a) noted that the dark colour is epistatic to both the pale and tinged colours.

Several modifier genes for flower colour have been reported. In 1945 Krishnaswamy et al. described a spreading factor, designated St, responsible for the full expression of colour on the dorsal surfaces of the standard petal. In 1970 Kolhe reported that yellow stripes on the dorsal surface of the standard are dominant over yellow, and he proposed the symbol $Ystp$ for the gene that conditions the trait. Kolhe (1970) also proposed the symbol Pf for a dominant gene responsible for purple flowers. However, in my opinion (Fery, 1980) the Pf gene is the general colour factor C or one of the complementary genes for seed-coat colour or pattern.

Pod

The pods of many cowpea cultivars contain anthocyanin and are either partially or wholly purple. Harland (1920) found that wholly purple pods are dominant to green pods, and he proposed the symbol P for a single gene that conditions the trait. He also observed that both the Bl gene for black seed coat and the E gene for the New Era seed-coat pattern condition purple pod tips; he speculated that both factors are allelic to P. Later, Mortensen and Brittingham (1952) showed that the P gene is allelic to the Bl gene and they redesignated the genes P^P and P^b, respectively. They noted that the P^P allele is dominant to the P^b allele in pod colour but recessive to it in seed-coat colour. They also speculated that it is impossible to obtain a true-breeding plant with both purple pod and black seed coat because such a plant would be

THE GENETICS OF COWPEAS

heterozygous at the P locus. Sen and Bhowal (1961) described three additional alleles at the P locus, p^t, p^v and p^g, and identified two additional pod-pigmentation genes, Ps and Pu. The p^t, p^v and p^g alleles govern green pod with purple tip, green pod with purple ventral suture, and purple pod with green sutures, respectively; except for a reversal of dominance in the p^v p^g heterozygote in unripe and semiripe pods, dominance was in accordance with increasing pigmentation. The Ps gene conditioned green pods with purple sutures, scattered small purple patches, and purple tips. The Pu gene conditioned wholly purple pods, and the Pu^+ allele conditioned green pods with faintly purple sutures and tips.

Capinpin (1935) found that a single dominant gene conditions black pods over white, and he proposed that the gene be symbolized B. However, I (Fery, 1980) redesignated the gene Bk because the symbol B had been assigned previously. Mortensen and Brittingham (1952) proposed the symbol k for a recessive gene that conditions drab pods. Saunders (1960a) proposed the symbol C for a gene that conditions cerise pods in the absence of gene P, the symbol S for a gene that conditions speckled pods, and the symbol Y for a gene that conditions brown or drab pods over the straw colour. Since the C, S and Y symbols had already been assigned previously, I (Fery, 1980) proposed that Saunders' (1960a) C, S and Y genes be redesignated Ce, Sk and Bp, respectively. Hare (1956) studied yellow pod colouring and found that the trait is monogenically dominant to both the golden and red pod colours.

Krishnaswamy et al. (1945) demonstrated that two complementary factors, designated I_1 and I_2, condition the straw yellow of dry pods. They also showed that a third factor, designated Cbr, converts the straw yellow to a cocoa brown. Since the symbol I had been used previously, I (Fery, 1980) redesignated I_1 and I_2 as $Sy-1$ and $Sy-2$ respectively. The homozygous recessive condition at one or both of the $Sy-1$ or $Sy-2$ loci results in ivory yellow pods. In 1961 Sen and Bhowal reported that the amber-straw and the brownish-straw colours of dry pods are monogenically dominant over the straw colour. Several published studies have addressed the genetics of colour in the immature pods: both Krishnaswamy et al. (1945) and Premsekar and Raman (1972) reported that light green is recessive to green and conditioned by a single gene. Krishnaswamy et al. (1945) proposed the symbol lg for their gene. Sen and Bhowal (1961) proposed the symbols G^D, G and g for an allelic series that conditions dark green, green, and yellow-green pods, respectively. As the symbol G had been used by Harland (1920) for tinged flower colour, I (Fery, 1980) redesignated Sen and Bhowal's (1961) G^D, G, g genes Gn^d, Gn and gn^y, respectively. Two studies, one by Kolhe (1970) and the other by Singh and Jindla (1971), reported that a single gene conditions green pods over cream or white pods, Kolhe (1970) symbolizing the gene as Gp and Singh and Jindla (1971) symbolizing it as G. In 1980, I redesignated the gene as Gnp.

Foliage

When the general colour factor C and the anthocyanin factor N are present, green plant organs other than pods can contain anthocyanin (Spillman, 1912). Harland (1919b, 1920) assigned the symbol X to a gene that conditions anthocyanin production in vegetative parts. He noted that the genes P, Bl, and E can cause anthocyanin development in the calyx and peduncle. Sen and Bhowal (1961) observed that the gene Pu conditions wholly purple stem and petiole, p^g conditions a partly coloured stem having reddish-purple patches on the nodes and internodes and purple streaks on the petioles, p^t conditions stems with purple spots at the internodes, Pb conditions purple petiole base, Pbr conditions purple branch base, and p^v conditions the spread of purple on the calyx. Venugopal and Goud (1977) reported that two independent genes govern pigmentation in the nodal regions of the main stem and the bases of tertiary branches, peduncles, and stalks of trifoliate leaves. They proposed the symbols P_1 and P_2 for these genes, but I (Fery, 1980) redesignated them Pp-1 and Pp-2. Krishnaswamy et al. (1945) found that the gene, $Pcot$, for purple cotyledon, is monogenically dominant over that for white cotyledon.

Saunders (1960a) reported that plants with pale-green foliage have a noticeable deficiency in chlorophyll, and he proposed the symbol g for the recessive gene that conditions this trait. As the symbol g had been used previously, I (Fery, 1980) redesignated Saunders' (1960a) g gene as pg. Kolhe (1970) found that light-green foliage is dominant to green foliage and is conditioned by two complementary genes that he designated Lg-a and Lg-b. Since Krishnaswamy et al. (1945) had already used the symbol lg for light-green pod, I (Fery, 1980) proposed that Kolhe's (1970) Lg-a and Lg-b be redesignated Lgf-1 and Lgf-2, respectively. Sen and Bhowal (1961) reported that the allelic series Gn^d, Gn, and gn^y for dark green, green, and yellow-green pod colours, respectively, condition similar colour responses in the leaf, calyx, and dorsal surface of the standard. Singh and Jindla (1971) assigned the symbol Gr to a gene that conditions green buds over white. Premsekar and Raman (1972) reported that dark-green calyx is dominant over pale green and the segregation pattern observed in an F_2 population suggests monogenic inheritance. Kolhe (1970) assigned the symbol Bcy to a gene that conditions brown calyx over green.

PLANT HABIT

Brittingham (1950), Premsekar and Raman (1972), and Krutman et al. (1975) reported that a single dominant gene conditions the climbing characteristic in cowpea, and Brittingham (1950) proposed that this gene be designated T (Table 4). Later, Norton (1961) and Kolhe (1970) both studied the indeterminate or vining characteristic, and each concluded that the trait is governed by two duplicate genes. Norton (1961) assumed that one of these genes was the T gene described by Brittingham (1950), and he proposed that the second

Table 4. Estimates of broad-sense heritability (%) for growth-habit traits

Reference	Heritability (%)						
	Plant height	Branches/ plant	Main branch length	Leaves/ plant	Stem girth	Vine length	Peduncle length
Angadi et al., 1978	—	68.4	—	—	—	—	—
Bapna and Joshi, 1973	45.3	48.8	—	—	—	—	—
	29.6	38.8	—	—	—	—	—
Erskine and Khan, 1978	51.0	—	—	—	—	—	—
	13.9	—	—	—	—	—	—
	5.9	—	—	—	—	—	—
	41.8	—	—	—	—	—	—
	49.8	—	—	—	—	—	—
Kheradnam and Niknejad, 1974	—	15.0	—	—	—	—	—
Kohli et al., 1971	34.6	44.4	58.6	63.7	48.4	—	—
	51.3	20.1	47.7	30.2	49.6	—	—
	42.8	43.7	43.3	28.9	51.6	—	—
	29.6	6.0	38.5	24.7	43.3	—	—
	17.5	33.5	41.0	33.7	33.6	—	—
Kumar and Mishra, 1981	15.5	—	—	—	—	—	—
Lakshmi and Goud, 1977	97.2	—	—	—	—	—	—
Sohoo et al., 1971	58.3	61.3	—	—	—	82.0	—
	72.5	94.4	—	—	—	94.9	—
	67.8	74.2	—	—	—	77.0	—
Tikka et al., 1977	66.5	40.3	—	—	—	—	—
Trehan et al., 1970	24.2	24.6	—	—	—	—	22.8
Veeraswamy et al., 1973	57.9	61.3	—	—	—	—	—

gene be designated *V*. Kolhe (1970) concluded that his results were in agreement with Norton's (1961) and he proposed that the genes be designated *V–1* and *V–2*. Since the symbol *T* had been used previously by Spillman and Sando (1930) and the symbol *V* had been used by Saunders (1959), I (Fery, 1980) proposed that the two vining genes be redesignated *Vi–1* and *Vi–2*.

In 1971 Singh and Jindla reported that the trailing growth habit is conditioned by two complementary genes and a third gene that is expressed only when both complementary genes are homozygous recessive. They proposed the symbols *T–1*, *T–2*, and *T–3*, but *T* had been assigned earlier and *T–1* and *T–2* were probably redesignations for *Vi–1* and *Vi–2*, respectively, so I (Fery, 1980) proposed that the *T–3* gene be redesignated *Vi–3*.

In recent years several researchers have used quantitative procedures to study the genetics of growth-habit traits such as plant height, branch number, main branch length, leaf number, stem girth, vine length, and peduncle length, and more than 50 heritability estimates have been published. Although there is considerable variation between studies, the results indicate that most growth-habit traits are at least moderately heritable. For example, the average heritability estimates for plant height and branch number are 43.6 per cent and 45.0 per cent, respectively.

Two genes that govern pod position within the plant canopy have been reported. Krishnaswamy *et al.* (1945) assigned the symbol *Pn* to a dominant gene that conditions long peduncles. Singh and Jindla (1971) found that erect pod attachement is dominant to drooping pod attachement and is governed by a single gene that they designated *E*. However, as the symbol *E* had been used previously by Harland (1920), I proposed (Fery, 1980) that the erect pod gene be redesignated *Er*.

Fawole and Afolabi (1983) have recently described a recessive gene that conditions branched peduncles capable of producing 2–10 pods/peduncle. The symbol they proposed for this gene, *bp*, had been used previously (Fery, 1980), so I propose that the branching peduncle gene be symbolized *bpd*.

DELETERIOUS GROWTH MUTANTS AND LETHAL FACTORS

In 1945 Krishnaswamy *et al.* identified a recessive gene, symbolized *ax*, that activates the usually dormant buds in the axils of cotyledonary leaves; plants homozygous for the *ax* gene exhibit axillary branches not seen in normal plants. Saunders (1960b) reported that spindly growth habit is governed by a single recessive gene, which I (Fery, 1980) proposed be symbolized *sh*. Sharma (1969a,b) postulated that a wide-range mutator gene is responsible for a 'late giant' mutant that he was able to induce repeatedly with chemical mutagens. Jones (1965) reported that a single recessive gene, *df*, conditions a dwarf mutant that he induced. Swollen stem base is dominant to normal stem base and is conditioned by the gene *Sw* (Sen and Bhowal, 1961; Ojomo, 1977).

Krishnaswamy *et al.* (1945) reported the first lethal gene in cowpea. They described the gene *xantha seedling*, *xn*, as a homozygous lethal. In 1952 Saunders reported that two lethal conditions were each controlled by two complementary genes. He found that when both genes are present, a delayed lethal reaction is observed in the F_1 seedling. He proposed the symbols L_1 and L_2 for the genes controlling the first lethal condition and the symbols L_3 and L_4 for those controlling the second lethal condition. However, L was first used by Harland (1919a), and, to avoid confusion, I proposed (Fery, 1980) that the four genes be redesignated *Le–1*, *Le–2*, *Le–3* and *Le–4*, respectively. Saunders (1960c) later reported two additional lethal conditions but speculated that both were modifier gene-induced variants of the *Le–3/Le–4* lethal condition.

LEAF TRAITS

Six different genes have been reported to condition leaf size or shape in cowpea. Small leaf, *ls*, is recessive to large leaf (Krishnaswamy *et al.*, 1945); long leaf, *Llf*, is dominant to short leaf (Kolhe, 1970); and the unifoliate leaf, *un*, is recessive to the trifoliate leaf (Rawal *et al.*, 1976). Krishnaswamy *et al.* (1945) proposed the symbol L for a dominant gene that conditions the lanceolate leaf shape, but rather than use this symbol, which had been assigned previously by Harland (1919a), I (Fery, 1980) proposed *La*. Both Saunders (1960b) and Kolhe (1970) reported that the narrow leaf shape is conditioned by a single dominant or partially dominant gene, the latter proposing the symbol *Nlf* for the gene. Jindla and Singh (1970) and Ojomo (1977) studied the hastate leaf shape and decided that the trait is inherited in a dominant fashion. However, Jindla and Singh (1970) concluded that the hastate leaf is conditioned by four genes, and Ojomo (1977) concluded that only one gene was involved. Jindla and Singh's (1970) theory must be considered preliminary because they examined a single F_2 population only. Ojomo (1977) proposed that the hastate leaf gene be symbolized *Ha*.

POD TRAITS

Size

Several studies indicate that pod length is moderately to highly heritable; published broad-sense heritability estimates average 75.2 per cent (Table 5). Additive gene action is important, and high heritability is often accompanied by high genetic advance (Veeraswamy *et al.*, 1973; Tikka *et al.*, 1977). However, non-additive gene action can be important in some instances (Singh and Jain, 1972). Long pod is usually dominant or partially dominant over short pod (Fennell, 1948; Roy and Richharia, 1948; Brittingham, 1950; Capinpin and Irabagon, 1950; Jindla and Singh, 1970), but there have been

Table 5.　Estimates of broad-sense heritability (%) for cowpea pod and seed traits

Reference	Heritability (%)					
	Pod length	Pod girth	Pod breadth	Seeds/pod	100-seed wt.	Testa wt.
Angadi et al., 1978	—	—	—	—	98.9	—
Aryeetey and Laing, 1973	60.7	—	—	47.8	63.0	—
	60.3*	—	—	37.8*	37.4*	—
Bapna and Joshi, 1973	—	—	—	78.9	80.6	—
	—	—	—	52.2	85.3	—
Bhowal, 1976	76.9	—	80.5	—	84.1	—
Bliss et al., 1973	—	—	—	53.0	84.0	—
Bordia et al., 1973	59.8	—	—	50.6	75.4	—
Chandrappa et al., 1974	—	95.3	—	94.0	96.6	—
Erskine and Khan, 1978	7.3	—	—	22.1	9.3	—
	20.9	—	—	23.1	50.4	—
	4.8	—	—	44.8	35.7	—
	33.7	—	—	23.9	54.4	—
	0.0	—	—	9.6	3.8	—

Reference							
Kheradnam and Niknejad, 1974	—	—	—	64.0	—	75.0	—
Lakshmi and Goud, 1977	—	95.4	—	70.1	—	93.8	—
Leleji, 1975	—	54.4†	—	41.7‡	—	68.3§	—
Singh and Mehndiratta, 1969	—	80.5	—	64.2	—	95.9	—
Sohoo et al., 1971	—	94.8	—	74.3	—	97.3	—
	—	75.9	—	98.9	—	71.8	—
	—	87.0	—	68.4	—	62.0	—
Tikka et al., 1977	—	95.0	—	86.3	—	—	96.0
Trehan et al., 1970	—	24.6	—	23.5	—	—	—
Veeraswamy et al., 1973	—	97.5	—	33.3	—	—	—

* Narrow-sense heritability estimate.
† Average of five crosses.
‡ Average of six crosses.
§ Average of seven crosses.

some reports of short pod being partially dominant (Leleji, 1975; Bhowal, 1976).

Although Aryeetey and Laing (1973) estimated that 0.44 effective factors condition pod length, most published reports indicate that at least two genes are involved, including that of Krishnaswamy *et al.* (1945) who used a cross between subspecies *sesquipedalis* and subspecies *cylindrica* to study pod length. They proposed the symbol *Sp* for a gene that conditions long pods (like those of subspecies *sesquipedalis*) and the symbol *Cy* for a gene that conditions short pods (like those of subspecies *cylindrica*). Brittingham (1950), Capinpin and Irabagon (1950), and Bhowal (1976) suggested that eight, two, and six genes respectively, condition pod length. Data published by Fennell (1948), Roy and Richharia (1948), and Jindla and Singh (1970) indicate that pod length is conditioned by multiple factors.

Bhowal (1976) observed that narrow pod is partially dominant over broad pod. He calculated a heritability estimate of 80.5 per cent for pod breadth and concluded that 10 genes were involved. Chandrappa *et al.* (1974) reported a heritability estimate of 95.3 per cent for pod girth.

Seed number and spacing

The number of seeds per pod is moderately heritable under most environmental conditions; published heritability estimates average 52.8 per cent. Non-additive gene action accounts for much of the variation. Singh and Jain (1972), for example, demonstrated that specific combining ability was important but that general combining ability was not. Also, Tikka *et al.* (1977) observed that the high heritability was accompanied by a comparatively low genetic advance. Kheradnam and Niknejad (1971), however, reported that both general and specific combining abilities were important. Leleji (1975) reported that a small number per pod is partially dominant over a larger number. Aryeetey and Laing (1973) reported that a single effective factor conditions seed number per pod.

Fennell (1948) observed that normal seed spacing within the pod is dominant over the wide spacing that is characteristic of the yard-long bean, and he concluded that the trait is multigenic. Kolhe (1970) reported that dense grain deposition is dominant to thin grain deposition, but he concluded that the trait is governed by two complementary genes, symbolized *Dgd–a* and *Dgd–b*.

Pod shape and texture

Spillman and Sando (1930) reported that the alfalfa-like curled pod is semidominant over the normal pod shape. They concluded that a single gene conditioned the alfalfa-like shape and they proposed that the gene be symbolized *A*.

Several inheritance studies of pod texture in crosses of subspecies *unguiculata* × subspecies *sesquipedalis* have been reported. Krishnaswamy *et al.* (1945) reported that the wrinkled appearance that is typical of dry *sesquipedalis*-type pods is conditioned by the two complementary recessive genes *pa–1* and *pa–2*. Kolhe (1970) concluded that the wrinkled pod is dominant to smooth pod and is conditioned by two complementary genes *Wp–a* and *Wp–b*. Premsekar and Raman (1972) found that the tough cowpea-type pod is dominant to the soft subspecies *sesquipedalis*-type pod and is conditioned by a single gene. They also found that the smooth pericarp of dry cowpea pods is dominant to the shrunken pericarp of dry *sesquipedalis*-type pods and is governed by a single gene. I (Fery, 1980) proposed that Premsekar and Raman's (1972) genes for soft subspecies *sesquipedalis*-type pod and shrunken pericarp be designated *stx* and *shp*, respectively.

SEED TRAITS

Morphology

Seed size, usually measured as 100-seed weight, is moderately to highly heritable, with published estimates averaging 67.8 per cent. Seed size of F_1 plants is normally intermediate between parents, but partial dominance for small seeds has been reported (Mackie, 1946; Leleji, 1975; Bhowal, 1976; Mak and Yap, 1980). Singh and Jain (1972) reported that both general and specific combining abilities are important, but Kheradnam and Niknejad (1971) concluded that general combining ability is the more important. Seed size is conditioned by several genes: Sene (1967), Aryeetey and Laing (1973), and Bhowal (1976) reported estimates of six, 10, and four genes, respectively.

The long reniform seed shape that is characteristic of the subspecies *sesquipedalis* is dominant or partially dominant to the shorter and rounder shape that is characteristic of the subspecies *unguiculata*. Fennell (1948) and Brittingham (1950) suggested that the reniform shape is conditioned by multiple factors, but Kolhe (1970) and Premsekar and Raman (1972) concluded that the trait was inherited monogenically. Kolhe (1970) assigned the symbol *Lg* to the gene conditioning long grain. However, this symbol was used previously by Krishnaswamy *et al.* (1945), so I proposed (Fery, 1980) that it can be redesignated *reniform shape*, symbolized *Rs*.

Inheritance of seed-coat structure has been studied by several researchers. Spillman and Sando (1930) assigned the symbol *L* to a dominant gene conditioning longitudinal furrowing; however, as Harland (1919a) had used this symbol earlier, I (Fery, 1980) proposed *Lf* as an alternative. Krishnaswamy *et al.* (1945) described a recessive gene that conditions loose testa, and proposed the symbol *ls* for the gene. Ojomo (1972) studied seed-coat thickness, and he concluded that two duplicate genes, symbolized *Th* and *Co*, condition the trait. Rajendra *et al.* (1979) described two anatomically different macrosclereid arrangements in the seed coat, perpendicular and parallel with respect to the cotyledon. They concluded that the perpendicular arrangement is responsible for smooth coat, and the parallel arrangement is responsible for rough seed coat, with the former being dominant and a single gene controlling the two macrosclereid traits. Rajendra *et al.* (1979) suggested the nomenclature designation PC for the perpendicular macrosclereid arrangement, and I propose that the gene conditioning this trait be symbolized *Pc*.

Krishnaswamy *et al.* (1945) studied hollow seeds, and they concluded that the trait is conditioned by a single recessive gene, symbolized *hs*.

Seed-coat pattern

Although seed-coat patterns are inherited independently of colour, the appearance of any pattern is dependent upon the presence of the general colour factor *C*. Spillman (1911) was the first to attempt an explanation of the inheritance of the eye patterns. He proposed the symbols *w* and *h* (Sen and Bhowal 1961 proposed that lower case letters be used) for genes conditioning the Watson eye (indefinite eye margin) and the Holstein patterns, respectively. Both patterns are recessive to solid pigmentation. The genotype for the ordinary or small eye is *w w h h*; the large eye is the heterozygote between the Holstein and small eye patterns *w w + h*. However, Spillman and Sando (1930) later concluded that the big eye genotype was homozygous at the *h* locus and heterozygous at either or both the *w* and *Na* loci. Harland (1919a, 1920, 1922) described three additional genes conditioning seed-coat patterns: *E*, the New Era pattern (dark-blue, irregular dots); *h–2*, a duplicate Holstein pattern gene; and *y*, very small eye. Additionally, he concluded that mottling is indicative of the heterozygous condition at the *Bl* locus. Sasaki (1922) studied an eye pattern intermediate between large eye and Holstein patterns and concluded that it is conditioned by a single recessive gene. Spillman and Sando (1930) assigned the following symbols to genes governing seed-coat pattern: *D*, dense speckling; *E*, narrow eye; *F*, fine and dense speckling (blue); *G*, dotting of Holstein seeds; *S*, spotting (patches of black pigment); *T*, Taylor pattern (thinly scattered bluish-purple dots), *W*, whippoorwill mottle pattern (irregular areas of dark shade separated by lighter areas); and *X*, Taylor inhibitor. As *D, E, G, W,* and *X* had been used previously, I (Fery, 1980) suggested that Spillman and Sando's (1930) genes for dense speckling, narrow eye, dotting, whippoorwill, and Taylor-inhibitor patterns be redesignated *De, Na, Dt, Wh,* and *In–T*, respectively. *De* is epistatic to *T*. Krishnaswamy *et al.* (1945) proposed the symbol *Sm* for the gene that conditions the smoky pattern (minute dots on maroon background). Saunders (1959, 1960b) described *E* for stippling, *o* for hilum ring, *S* for speckled, and *V* for mottling, but, to avoid confusion, I (Fery, 1980) proposed that the genes for stippling and speckled patterns be redesignated *Stp* and *Spk*, respectively. The *V* and *Stp* combination results in the mottling and stippling seed-coat pattern characteristic of the cultivar Victor. Several of the genes governing seed-coat patterns, e.g. *E*, *De*, and *Stp*, may be allelic, but their identities probably should be kept separate until allelism is proved (Fery, 1980).

Franckowiak and Barker (1974) concluded that several of the genes are allelic to the gene governing black seed coat (*Bl*) (they used Harland's (1919a,b, 1920) symbol *B*), and they proposed the symbols: B^c for the New Era pattern; Bl^f, blue; Bl^l, black with light spots; Bl^s, black spots; Bl^t, Taylor pattern; and Bl^x, modifier of Bl^f. The dominant gene *Mu* conditions mottling in the seed coat. Several of these *Bl* alleles are likely genes described by earlier researchers (Fery, 1980) but are impossible to reconcile. Franckowiak and Barker's (1974) *Bl* alleles should be kept separate from the genes described earlier (Fery, 1980).

Protein

Significant variability in seed protein and some of the essential amino acids has been identified in cowpea accessions (Bliss *et al.*, 1973; Bliss, 1975). Bliss *et al.* (1973) estimated heritability of 29 per cent for protein, 27 per cent and 34 per cent for cystine, 46 per cent and 54 per cent for methionine, and 4 per cent and 39 per cent for tryptophan. However, they noted relatively high and negative genotypic and phenotypic correlations of yield with percentage of protein, yield with methionine content, and 50-seed weight with tryptophan content. Dessauer and Hannah (1978) concluded that high-methionine content in cowpea seed is under genetic control and associated with recessive alleles. Mak and Yap (1980) reported that high protein is associated with recessive genes and noted that high-yielding parents tend to carry dominant genes.

YIELD

The yields of both the reproductive and the vegetative portions of the cowpea plant are moderately heritable under most environmental conditions (Table 6). Heritability estimates for pod number, seed yield, and fresh fodder yield, for example, average 53.1 per cent, 45.0 per cent and 54.7 per cent, respectively. Kheradnam and Niknejad (1971) and Singh and Jain (1972) reported that both general combining ability and specific combining ability are important for grain yield; however the high potential for genetic gains reported by Singh and Mehndiratta (1969), Trehan *et al.* (1970), Veeraswamy *et al.* (1973), and Tikka *et al.* (1977) suggest that the expression of reproductive yield is conditioned largely by additive gene action, as is that for fodder yield (Kohli *et al.*, 1971).

Table 6. Estimates of broad-sense heritability (%) for various yield parameters

Reference	Clusters/plant	Pods/cluster	Pods/plant	Heritability (%) Pod wt./plant	Seeds/plant	Seed wt./plant	Fresh fodder yield	Dry fodder yield
Aryeetey and Laing, 1973	—	—	28.2 / 19.8*	—	—	—	—	—
Bapna and Joshi, 1973	—	—	62.6 / 44.3	—	55.3 / 35.8	26.2 / 54.0	—	—
Bliss et al., 1973	—	—	7.0	—	—	12.0	—	—
Bordia et al., 1973	—	—	36.8	—	—	23.9	—	—
Chandrappa et al., 1974	—	89.5	68.0	86.8	—	73.0	—	—
Erskine and Khan, 1978	—	—	44.6 / 51.5 / 47.0 / 62.8 / 46.8	—	—	57.6 / 57.1 / 50.5 / 80.8 / 53.4	—	—

Reference								
Kheradnam and Niknejad, 1974	47.0	—	—	44.0	—	35.0	—	—
Kohli et al., 1971	—	—	—	—	—	—	64.4	59.2
	—	—	—	—	—	—	28.9	32.0
	—	—	—	—	—	—	23.3	13.6
	—	—	—	—	—	—	32.3	23.6
	—	—	—	—	—	—	24.6	14.9
Kumar and Mishra, 1981	—	—	—	—	—	27.2	34.6	21.3
Lakshmi and Goud, 1977	—	—	—	41.9	—	47.5	—	—
Singh and Mehndiratta, 1969	50.5	—	—	56.1	—	35.6	—	—
Sohoo et al., 1971	—	—	—	97.4	—	—	90.0	—
	—	—	—	94.8	—	—	98.6	—
	—	—	—	83.2	—	—	95.6	—
Tikka et al., 1977	—	—	—	94.5	—	70.0	—	—
Trehan et al., 1970	—	—	—	21.4	—	19.6	—	—
Veeraswamy et al., 1973	48.7	—	70.0	62.8	—	41.9	—	—

* Narrow-sense heritability estimate.

RESISTANCE TO BACTERIAL DISEASES

Bacterial diseases are not as widespread on cowpeas as are fungal and viral diseases. However, two bacterial diseases that cause economic damage have been the focus of investigations into the genetics of resistance: bacterial blight and bacterial pustule.

Bacterial blight, also known as bacterial canker, is caused by *Xanthomonas vignicola*. Lefebvre and Sherwin (1950) reported that resistance is conditioned by a single dominant gene. However, in later studies by Singh and Patel (1977), the resistance in Iron was found to be conditioned by a single recessive gene, *bc-2* (Fery, 1980), and the resistance in P-309, P-426 and P-910 by a single dominant gene, *Bc-1* (Fery, 1980). Raj and Patel (1978) confirmed these findings, although resistance in some cowpea lines appears to be controlled by two genes (IITA, 1976). Prakash (1980) used biometrical techniques to study the resistance in the line 779, and he concluded that the trait is moderately to highly heritable.

Bacterial pustule is incited by *X. campestris* pv. *vignaeunguiculatae*. Patel (1982a) studied the inheritance of three types of reactions, brown hypersensitive resistant (BHR), non-hypersensitive resistant (R), and susceptible (S), produced by three races of the pathogen. He found that the BHR reaction was dominant over the R and S reactions and that the R reaction was recessive to the S reaction. He concluded that two genes are involved in the BHR reaction, one gene conditioning resistance to race 1 and the other conditioning resistance to both race 1 and 2. The R reaction, which is effective against all three races of the pathogen, is conditioned by one, two, or three recessive genes. Patel (1982a) proposed the gene symbols *Bp-1* and *Bp-2* for the dominant BHR genes and the symbols *bp-3*, *bp-4* and *bp-5* for the recessive *R* genes. However, as I (Fery, 1980) had previously used the gene symbol *Bp*, I suggest the symbols *Bpl-1*, *Bpl-2*, *bpl-3*, *bpl-4* and *bpl-5*, respectively.

RESISTANCE TO FUNGAL DISEASES

Breeding cowpea varieties with resistance to fungal pathogens has been a major goal of many improvement programmes since the early part of the century. Results of inheritance studies have been published for nine fungal diseases: anthracnose, Cercospora leaf spot, charcoal rot, *Fusarium* wilt, powdery mildew, rust, stem rot, target leaf spot and *Verticillium* wilt.

For example, immunity, a hypersensitivity-type resistance, and field resistance to anthracnose have been reported (IITA, 1976), and evidence is that both dominant and recessive genes condition the resistance (University of Ife, 1972; IITA, 1974, 1976).

My colleagues and I (Fery *et al.*, 1976) studied the inheritance of resistance to leaf spot caused by *Cercospora cruenta*, and we identified two genes—one dominant *Cls-1* and the other recessive *cls-2*, each of which conditions a high level of resistance. In a subsequent study (Fery and Dukes, 1977a), *Cls-1* was completely dominant in all crosses examined, and we found no evidence that either modifier genes or environment had any impact on its expression. The *cls-2* gene, however, was found not to be a complete recessive during weak epiphytotics but rather an incomplete dominant that is highly influenced by the environment when in the heterozygous condition. The *Cls-1* and *cls-2* genes are not allelic or linked.

Resistance to charcoal rot, sometimes called ashy stem blight, is a dominant trait and is found in Brabham, Iron and Victor (Mackie, 1934, 1939).

Likewise, the cultivar Iron has been reported to be resistant to all three races of *Fusarium oxysporum* f. sp. *tracheiphilum* (Mackie, 1934). Detailed genetic data have not been published, but a single dominant gene probably conditions resistance to race 1, and two dominant genes probably condition resistance to races 2 and 3 (Hawthorne, 1943; Hare, 1957a).

Dundas (1939) reported that resistance to powdery mildew is conditioned by a single dominant gene; however, Fennell (1948) found resistance to be a recessive trait conditioned by multiple genes. Fennell (1948) speculated that he might have been working with an *Erysiphe polygoni* race different from Dundas (1939).

Available evidence suggests that resistance to cowpea rust caused by *Uromyces phaseoli*, is conditioned by a single dominant gene (IITA, 1976).

Purss (1958) identified a single, dominant gene in the cultivar California Blackeye 5 that conditions an almost immune reaction to cowpea stem rot. Partridge and Keen (1976) have reported an association between this gene and the phytoalexin kievitone, and I (Fery, 1980) assigned the symbol *Sr* to the gene for stem-rot resistance.

The cultivar VITA 3 has been reported as immune to target leaf spot (IITA, 1976), which is caused by *Corynespora cassiicola*, and the available evidence suggests that the immunity is conditioned by a single recessive gene.

Moore (1974) reported that resistance to *Verticillium* wilt exhibited by the line M62 is governed by a single dominant gene, symbolized *Vw* (Fery, 1980).

RESISTANCE TO VIRAL DISEASES

Plant resistance is the only feasible method of control of viral diseases. Information on the inheritance of resistance is available for eight cowpea viruses—bean yellow mosaic, blackeye cowpea mosaic, cowpea chlorotic mottle, cowpea mosaic, cowpea mottle, cucumber mosaic, southern bean mosaic and tobacco ring spot—as well as cowpea stunt, a viral disease caused by a synergistic effect of two viruses.

Reeder *et al.* (1972) identified a single recessive gene in PI 297562 that conditions resistance to the cowpea strain of bean yellow mosaic virus

(BYMV-CS), the symbol *by* being assigned to the BYMV-CS resistance gene (Fery, 1980).

Taiwo *et al.* (1981) and Walker and Chambliss (1981) identified single recessive genes that condition resistance to blackeye cowpea mosaic virus, the symbol *bcm* being assigned to the gene by Taiwo *et al.* (1981) and *blc* being assigned by Walker and Chambliss (1981). The *cbm* and *blc* genes are probably indistinguishable.

Rogers *et al.* (1973) reported that resistance to chlorotic mottle virus is exhibited by PI 255811 and is conditioned by a single recessive gene, which I later (Fery, 1980) symbolized *cc*. Subsequent work by Kuhn *et al.* (1981) showed that viral movement is conditioned by a single dominant gene, and they proposed the symbol *Mv* for this gene. They concluded that the recessive allele at this locus is probably a synonym of Rogers *et al.*'s (1973) *cc* gene. Since the symbol *Mv* was used previously by Franckowiak and Barker (1974), I propose that Kuhn *et al.*'s (1981) *Mv* gene be redesignated *Mvi*. Additionally, Kuhn *et al.* (1981) showed that the total accumulation of virus in infected plants and the rate of virus replication is probably conditioned by several genes. They also found that the symptoms, i.e. intensity of chlorosis, are not related to the accumulation of virus, although pod and seed yields are.

Bliss and Robertson (1971) studied both tolerance and resistance to cowpea yellow mosaic virus, reporting that three additive loci condition the tolerance exhibited by the cultivar Alabunch and that an unrelated dominant gene *Ymr* conditions the resistance exhibited by the cultivar Dixielee. Lines with immunity to cowpea mosaic virus (which includes cowpea yellow mosaic) have been identified in the IITA collection of germ plasm (IITA, 1974). Data of Raj and Patel (1978) suggest that a recessive gene conditions the resistance exhibited by the cultivars Cream Pea and Iron; however, their data indicate that minor genes, gene interactions such as epistasis, and environment strongly influence the resistance. De Jager and Wesseling (1981) studied a hypersensitive type of resistance, but they concluded that it had limited value against field isolates of cowpea mosaic.

Patel (1982b) studied the inheritance of cowpea reactions to two strains of cowpea mosaic virus from Tanzania, and he concluded that dominant alleles at two loci interact to condition four types of host reactions: resistant or immune (R), susceptible (S), necrotic local lesion (NLLR), and lethal susceptible (LS). The gene *MVN* controls the necrotic (hypersensitive) reactions, NLLR and LS, and the gene *MVS* controls the S reactions; the recessive *mvs* allele controls the R reaction. The *MVN* allele is epistatic to the *MVS* allele, and hypostatic to the *mvs* allele. Patel (1982b) identified three alleles, designated *MVN-1*, *MVN-2* and *MVN-3*, at the *MVN* locus, *MVN-1* being dominant over *MVN-2* and *MVN-3*, and *MVN-3* probably being dominant over *MVN-2*. To bring Patel's (1982b) symbolization into compliance with the *Vigna* gene nomenclature rules (Fery, 1980), I suggest his symbols be changed to Mvn^1, Mvn^2, Mvn^3 and *Mvs*.

Bliss and Robertson (1971) also reported that tolerance to cowpea mottle virus is dominant to susceptibility and their data suggest that the tolerance is conditioned by one or two genes.

Pio-Ribeiro *et al.* (1980) studied the inheritance of the synergistic necrotic reaction associated with cowpea stunt using plant populations inoculated with both blackeye cowpea mosaic virus and cucumber mosaic virus, and they concluded that the reaction is conditioned by an incompletely dominant gene, symbolized *Nv*.

Resistance to cucumber mosaic virus is conditioned by a single dominant gene (Sinclair and Walker, 1955; deZeeuw and Crum, 1963; Khalf-Allah *et al.*, 1973). DeZeeuw and Crum (1963) assigned the symbol *C* to this gene, but since *C* is the symbol for the general colour factor, I (Fery, 1980) proposed the gene for resistance be symbolized *Cm*.

Brantley and Kuhn (1970) identified a single dominant gene that conditions a hypersensitive-type resistance to southern bean mosaic virus, and I (Fery, 1980) proposed the gene be symbolized *Sbm*.

DeZeeuw and Ballard (1959) studied a hypersensitive-type of resistance to tobacco ring spot (TRSV), and they concluded that the resistance is conditioned by a single dominant gene. Subsequent work by deZeeuw and Crum (1963) indicated that TRSV susceptibility is affected by an epistatic inhibitor. The symbol *T* was proposed for the resistance gene, but I (Fery, 1980) suggested the symbol *T*, which had already been assigned, be changed to *Tr*.

RESISTANCE TO ROOT-KNOT NEMATODES

Several species of the root-knot nematode genus *Meloidogyne* are pathogenic to cowpeas. Root knot, the disease caused by the nematode, is serious in many parts of the world. Webber and Orton (1902) were the first to recognize the existence of natural resistance, and Orton (1913) reported that the resistance is inherited as a dominant trait. Hare (1959) examined 80 F_3 progenies of a resistant by susceptible cross, and he speculated that resistance to *M. incognita*, the most common *Meloidogyne* species pathogenic to cowpeas, is conditioned by a single, dominant gene. Amosu and Franckowiak (1974) conducted similar studies using several F_2 progenies of resistant by susceptible crosses, and their data support Hare's hypothesis. Raj and Patel (1978) examined the F_1 and F_2 progenies of several resistant by susceptible crosses, and they observed patterns of both dominant and recessive monofactorial inheritance for resistance to *M. incognita*. Dukes and I (Fery and Dukes, 1980) studied the parental populations and the F_1, F_2 and backcross progenies of a resistant by susceptible cross and found that the resistance to *M. incognita* exhibited by Mississippi Silver is controlled by a single, dominant gene. Allelism tests confirmed that the same gene conditions resistance in several other cultivars. Additionally, we showed that the gene conditions resistance to *M. javanica* and *M. hapla*, proposing the symbol *Rk* to denote the gene. In subsequent work (Fery and Dukes, 1982), we identified a single, recessive gene that conditions an intermediate type of resistance to *M. incognita*; this new gene is allelic to the *Rk* gene and is designated root-knot resistance—intermediate (symbolized rk^i) (Fery and Dukes, 1982).

RESISTANCE TO INSECTS

Resistance may prove valuable as a means of insect control, or as an adjunct to other control measures, but there is not yet an extensive amount of information on the inheritance.

Kolhe (1970) studied the inheritance of resistance to an unidentified storage pest in parental, F_1 and F_2 generations of a subspecies *unguiculata* by subspecies *sesquipedalis* cross. He found that the resistance exhibited by the *sesquipedalis* parent is conditioned by a single, recessive gene that he symbolized *Rh*. Since the gene is recessive, I (Fery, 1980) used *rh* rather than *Rh* as the preferred symbol.

Youngblood and Chambliss (1973) examined resistance to cowpea curculio, *Chalcodermus aeneus*, in F_2 and backcross progenies of a resistant by susceptible cross and concluded that the trait is inherited quantitatively. Cuthbert *et al.* (1974) demonstrated that each of three types of resistance—non-preference, a pod factor inhibiting penetration through the pod wall by the adult, and antibiosis—is under genetic control and can be transmitted to progeny. Cuthbert and I (Fery and Cuthbert, 1978) reported that non-preference is inherited in a partially dominant manner; broad-sense heritability estimates ranged from 0 per cent to 19 per cent. We studied the pod-factor resistance (Fery and Cuthbert, 1975) and calculated the narrow- and broad-sense heritability estimates of 45 per cent and 49 per cent, respectively. We found that the gene action was primarily additive, and the estimated number of effective factors conditioning the trait was one pair. Chambliss *et al.* (1982) investigated the genetic nature of the pod factor by analysing the pod-wall strength of dry pods in parental, F_1, F_2 and backcross generations of a resistant by susceptible cross. They echoed our findings, with narrow- and broad-sense heritability estimates of 45 per cent and 49 per cent, noting the gene action was mostly additive. The estimated number of effective factors was 1.37. These results suggested that pod-wall strength is closely associated with pod-factor resistance. In fact, Hossain and Chambliss (1983) later demonstrated that selection of increased pod-wall strength does result in increased levels of the pod-factor-type resistance.

Resistance to the pod borer, *Maruca testulalis*, is dominant (Woolley, 1976), and the trait is probably controlled by several genes.

Studies at IITA (1981–83) have indicated that resistance to aphids, *Aphis craccivora*, is controlled by a single dominant gene; resistance to thrips

(*Megalurothrips sjostedti*) and storage weevil (*Callosobruchus maculatus*) is controlled by recessive genes.

GENE LINKAGE

Research on gene linkage in cowpeas has been minimal, and many of the reported links (Table 7) need further study and verification. Harland (1920) observed that *Bl, P* and *E* are so tightly linked that these factors might be allelic rather than linked. Spillman and Sando (1930) suggested linkage as the probable explanation for the association between traits conditioned by *Pr, Br, De, T, F* and probably *S* genes. Brittingham (1950) reported the qualitative–quantitative links for several genes: *Vi-1* and the genes conditioning early pod maturity; *C* and the genes conditioning pod length; *Bu* and the genes conditioning pod length; and *C* and the genes conditioning seed size. However, Saunders (1960b) suggested that the association between the colour factor *C*, pod length, and seed size reflects the multiple effects of the *C*

Table 7.   Genetic linkages in the cowpea

Observed link*		Cross-over frequency (%)	Reference
Bf	*— V*	25.5	Saunders, 1959
Bg	*— Pf*	20.3	Kolhe, 1970
Bg	*— Ystp*	39.5	Kolhe, 1970
Bl	*— Bp*	—	Saunders, 1960b
Bl	*— Ce*	34.4	Saunders, 1960b
Bl	*— E*	—†	Harland, 1920
Bl	*— P*	—†	Harland, 1920
Bp	*— Spk*	20.0	Saunders, 1960a,b
C	*— V*	20.0	Saunders, 1960b
Cm	*— Tr*	25.0	DeZeeuw and Crum, 1963
E	*— P*	—†	Harland, 1920
F	*— T*	—†	Spillman and Sando, 1930
o	*— P*	15.4	Saunders, 1960b
P	*— sh*	34.6	Saunders, 1960b
p'	*— Pb*	19.9	Sen and Bhowal, 1961
p'	*— Pbr*	13.5	Sen and Bhowal, 1961
Pf	*— Ystp*	20.4	Kolhe, 1970
pg	*— w*	23.4	Saunders, 1960b
Sk	*— Spk*	—†	Saunders, 1960a
Wp-a	*— Dgd-a*	5.8	Kolhe, 1970

* Preferred gene symbols used (see Table 1).
† Tightly linked.

gene rather than linkage. Additionally, he suggested that seed-coat colour and date of maturity are associated in a quantitative–quantitative linkage. Similarly, Roy and Richharia (1948) suggested that the gene systems conditioning pod length and the fibre in pod wall are linked. Kuhn *et al.* (1981) speculated that the *Mvi* gene that conditions the movement of cowpea chlorotic mottle virus within the plant is closely linked to one of the genes that controls the replication of the virus.

PART 3

Physiology

Cowpea Research, Production and Utilization
Edited by S. R. Singh and K. O. Rachie
© 1985 John Wiley & Sons Ltd

4

The Physiology of Cowpeas

R. J. Summerfield*, J. S. Pate†, E. H. Roberts* and H. C. Wien‡
* *Department of Agriculture and Horticulture, Plant Environment Labora-*
tory, University of Reading, England; † Department of Botany, University of
Western Australia, Nedlands, Australia; ‡ Department of Vegetable Crops,
New York State College of Agriculture and Life Sciences, Cornell University,
Ithaca, USA

Since the pioneering studies of Njoku (1959), Wienk (1963), Ezedinma (1966) and Ojehomon (1968a,b) who, collectively, investigated the growth and dry-matter partitioning and the flowering behaviour of cowpea, much has been learned about the physiology of the crop: its growth (dry-matter production and distribution, carbon metabolism and nitrogen nutrition), development (the modulation of and genetic variation in rates of progress toward flowering, maturity and death), adaptation (to both aerial and edaphic environments), and its symbiosis with strains of *Rhizobium*. We deal with these aspects only briefly here but in a manner that complements three reviews of the subject matter published recently (Summerfield *et al.*, 1983; Summerland and Roberts, 1985, Wien and Summerfield, 1984).

SEEDS, SEED GERMINATION AND SEEDLING EMERGENCE

Selection during and after domestication of cowpea has led to considerable change in reproductive strategy and structure. Some changes are pertinent to current efforts to increase not only the magnitude of economic yield but also the rate and synchrony of seed production:

○ The life forms of cowpea cultivars have tended to greater 'annual-ness', i.e., in physiologic terms, after flowering, plants are programmed to allocate carbon and nitrogen in larger proportions to fruits than to vegetative components;

○ Individual fruit and seed sizes have increased;

○ Fruits are less prone to dehisce;
○ Seed dormancy has been reduced or even lost (Rawal, 1975b; Smartt, 1978).

Seed dormancy, conditioned by impermeable testae, is common in wild and weedy subspecies but is confined to a few smooth-seeded cultivated lines (Lush *et al.*, 1980). When seeds are sown at conventional depths (2–3 cm) at 28°C, germination is epigeal, and seedlings emerge within 2–3 days (Lush and Wien, 1980). The base temperature for germination of the improved line TVx 3236 (one of IITA's most productive, thrips-resistant releases, which has become popular among African farmers) is 8.5°C (Covell, 1984). Temperatures over about 40°C are detrimental for hypocotyl elongation (Onwueme, 1974; Ndunguru and Summerfield, 1975a).

Although the size of the seedlings at emergence is directly and positively related to the weight of cotyledons of the seeds sown (Lush and Wien, 1980), subsequent seedling growth is affected little since cotyledonary dry matter is mobilized rapidly during hypocotyl elongation and the cotyledons abscise within about 5 days of emergence (Ndunguru and Summerfield, 1975b). We postulated elsewhere that the total commitment of cotyledonary reserves early in crop life could invoke a penalty later (Summerfield *et al.*, 1983). Sprent and Thomas (1984) agreed: when seed reserves of nitrogen are exhausted, symptoms of nitrogen deficiency develop if nitrogen fixation has not begun.

ACCUMULATION AND PARTITIONING OF DRY MATTER

The rate at which a crop increases in dry weight depends on leaf area index (LAI) and net assimilation rate (NAR). Changes in LAI depend on growth in leaf area and senescence, whereas NAR reflects the balance between photosynthetic gain and respiratory loss of carbon. Knowledge of these four processes and how they are affected by environment is incomplete (Elston *et al.*, 1975); they are all under genetic control (Wallace *et al.*, 1972) but unlikely to be equally amenable to exploitation by breeding (Evans, 1974).

Leaves are initiated about twice as rapidly at 30°C than at 20°C. Minimal temperatures at night limit expansion, with the base temperature for leaf appearance and expansion being estimated at 16°C and 21°C, respectively (Littleton *et al.*, 1979a).

In general, LAI of both determinate and indeterminate cultivars is larger at close (e.g. 25 cm × 25 cm) than at wider (e.g. 25 cm × 125 cm) spacings. Maximal light interception is achieved with an LAI of 3 under humid tropical conditions (Figure 1) but larger values of LAI may be required in regions of greater insolation (Wien, 1982).

Many more or less determinate cultivars have a maximum LAI of 3 but vary considerably in the time taken to achieve this: from about 30 days in

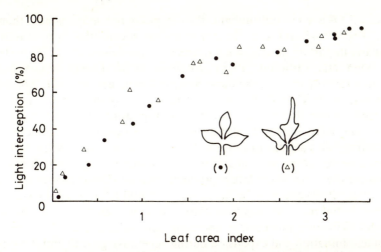

Figure 1. Relation between leaf area index and light interception (percentage) for two cowpea cultivars differing in leaf shape (second rains, 1978; from IITA, 1979)

Delhi (24 plants/m^2) to 60 days for crops (16 plants/m^2) at 24.8°C, the average temperature at Ibadan, Nigeria (Littleton *et al.*, 1979a; Wien *et al.*, 1979b; Chaturvedi *et al.*, 1980).

Maximal values of LAI, whether close to 3 or much larger and typical of values for indeterminate cultivars (>5), usually coincide with the appearance of the first flowers in determinate types. The onset and rate of leaf death are hastened by hot, arid conditions, but effects of weather at this time are difficult to separate from the effects of reproductive load and synchrony of seed filling, and both hormonal and nutritional factors may be involved.

Seed yield in cowpeas appears to rely on 'current (carbon) account'. In carbon-14 labelling experiments, only 8 per cent of the carbon fixed 1 week before anthesis could be recovered in seeds at harvest, and just 15 per cent of carbon taken up at flowering ended in the seeds (IITA, 1976). Recent findings by Atkins *et al.* (1982) have indicated that about 20 per cent of the carbon needed by the fruits and nodulated roots during active seed filling may originate from breakdown of structural components of stems and petioles. The protein-rich seeds need to accumulate nitrogen, too, and about 50 per cent of total N in seeds originates in vegetative organs, especially leaves (Eaglesham *et al.*, 1977; IITA, 1977).

Carbon exchange rate

Cowpea leaves have maximal rates of (C3) photosynthesis at full expansion and for about 20 days thereafter (in cultivar TVu 4552). Green fruits have a net loss of CO_2 even in full sunlight; they respire much more rapidly than do leaves. These high rates of fruit respiration mean that temperature is particularly important at irradiance values less than full sunlight (IITA, 1975, 1976; Wien and Littleton, 1975). Littleton et al. (1981) reported that respiration rates of fruits in the dark increased by 62 per cent as temperature increased from 28°C to 40°C but in full light were reduced by 81 per cent. Thus, fruits above the cowpea canopy may represent a smaller respiratory burden than those under it, although the peduncles elevating them have gas-exchange rates similar to those of fruits (Littleton et al., 1981). Still, the peduncles serve, along with the stems, as a source of carbohydrates and nitrogen for the seeds in the late reproductive phase (Eaglesham et al., 1977; Atkins et al., 1982) and may be far less detrimental to large seed yields than suggested in the past. No positive relation has been demonstrated between maximum C fixation per hectare and economic yield in grain legumes; the rates of leaf initiation and expansion during vegetative growth and the rate of leaf area decline during the reproductive period are probably more important.

Others have suggested that selection for photosynthetic rate or some related attribute such as RuBP carboxylase activity is worthwhile, but we disagree because of the ignorance about how rates of respiration are related to those of photosynthesis and to the accumulated dry weight of a crop stand (Monteith, 1972); the large variations in respiratory activity between organs at different times (Evans, 1974; Elston et al., 1975; Tanaka, 1977); and the paucity of information on differences in carbon and nitrogen economy attributable to the host, to strain of *Rhizobium*, or to the environment (Pate and Minchin, 1980; Schubert and Ryle, 1980). Also, the number of biochemical processes (and hence genes) involved in variations of respiration rates or carbohydrate consumption in nodular fixation could be large, with several compensatory processes being involved.

Dry-matter production and partitioning

Cowpea crops in the vegetative and early reproductive stages of development produce dry matter at rates comparable with those recorded for soybean (IITA, 1973; Summerfield et al., 1983) but smaller than the largest values recorded for other broadleaf crops with C3 metabolism (a daily rate of 34–39 g/m^2; Monteith, 1978).

The limited data available (Grancher and Bonhomme, 1974; Littleton et al., 1979b) indicate that vegetative cowpea crops intercept about 50 per cent of total incoming solar radiation during the linear period of vegetative growth and assimilate dry matter with a conversion efficiency of 1.7–1.9 per cent. Large amounts of dry matter are produced when crops of more or less determinate cultivars maintain relatively large and healthy leaf areas for prolonged periods: in the crops that produced the most dry matter in the trials of Littleton *et al.* (1979b), an early cool period delayed the onset of leaf death, and warm conditions after 20 days produced a large leaf area at flowering so that relatively young leaves were present when plants became reproductive. For large economic yields, leaf area indices between 1 and 2 are required for as long as possible after flowering but need to be coupled with efficient partitioning of dry matter into fruits (IITA, 1975).

Indirect evidence of the benefits of maintaining relatively large photosynthetic rates (or, in other words, prolonging photosynthesis) into the reproductive period was obtained from experiments with reflectors placed near the base of plants at flowering to increase irradiance (by 22 per cent). Seed yields increased by the same magnitude (Wien, 1982). In an experiment involving 16 genotypes of nine species of grain legume, Laing *et al.* (1983) showed that leaf area duration (LAD; the integral of the time course of LAI) alone explained a remarkably large proportion ($r^2 = 0.99$) of the variation in seed yield (Figure 2).

Although in the savanna zone of West Africa the values for monthly solar radiation during the growing season (May–October) average 42 per cent greater than those in the growing season (March–November) in the forest

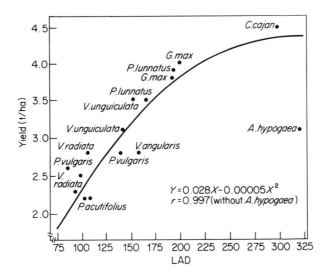

Figure 2. Relation between seed yield (14 per cent moisture content) and leaf area duration (LAD; the integral of LAI from emergence to physiologic maturity) for fifteen genotypes from seven species of grain legume (data for *A. hypogaea*, a special case, were omitted from the regression analysis). Reproduced from Laing *et al.*, 1983 with permission of the authors

zone, there is no evidence that cowpeas have larger crop growth rates in the former location (Enyi, 1973b; Kassam and Kowal, 1973; Wien, 1973; Monteith, 1978; Littleton et al., 1979b).

Cowpea yields seem to be limited by the poor ability of genotypes to assimilate carbon and nitrogen during the reproductive period and to partition large amounts of their daily gains of these two elements into fruits (Wien, 1973; IITA, 1977; Pate and Minchin, 1980). Thus, information on not only economic yield but also harvest index (both for seed dry weight and for nitrogen) is needed if one is to identify potentially productive symbiotic associations. These data are seldom available.

We can make few generalizations except that sole crop cowpeas grown with adequate crop protection can produce excellent yields of protein-rich food by any standards. Short-duration crops and more or less determinate growth habits may be correlated with large harvest indices (although cause and effect are difficult to identify), but the long duration of some indeterminate types seems to compensate for smaller rates of yield production. Cultivars may contain large amounts of nitrogen but yield poorly, and vice-versa. High-yielding cultivars may or may not apportion most of their nitrogen to fruits (compare IITA, 1977 with Chaturvedi et al., 1980), and if they do (IITA, 1977), large seed yields could be incompatible with benefits to succeeding crops in terms of nitrogen in crop residues or minimized depletion of inorganic nitrogen from soils (Nair et al., 1979).

Clearly, much remains to be done before the interactions between carbon metabolism and nitrogen fixation are understood well enough for breeders to select nodulated cowpeas that utilize both elements efficiently to produce seeds.

FLOWERING AND FRUITING

Cowpeas show extreme variation in the start and end of the reproductive period. Some cultivars flower within 30 days from sowing and are ready for dry-seed harvest 25 days later; others take more than 100 days to flower and take between 210 and 240 days to mature.

The floral biology and environmental regulation of flowering in cowpeas have been reviewed in considerable detail only recently (Summerfield and Roberts, 1985).

In general, genotypes that flower early have shorter blooming periods (the number of days for which new flowers continue to open) than do later-flowering ones (about 18 and >30 days, respectively). Flowers are produced in inflorescences that are compound racemes of several modified simple racemes borne on peduncles usually between 5 and 60 cm long. The majority of peduncles arise from axillary nodes in which only one of the three buds present normally develops. Thus, individual nodes generally produce just a single peduncle (or major branch).

Most flowers are self-pollinated, and so the discovery of genetic male sterility (Rachie *et al.*, 1975b), governed by a single recessive gene, has greatly increased the rate at which crosses can be made, and has made possible the use of population improvement and recurrent selection methods in breeding programmes (Smithson *et al.*, 1980).

Reproductive development, yield potential and seed yield in cowpeas are notoriously sensitive to the vagaries of weather. Numerous studies have shown that most cowpea genotypes respond to photoperiod in a manner typical of quantitative short-day plants; that some genotypes are insensitive to a wide range of photoperiods; and that warmer temperatures can hasten the appearance of flowers in both photoperiod-sensitive and insensitive genotypes.

Local populations of cowpeas grown by farmers in West Africa are well adapted so that they start to flower at the end of the rains at a particular location (Wien and Summerfield, 1980). The initiation of reproductive primordia and the expansion of visible buds into open flowers require two distinct photoperiod triggers, longer and shorter, respectively. Lush and Evans (1980) have proposed an alternative regulatory mechanism: they noted that only two to four inductive cycles were required for floral initiation compared with about 20 for development to open flowers. Of course, either mechanism will result in the timely adaptive development shown by cowpeas in West Africa, and both will depend on the sensitivity of individual genotypes (rate of advance in flowering per unit change in photoperiod) and on the difference between the experienced and critical daylengths.

After extensive research in controlled environments in which a diverse range of genotypes was grown in large numbers of photothermal regimens (Hadley *et al.*, 1983), we now believe that the flowering responses of cowpeas to wide ranges of photoperiod and temperature are amenable to quantitative description by several general equations. These data have exciting implications for plant breeders who wish to characterize their parental germ plasm into 'response classes' of known sensitivity to photoperiod and temperature. The principle is that certain parts of the response surface are affected exclusively by either mean temperature or photoperiod, and that the response to either, in terms of the reciprocal of the time taken to flower, is linear. Thus, from the results of a few treatments, one can determine the values of the relevant constants applicable in these equations and so predict the entire response surface (Figures 3 and 4).

For example, for photoperiod-insensitive TVu 1009 at any daylength and for photoperiod-sensitive TVu 1188 in daylengths shorter than the critical value (p_c), rate of progress toward flowering $(1/f)$ is a simple, linear function of the mean temperature ($\bar{t}$):

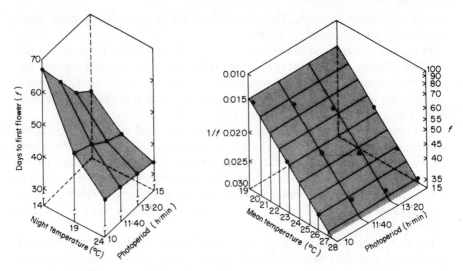

Figure 3.   Effects of photothermal regimen on flowering in cowpea TVu 1009. Response surfaces show (a) effects of photoperiod (h) and night temperatures (°C) on days from sowing to flowering (f) and (b) effects of photoperiod and mean diurnal temperature ($\bar{t}$°C) on rate of progress towards flowering ($1/f$) and days to flowering (f) on left and right-hand ordinates, respectively [where $r^2 = 0.963$ for the fitted response surface in (b)]

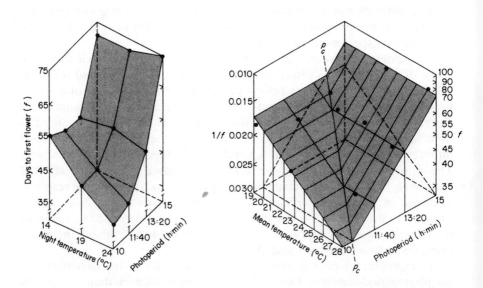

Figure 4.   As in Figure 3 for cowpea TVu 1188 [where $r^2 = 0.979$ for the fitted response surface in (b)]

$$1/f = a + b\bar{t},$$

where a and b are constants.

If this equation applies, a base temperature, t_0, can be defined as the temperature at which the rate of progress toward flowering is zero (i.e. $1/f = 0$), and its value calculated from the constants of the first regression equation:

$$t_0 = -a/b.$$

Furthermore, the thermal sum, T_f (in day degrees above the base temperature), required for flowering is:

$$T_f = 1/b.$$

For photoperiod-sensitive genotypes such as TVu 1188, a second response, superimposed on the 'basic' temperature response, simply and accurately describes the rate of progress toward flowering in photoperiods longer than the critical value:

$$1/f = a' + b'p,$$

where a' and b' are constants (b' is always negative for short-day species such as cowpeas) and p is photoperiod (where $p > p_c$).

Thus, time to flowering is determined exclusively by either mean temperature or photoperiod—whichever causes the greater delay. In any photothermal regimen, f can be predicted from the equations, and the correct solution is that which indicates the longer period to flowering.

The critical photoperiod, p_c, decreases at warm temperatures and its value is represented by the line of intersection of the response surfaces for photoperiod and mean temperatures ($\bar{t}$). Thus, algebraically, p_c can be determined when the values for f given by the equations are identical:

$$p = p_c \text{ when } a + b\bar{t} = a' = b'p.$$

So:

$$p_c = (a - a' + b\bar{t})/b'.$$

Now that the dependence of critical photoperiod on mean temperature has been identified and quantified, the anomalous flowering responses of some genotypes at different latitudes and sowing dates can be better appreciated. For example, a photoperiod-sensitive genotype may flower later than expected in short days if it is relatively insensitive to temperature and the shorter days coincide with relatively warm conditions compared with those experienced when days are longer. In these circumstances the difference between the shorter and the critical photoperiod could still be greater than that between the longer and the critical photoperiod (i.e. plants would be less strongly induced to flower in short days).

Cowpea breeders have attempted to select (albeit indirectly) for insensitivity to photoperiod so that, in theory, production is not restricted by latitude or date of planting. What is now clear is that screening for effects of photoperiod but neglecting those of temperature leads to inaccurate conclusions about the invariance of particular genotypes. Of the first five improved cowpea lines released from IITA (the so-called VITA material), two (VITA 2 and 5) are insensitive to a modest difference of 100 min in photoperiod between 11 h 40 min and 13 h 20 min, one (VITA 4) is moderately sensitive to this difference, and two (VITA 1 and 3) are acutely sensitive. Furthermore, photothermal-sensitive cultivars that maximize vegetative growth under the restrictive conditions of a cereal intercrop may be particularly suitable for the West African savanna zone (Wien and Smithson, 1979).

Plant size at flowering

Plant size at flowering (and so the number of nodes produced) has important consequences for subsequent economic yield in determinate genotypes, which have only a limited capacity for continued growth and leaf production once the first flush of fruits has been set. Adverse conditions during the vegetative period (IITA, 1975; Summerfield et al., 1976) may stunt plants sufficiently to prevent recovery during the reproductive period and so yields are small. However, the large majority of cowpeas are indeterminate botanically (Porter et al., 1974), and plant size at flowering in these types has relatively little effect on economic yield.

Shedding immature reproductive structures

As in other grain legumes (Summerfield and Wien, 1980), in cowpeas the period of anthesis is characterized by a profligate loss of flower buds and open flowers and, thereafter, of immature fruits. Numerous factors have been suggested for this loss in reproductive potential (Sinha, 1977; Summerfield and Wien, 1980).

In general, the first two flowers to open in any cowpea inflorescence are likely to produce fruits, whereas the third pair, if they open, have little chance of doing so. We have calculated from the detailed observations of Ojehomon (1968b) that the respective contributions of racemes 1, 2 and 3 (in acropetal order within any inflorescence) to the total number of open flowers is in the ratio of 6:3:1. In environments conducive to either small, intermediate or large seed yields (43–92 g/plant), the likelihood of premature abscission of flowers from specific racemes varies little: about 10, 34 and 52 per cent failed to produce immature fruits at racemes 1, 2 and 3, respectively (Stewart *et al.*, 1980).

If data from Stewart *et al.* (1980) are typical of cowpea crops, the premature abscission of immature reproductive structures may be less important in determining yield potential than earlier thought. If, in the absence of pests, the first pairs of flowers in cowpea inflorescences contribute at least 60 per cent of the total that reaches anthesis and the chance of each of these flowers failing to set fruits is only 1 in 10, then the minority of later flowers that are the ones most likely to abscise should not have a dramatic effect on yield potential; the number of inflorescences (reproductive nodes) is likely to be more important. Although maximum yield was more than double the minimal value in studies by Stewart *et al.* (1980), the relative loss in yield potential because of premature fruit abscission was remarkably invariant (35.7–41.4 per cent). Indeed, if the total number of open flowers per plant recorded in this investigation is adjusted to give the estimated numbers at particular racemes, the consequences of a small number of reproductive nodes on the number of potentially fruitful flowers is clear: the largest yielding plants opened 67 per cent more flowers at raceme 1 than those that produced the smallest yields. We interpret these data to show that variations in phenologic potential (the number of reproductive nodes per plant) have large effects on seed yield, whereas only in rather specific situations will the premature abscission of potentially fruitful flowers become of major significance.

Increased nodes inevitably mean increased numbers of leaves and, depending on the length of internodes, a longer main stem or increased branching: leaves, main stem and branches together often represent at least 50 per cent of maximum total plant dry weight so economic yields might also be expected to show a close positive relation with total plant dry weight.

We do not offer variations in phenologic potential as the only cause of variation in potential yield (under tropical field conditions, for example, 'stress' induced by various biological or physical agents may lead to the premature abscission of large numbers of flower buds); rather we seek to redress what seems to be an unjustified bias in research activity focusing on a loss in 'theoretical yield potential', which has only limited practical significance.

In studies with a wide range of cultivars of diverse habit and origin, Wien and Roesingh (1980) found that inherent differences in flower bud set between cultivars were related to resistance to the abscission-causing insect *Megalurothrips*. In those cultivars having some resistance to thrips, 76 per cent of the flower buds on the five lowermost main stem peduncles set fruits compared with an average of only 50 per cent for the thrips-susceptible genotypes when both were tested under conditions of total insect control. These results suggested that ethephon sprays could be used as a screening tool to select cultivars less susceptible to flower abscission, and the technique is now used routinely to select cultivars for tolerance to flower thrips (IITA, 1980).

Dense plantings and poor irradiance, luxurious applications of phosphorus; drought; high temperatures, especially at night; unfavourable photoperiods; and insect depradation are detrimental to fruit set in cowpeas, although appreciable genetic diversity is available for most, if not all, of these characters (IITA, 1973; Lush and Evans, 1980; Roesingh, 1980; Turk *et al.*, 1980; Hall and Grantz, 1981; Warrag and Hall, 1983, 1984a,b; Pulver, Brockman and Wien, unpublished).

We urge that future studies be broadened to include quantitative estimates of susceptibility to abscission at successive stages of reproductive development. Only when these data become available will the relative importance of 'numbers' and 'proportions' of different reproductive structures be established precisely.

Duration of reproductive period

The reproductive period of cowpeas is composed of overlapping periods of development of individual fruits, each lasting about 19 days (Wien and Ackah, 1978). The longer the reproductive period, the greater the number of fruits that mature and the larger the yield (Wien and Summerfield, 1984).

Cultivars that have a small percentage of bud set under field conditions are not restricted vegetatively by the demands of fruit growth (Huxley and Summerfield, 1976; Wien and Tayo, 1978). Consequently, the morphology of such plants consists of numerous branches with many nodes compared with the more confined growth of a cultivar with good fruit set characteristics.

Genetic differences in the duration of the reproductive period are related to growth habit, with determinate cultivars of limited leaf areas senescing as early as 20 days after the onset of flowering and indeterminate cultivars requiring 45 days after flowering to senesce and die during the same growing season (IITA, 1974). In one study (Summerfield *et al.*, 1978) testing responses to increasing day temperatures from 27°C to 33°C, all but 1 of 22 cultivars experienced a 20–60 per cent reduction in reproductive period, culminating in marked reductions in yield. However, TVu 662, a cultivar from northern Nigeria, contrasted sharply with the rest; it senesced and died after 53 days at 33°C but after 30 days at the 27°C day temperature. This cultivar also performed well under drought stress in the field. The end of the reproductive period is brought about by leaf senescence in this monocarpic annual species; the interrelations between leaf longevity, the environmental conditions that shorten the reproductive period and the growth of cowpea fruits are noteworthy.

FRUIT PHYSIOLOGY AND METABOLISM

Postanthesis distribution of photosynthate in cowpeas has been studied in detail in a photoperiod-insensitive line of the improved cultivar TVu 1190-E1, released from IITA as VITA 3 (Pate *et al.*, 1983). Budgets of transfer of carbon from individual leaves and from the main stem, petioles and peduncles to fruits have been constructed for unbranched plants: researchers assess the exportable surpluses of carbon from the source organs, measure the net exchanges of CO_2 and changes in carbon content of dry matter and allocate the surplus carbon to fruits and other sink organs on the basis of their relative ability to attract photosynthate from each source, as demonstrated by (^{14}C) carbon dioxide feeding. The resulting picture of carbon flow (Figure 5) bears evidence of a non-stratified, fully integrated pattern of assimilate allocation in which any currently active sink is able to attract carbon from all sources. There is only slight evidence of blossom leaves being especially committed to the nourishment of their subtended fruits—a situation contrasting markedly with that of other axillary-flowering grain legumes such as peas and field beans (Pate, in press a).

An unusual feature of the reproductive development of VITA 3 and certain other large-fruited cultivars of the subspecies *unguiculata* is premature yellowing and abscission of leaves that occur at reproductive nodes early during the fruiting period. A sharp decline in photosynthesis occurs first in the leaf subtending the older fruit—often as early as 10 days after anthesis and a week or more before this fruit has reached its peak demand for carbon (Pate *et al.*, 1983). The fruit is then forced to draw heavily on other photosynthetic surfaces for the bulk of its assimilate requirements during pod filling.

Early senescence also occurs in the blossom leaves at other reproductive nodes so that fruits are forced to draw increasingly upon assimilate sources outside their own nodes. Subdominant sinks suffer, nitrogen fixation declines sharply, root and shoot growth ceases and young flowers and fruits at upper nodes of the shoot absciss. 'Physiologic determinancy' is enforced so that lower leaves at vegetative nodes eventually bear almost full responsibility for photosynthetic gains of carbon.

Fruit-induced senescence of blossom leaves has been well authenticated and studied experimentally in soybean (Nooden and Murray, 1982; Neumann and Nooden, 1984) and is suggested to result primarily from hormones released by adjacent fruits. In line with this suggestion, blossom leaf senescence of VITA 3 can be prevented by removal of fruits and delayed significantly if vascular strands are cut either at the dorsal vasculature of the fruit or at the node connecting peduncle and petiole. Reducing the number of seeds that normally develop at a node by artificially removing them from fruits or cutting off distal portions of the fruit also prevents or delays leaf yellowing (J. S. Pate and M. B. Peoples, unpublished). With 6 or fewer seeds remaining in a fruit normally containing 16–18 seeds, the blossom leaf

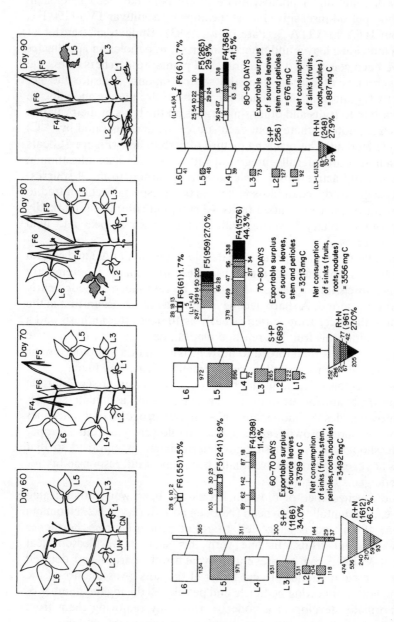

Figure 5. Post-anthesis carbon flow to fruits of cowpea (cv. Vita 3). The top row of figures denotes the morphological condition of plants at the four successive times of harvest, the lower row the net contributions of carbon from individual leaves (L_1–L_6) and stem + petioles (S + P) to nodulated root (R + N) and individual fruits (F_4–F_6) during the intervals between these harvests. Donor organs are hatch-coded so that their contributions to sink organs can be readily recognized. Amounts (mg) of carbon donated and received are given and percentage of total exported C received by each fruit are indicated. (From Pate, 1984.)

remains green until the end of the growth cycle; with 6–12 seeds/fruit, senescence occurs later than in control fruits containing a full seed complement.

Small-seeded genotypes (e.g. Caloona and TVu 354) bearing four to six sequentially filling fruits at each reproductive node seldom show premature yellowing of blossom leaves. Instead, they exhibit mass leaf senescence in late flowering, lower stem leaves at vegetative nodes senescing synchronously with or slightly before the leaves at upper reproductive nodes.

Premature senescence of foliage at reproductive nodes is clearly counterproductive to yield of a large-fruited cultivar such as VITA 3. For example, the yield of seed and fixed N in this cultivar can be increased by as much as 15–20 per cent simply by the removal of the first pair of flowers to form on the inflorescences at the first two reproductive nodes (J. S. Pate, M. B. Peoples and C. A. Atkins, unpublished). This effects a 2- or 3-day delay in fruit set at each node and a comparable delay in senescence of blossom leaves; photosynthesis and nitrogen fixation are able to continue for a few crucial days—hence the significant yield gains.

The overall effects of early senescence of blossom leaves in VITA 3 are evident when one assesses the sources of carbon available to those fruits that form at the first reproductive node (Figure 5). Of a total of 2342 mg C imported by these fruits, only 189 mg (8 per cent) is donated by the subtending lead, compared with 555 mg (24 per cent) from the leaves of the next reproductive node, 500 mg (21 per cent) from the reproductive node above, 476 mg (20 per cent) from starch and sugar mobilized from stems and penduncles, and the remaining 622 mg (27 per cent) from lower leaves at non-reproductive nodes.

Sources of nitrogen to fruits

Data directly comparable with those for postanthesis economy of carbon are also available for mobilization of nitrogen during the fruiting of VITA 3 (Peoples et al., 1983). Nitrogen fixed before flowering contributes approximately three-fifths of the fruits' total requirement for N and, within this amount, leaves, nodulated roots, stems plus petioles, and peduncles donate mobilized nitrogen in the relative proportions of 5:2:1:1, respectively. The blossom leaves lose 70–77 per cent of their N before abscission, whereas lower leaves at non-reproductive nodes have mobilized only 44–57 per cent of their N by fruit maturity and, in most cases, have not abscised (Peoples et al., 1983). The remaining 40 per cent of the fruits' N budget is met by fixation after flowering. In sharp contrast, nitrogen mobilization profiles of other genotypes of cowpeas have attributed more than three-quarters of the fruits' N to fixation after flowering (Eaglesham et al., 1977; Neves et al., 1981).

In studies of VITA 3, Peoples *et al.* (1983) found that peptide hydrolase activities of extracts of leaves, roots, and nodules during fruiting were related to N loss from these organs, and hydrolase activities of leaf extracts against RuBP carboxylase were strongly and positively correlated with measured *in-vivo* loss of this protein from the leaflets. Since preferential degradation of RuBP carboxylase occurs during leaf senescence, especially in blossom leaves, the regulation of hydrolases specific to its breakdown by the senescence signal from fruits merits further examination. As with carbon for fruits at the first-formed node, the subtending leaf provides its fruits with less than one-tenth of their total N intake.

Organic solutes imported by the fruit

Root bleeding sap or, better, tracheal sap obtained by mild vacuum extraction of peduncles, offers a means of sampling the xylem stream available to the fruit (Peoples *et al.*, in press a,b), while the recently discovered technique of cryopuncturing of the dorsal suture of fruits (Pate *et al.*, 1984) enables the phloem stream to be sampled throughout development. Analyses of these sap samples show that when plants rely completely on nitrogen fixed by their nodules, the xylem stream carries N mainly as ureide (77 per cent of total N), the rest as amino acids, principally glutamine (Pate *et al.*, 1984; Peoples *et al.*, in press a,b). Fruit phloem sap of the symbiotic plants, however, has only 10–15 per cent of its N as ureide, the remainder as amino acids. Thus, much of the ureide synthesized in root nodules is probably broken down in the leaves to form amides and amino acids—the principal solutes to be loaded onto phloem streams serving the fruits. When plants have access to nitrate and are not nodulated, the xylem sap has a ureide:amino acid plus amide:nitrate balance of 2:20:78 (N by weight basis), the corresponding fruit phloem sap having a ratio of 10:90:0.1. Feeding roots with (^{15}N) nitrate results in heavy enrichment of the N of amides and certain amino acids, but not of ureides, confirming that nitrate reduction in the shoot leads to the production of mobilizable amino compounds, just as in the parallel case of metabolism of root-derived ureides into amino acids and amides before phloem translocation in the shoots of symbiotic plants (Pate *et al.*, 1984).

Transpiration

During its postanthesis development a VITA 3 cowpea plant transpires on average 280 mℓ water/g of dry matter accumulated, a transpiration ratio within the range normally associated with mesophytic C3 plants. In plants well supplied with water, transpiration rates at night are usually less than 5 per cent of those in the day (Pate *et al.*, in press). Studies of water loss of attached fruits show them to transpire at less than 20 per cent of the rate per unit surface area as their supporting peduncle. The transpiration ratio of the fruit is extremely small (6–9 mℓ water/g dry matter) in comparison with other legume fruits (e.g. peas, lupins and soybeans) where values of 18–28 mℓ/g have been recorded (Pate, in press a,b). Fruits and peduncles exhibit only twofold or threefold diurnal amplitudes in their transpiration rates (Pate *et al.*, in press).

As with fruits of other legumes (Pate and Kuo, 1981) small rates of water loss and CO_2 exchange by cowpea fruits may be directly attributed to the sparse stomatal densities on, and heavy cutinization of, the pod-wall surface. Surface area:mass ratios of legume fruits are also not generally conducive to exchange, especially in cylindrical fruits such as those of cowpea. Primarily as a result of poor ventilation, cowpea fruits, like those of other legumes, have large concentrations of CO_2 (up to 2.5 per cent weight/volume) in their gas cavity. Internal CO_2 tends to be smaller during the day than at night, consistent with the pod wall being capable of photosynthesizing at least some of its own respired CO_2 and that of its seeds. Photosynthetic fixation by the pod wall is also indicated from the fruit's net loss of CO_2, which is smaller during the day (when respiration is likely to be proceeding quickly) than at night. Peoples *et al.* (in press) have reported that the pod wall of VITA 3 photosynthetically fixes 119 mg C over its growth cycle, an amount equivalent to a conservation of 7 per cent of the total intake of C through the peduncle.

Nevertheless, the CO_2 economy of cowpea fruits is relatively poor in comparison with those of other legumes. Fruits of lupins and peas, for example, photosynthetically derive from the atmosphere an amount of carbon equivalent to 15–20 per cent of their total budget. Unlike cowpea, both these species possess a photosynthetic inner epidermis to their pod walls and show significant photosynthetic CO_2 fixation in their inner pod tissues (Pate and Kuo, 1981; Pate, in press a,b). In contrast, the non-chlorophyllous endocarp of cowpea's pod wall functions as a temporary reservoir for nitrogen during fruit development (M. B. Peoples and J. S. Pate, unpublished).

Carbon and nitrogen economy of fruits

Budgets have been drawn up to depict the fate of 100 units by weight of carbon entering the fruit from the parent plant (Figure 6; Peoples *et al.*, in press). Net losses of carbon to the atmosphere are equivalent to 19 per cent of the fruit's intake of translocated carbon; the equivalent of 10 per cent of the carbon intake remains as non-mobilizable material in the dry pod wall; the remaining 71 per cent is recoverable as seed dry matter. This last value compares favourably with the values of 69 per cent and 50 per cent recorded, respectively, for conversion of imported carbon into seed dry matter of pea (cultivar Greenfeast) and lupin (cv. Ultra) by Pate (1984).

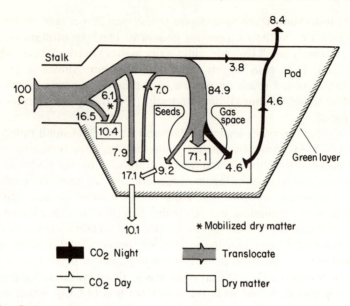

Figure 6.   Carbon economy for the complete maturation period of the VITA 3 cowpea fruit. Items of the budget of the fruit are expressed relative to a net intake of 100 units by weight of carbon through the fruit stalk. The contribution of mobilized dry matter (asterisks) from pod to seed during late fruit development is indicated, and details given of the day and night exchange of CO_2 by pod and seed. (From Pate, 1984.)

The fully ripened VITA 3 fruit carries 96 per cent of its total N in its seeds, an efficiency rating identical to the values recorded for lupin and pea (Pate *et al.*, in press). However, mobilization of nitrogen from the pod wall of VITA 3 during late fruit fill furnishes the equivalent of only 9 per cent of the seed's total requirement for N, compared with values of 15 per cent in pea and 23 per cent in white lupin (Pate, in press a,b). Nevertheless, the fruit emerges as an extremely efficient converter of imported organic materials into seed reserves.

Water economy of the fruit

The empirical modelling technique quantifying exchanges of C, N and H_2O between fruit and parent plant (Peoples *et al.*, in press a,b) was first devised by Pate *et al.* (1977) for fruits of white lupin, and more recently by Layzell and La Rue (1982) for soybean. The model is constructed from information on C and N incorporated into fruit dry matter, net atmospheric exchanges of carbon by the fruit, the C:N weight ratios, and concentrations of the organic solutes delivered to the fruit in xylem and phloem. It is assumed that the mass flow into the fruit is unidirectional from both xylem and phloem. The resulting model is then tested on the basis of whether the mixture of xylem and phloem streams estimated to meet the recorded C and N intake of the fruit supplies an amount of water matching precisely the water consumption of the fruit, as determined by transpiration loss and net change in tissue water.

When applied to lupin (Pate *et al.*, 1977; Pate, in press a,b), the model produces data that fit nearly perfectly with transpiration and tissue water losses, which account for 90 per cent of the water estimated to be carrying organic solutes into the fruit through vascular inputs. Xylem supplies the bulk (60 per cent) of the water entering the fruit, phloem virtually all of the carbon (98 per cent) and the major fraction of the nitrogen (89 per cent).

Applied to cowpea (Peoples *et al.*, in press a,b), the model indicates gross anomalies in the fruit's water budget: to meet the observed N increment of the fruit, much more water must enter through xylem and phloem than can be accounted for in tissue-water increments or transpirational loss. The discrepancy is resolved if one assumes that excess water delivered to the fruit by mass flow in xylem and phloem returns in some manner to the parent plant (recycled water, Figure 7). It is seen to represent, during the second half of

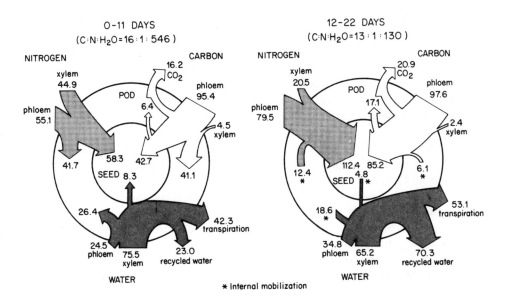

Figure 7. Proportional intake of carbon, nitrogen and water through xylem and phloem during (A) early (0–11 days) and (B) late (12–22 days) development of the Vita 3 cowpea fruit. All components of the fruit's budget are expressed relative to a net intake of 100 units of C, N or H_2O through xylem and phloem. Values for mobilization of C, N or H_2O (asterisks) are given for the second half of development and estimates of the hypothetical 'recycled' components of the water budget given for both stages of fruit growth. Ratios of absolute amounts by weight of C, N and H_2O consumed by fruits are given. (From Peoples *et al.* (in press a,b).)

fruit development, an amount of water greater than that lost in transpiration and equivalent to a return to the plant of up to 2–3 mℓ/day (Peoples *et al.*, in press a,b).

While the anatomical arrangement of the vasculature of cowpea fruits is compatible with fully reversible exchanges of water between seeds, pod and parent plant through the xylem (Pate *et al.*, in press; Kuo and Pate, submitted for publication), fruits of other varieties may behave differently from VITA 3. In fact, preliminary experiments on the small-fruited cultivar Caloona indicate that the fruit transpires sufficiently fast to require additional intake of water through the xylem day and night throughout most of fruit development.

The alleged senescence-inducing factor produced by fruits of large-fruited cultivars may travel from fruit to blossom leaf during the day via the xylem. This would explain why large fruits, with rapid rates of phloem intake relative to their surface area, are especially potent in inducing blossom leaf senescence, while small-fruited cultivars, not regularly generating an excess of water through phloem intake, bear little or no evidence of early loss of leaves at their reproductive nodes.

Metabolic transformation

Using data on the economy of total N within the VITA 3 fruit, and inventories of the amounts of different nitrogenous compounds supplied in the xylem and phloem streams and in the soluble and protein pools of pod wall, seed coat and embryo, one can construct budgets for import and utilization of ureides, amides and the common translocated amino acids by the fruit at different stages of its development (Peoples *et al.*, in press a,b). This information can then be considered alongside data on *in-vitro* activities in different parts of the fruit of certain key enzymes of nitrogen metabolism (e.g. urease, allantoinase, asparaginase, various ammonia assimilating enzymes and aminotransferases). Added to this, one may include information on the pools of nitrogenous solutes in embryo sac fluid, thus determining the key solutes in early embryo nutrition. Similarly, analyses of the leakage products of seed coats allow one to assess which solutes are traversing the apoplastic junction between seed coat and embryo. The flow pathways and metabolic routes for nitrogenous compounds to seeds of symbiotically dependent cowpea can then be assembled (Figure 8).

Fruit metabolism and seed composition

Study of the nitrogen metabolism of fruits of nodulated plants fixing nitrogen throughout their growth logically extends to inquiries into plants whose fixation has been impeded during postanthesis development and into non-nodulated plants relying entirely on nitrate for their nitrogen nutrition

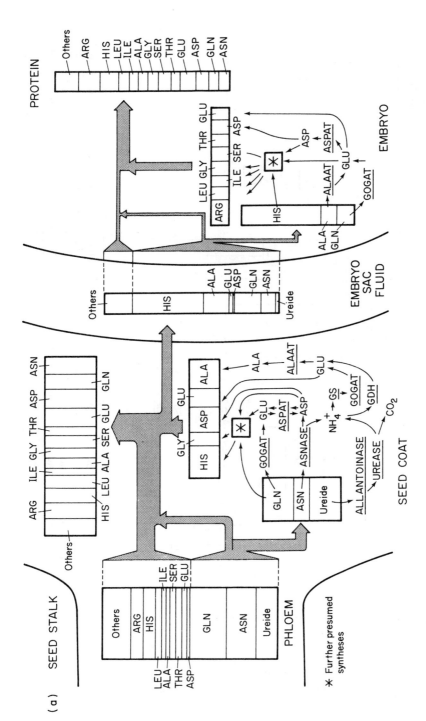

Figure 8. Suggested flow pathways and metabolic routings for nitrogenous compounds into and within seeds of symbiotically dependent VITA-3 cowpea during (a) the embryo sac fluid stage of development (4–9 days after anthesis) and

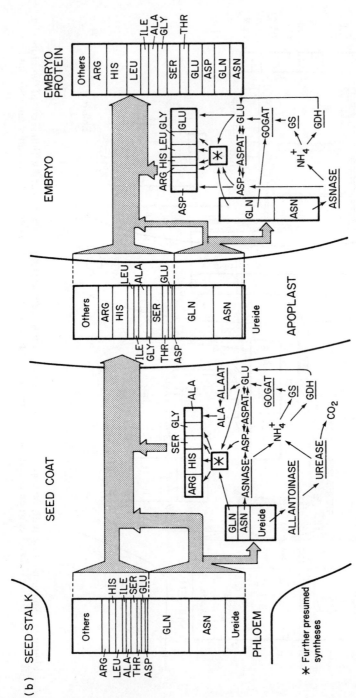

Figure 8. (b) the main seed-filling stage (10–18 days after anthesis). The schemes incorporate data for composition of phloem cryopuncture sap, embryo sac fluid and apoplast leakage products of seed coats. Each scheme depicts amounts of total N and the N of specific solutes entering the fruit compartments or bound into protein of seed coat and embryo. Arrow breadths and areas of rectangles depicting N solutes are drawn proportional to net amounts of N transported or consumed. It is assumed that the seed is fed exclusively through phloem and that analysis of embryo sac fluid indicates the spectrum of compounds available to the embryo. Enzymes whose *in vitro* activities were measured in the fruit: ASNASE—asparaginase; ALAAT—alanine aminotransferase; ASPAT—aspartate aminotransferase; GS—glutamine synthetase; GOGAT—glutamate synthase; GDH—glutamate oxidoreductase

(Peoples *et al.*, in press a,b). For example, when roots were flushed with 100 per cent oxygen (20 min/day) as a means of inhibiting nitrogenase activity during flowering and fruiting, the resulting nitrogen deficiency caused seed abortion and, for seeds that filled successfully, small concentrations of total N in the dry matter. Nitrogen stress also led to a more efficient mobilization of nitrogen from vegetative parts to fruits than in either the control symbiotic plants or NO_3-fed plants.

When examined in further detail in relation to nitrogen transport to fruits, the plants differed significantly in respect to the balance between ureides, amides and nitrate in xylem streams to the fruits but differed little in the nitrogenous solutes in the phloem stream (Figure 9). Analyses of this kind should provide answers on how best to manipulate the nitrogen nutrition of cowpea plants.

Seed yields

When grown by subsistence farmers in the lowland tropics of West Africa, cowpeas yield about 88 kg/ha (Slade, 1977). They are usually intercropped with cereals and are grown at populations of 1000 plants/ha, or less, without fertilizer or protection from insects and diseases. When grown as a monocrop with good management, cowpea crops can produce yields ranging from about 1000 to 4000 kg/ha (Table 1). Largest yields have been achieved by crops relatively late to flower and mature (grown at relatively cool temperatures), confirming results from controlled environments (Huxley and Summerfield, 1976). Unfortunately, the climatic conditions under which large yields were obtained have not been documented in most cases, so the main factors conducive to large yields cannot be quantified. In several instances local, unimproved cultivars achieved yields as large as those genotypes produced as a result of intensive breeding. This sobering fact emphasizes the importance of environment in cowpea yields (Erskine and Khan, 1977; Smithson *et al.*, 1980; Summerfield *et al.*, 1983). Thus, breeding and selection might be more usefully directed to overcoming the constraints imposed by adverse environments and pests than to aiming for yield improvements in situations conducive to large yields. This has, in fact, been the recent trend in cowpea-breeding programmes (Smithson *et al.*, 1980).

ENVIRONMENTAL STRESSES AND THEIR EFFECTS ON YIELD

Drought

Most cowpea crops are rainfed, a small proportion are irrigated (e.g. in the southern USA and in Iraq), and others utilize mainly water residual in the soil after a crop of rice has been harvested (Rachie and Roberts, 1974).

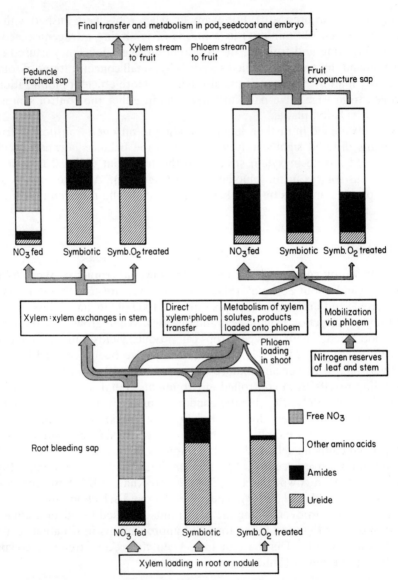

Figure 9. Scheme summarizing the principal effects of nitrogen regimen on proportional rates of transport of different classes of nitrogenous solutes in xylem and phloem of fruiting plants of cowpea (cv. VITA 3). The thicknesses of the arrows depicting flow of nitrogen are drawn in proportion to relative rates of N transfer between root, shoot and fruit. Processes involved in loading of phloem in the shoot are indicated. NO$_3$-fed—non-nodulated plants fed continuously with 10 mM NO$_3$; symbiotic—effectively nodulated plants fed minus nitrogen culture solution; *Symb.* O$_2$, treated—as symbiotic plants but with their rhizosphere flushed daily from 12 days before anthesis onwards with 100 per cent oxygen, thereby greatly inhibiting nitrogen fixation during post-flowering

Table 1. Some examples of the seed yields (kg/ha) produced by cowpea crops in well-managed trials (days to flowering and maturity of the crops and the location where they were grown are also shown)

Location	Cultivar	Seed yield	Days to		Reference
			First flower	Maturity	
Kobbo, Ethiopia	TVu 1977-OD*	4201	68	106	Nangju et al., 1977
Kobbo, Ethiopia	Local check	4072	71	97	Nangju et al., 1977
Riverside, Calif., USA	Calif. Blackeye (CB5)	3916	—	83–113	Turk et al., 1980
Shika, N. Nigeria	TVu 1283	3300	—	—	Akinola and Davies, 1978
Morogoro, Tanzania	Morod	3086	42	110	Enyi, 1973b
Ibadan, Nigeria	TVu 1977-OD*	1357	42	93	Nangju et al., 1977

* Released from IITA as VITA 4.

Dancette and Hall (1979) have stressed that maximal water requirements of annual crops in one region depend on crop duration; indeed, the two variables may be linearly correlated. Cowpeas have a wide range of adaptation in the tropics: from short-duration cultivars in semiarid regions (<600 mm annual rainfall) through the subhumid to the humid, forest belt (1000–1500 mm). Rachie (1977, 1978) has emphasized that, without fertilizer applications or irrigation, but with all major pests controlled, a cowpea crop in the semiarid or subhumid tropics can be expected to produce seed yields between 1600 (good) and 3000 (maximum) kg/ha within 85 days from sowing; the crop is said to need 'adequate water' for only 55 days (65 per cent) of its life. The question is not only how much is adequate but also on which days should it be available.

Based largely on the efforts of researchers at IITA (Wien *et al.*, 1979b) and others at the University of California, Riverside (Turk *et al.*, 1980; Turk and Hall, 1980a,b,c), a comprehensive picture has now emerged on the responses of cowpeas to water stress. The plants avoid drought by reductions in leaf areas, decreases in stomatal conductances, and changes in leaflet orientation (Shackel and Hall, 1979). Under moderate water stress (Wien *et al.*, 1979b), determinate cultivars produced close-to-normal seed yields in spite of stress imposed at flowering because the plants had matured their fruits before the stress became severe. When more severe stress was imposed at different times of the 35–60-day reproductive period in California, closure of stomata and reductions in leaf area combined to limit dry-matter production and led to substantial decreases in seed yield (Turk *et al.*, 1980).

Turk *et al.* (1980) found that California Blackeye 5 recovered more fully from stress at flowering than did the cultivar Chino-3. Hall and Grantz (1981) made selections from these cultivars for lines in which drought hastened maturity. The selections gave 36 per cent greater yields than did the parent varieties (512 kg/ha) when grown to maturity relying solely on stored soil moisture.

Withholding water during vegetative growth decreased the rate of initiation of main stem leaves, but Wien *et al.* (1979b) were not able to show a significant relation between rate of leaf appearance and leaf-water potential (range −0.7 to 1.2 MPa). On the other hand, Cutler (1979) described the seasonal changes in midday leaf-water potentials for rainfed crops at Niomo, Mali, and noted changes in leaf temperature and leaflet orientation when values were between −0.81 and −0.82 MPa for all five cultivars; he concluded that a midday value of about −0.82 MPa was critical for leaf-water potential, postulating sensitivity to 'mild (−1.0 MPa) midday deficits'. In his studies, changes in leaf-water potential were compared with the rate of elongation of terminal leaflets, and he concluded that leaflet elongation was a sensitive indicator of detrimental water deficits in vegetative plants (and see Elston and Bunting, 1980). The five genotypes received only 290 mm rain and experienced an average maximum temperature of 40°C throughout the reproductive period. One cultivar, TVu 662 (an indeterminate type classified as intermediate between semierect and semiprostrate and originating from southern Nigeria; Porter *et al.*, 1974), was by far the best 'drought avoider', producing yields of 1500 kg/ha. TVu 37 (perhaps better known colloquially as Pale Green, an erect, indeterminate vigorous cultivar with a tendency to twine, originating from South Africa; Porter *et al.*, 1974) was the most sensitive to drought, producing only 226 kg seed/ha. Earlier, in glasshouse studies of 22 cultivars (Summerfield *et al.*, 1978), TVu 662 had been the only one to produce dramatically larger yields under warm days than under cool days, and TVu 37 was one of the poorest yielders in the warm regimen (Tables 2 and 3).

Table 2. Relative changes (%) in seed yield (g/plant) of selected cowpea cultivars grown in factorial combinations of warm and cool days (33 and 27°C) with warm and cool nights (24 and 19°C) in a glasshouse (from Summerfield *et al.*, 1978)

Cultivar	Mean seed yield per plant in		Relative effect (%) of warmer temperatures on yield
	Warm days (33/19–24°C)	Cool days (27/19–24°C)	
Overall average of 22 cvs tested	17.9	31.0	−42.3
TVu 662	42.3	27.9	+51.6
TVu 37	19.2	55.2	−65.2

Wien and his colleagues (1979b) have urged breeders to use field conditions when screening for drought tolerance because plants in pots develop water deficits at abnormally rapid rates that preclude the acclimation typical of field-grown plants (Hall *et al.*, 1979). Whether severe drought stress primarily limits phenologic potential (Summerfield *et al.*, 1976) or the subsequent realization of this potential (Hall *et al.*, 1979) probably depends on previous and subsequent environmental conditions (e.g. the effects will be less severe when postdrought temperatures and vapour pressure deficits favour recovery), on the permanence of changes in processes such as leaf initiation and leaf expansion (Monteith, 1977), and on the cultivar's capacity to nodulate and supply its own nitrogen (Sprent, 1976).

Waterlogging

Even in the semiarid regions, intense rain storms can lead to ephemeral waterlogging of soils, especially during periods of vegetative growth. When

Table 3. Comparison of visual estimates of relative heat-tolerance (degree of fruit-set) of cowpea genotypes in the field at Imperial Valley, California with the relative yield (g seed/plant) of the same genotypes grown in pots in different thermal regimens in a glasshouse (from Warrag and Hall, 1983 and Summerfield et al., 1978, respectively)

		Yield of pot-grown plants in a glasshouse		
Genotype	Relative heat tolerance (field observations)	Cooler regimen (a)	Warmer regimen (b)	Relative yield (%) in warmer regimen*
TVu 76†	Extremely	14.2 (36)	20.3 (34)	+43.0
TVu 4552	tolerant	12.6 (37)	14.0 (36)	+11.1
TVu 354	Poor to moderately	45.6 (39)	16.9 (41)	−62.9
TVx 12-01E	tolerant	21.6 (34)	7.2 (34)	−66.7
TVu 4557‡	Poorly tolerant	16.5 (43)	8.8 (45)	−46.7

* Calculated as $[(a)-(b)/(a)] \times 100$ where (a) = seed yield in 27–24 °C day–night temperature and (b) = seed yield in a 33–24 °C regimen (both in combination with a daylength of 12 h). Figures in parentheses within body of table are days from sowing to first flowering.
† Known colloquially as 'Prima'.
‡ Released from IITA as VITA-5.

cowpeas rely on symbiotic nitrogen fixation rather than inorganic nitrogen (mineralized or applied), as they often do, distinct populations of primary, secondary and tertiary nodules (classified according to their relative position within the root system) supply the host with nitrogen in relative proportions depending on the strain of *Rhizobium*, the stage of reproduction of the host and the edaphic and aerial environments in which the symbiosis occurs. Differences in the relative rate and duration of nitrogen fixation by successive populations of nodules during crop ontogeny can have important consequences for the realization of yield potential (e.g. Minchin and Summerfield, 1978; Summerfield, 1980).

The morphology, vascular anatomy, principal export products and water relations of cowpea nodules can each be considered as components of a convenient strategy for coping with warm, dry environments (Sprent, 1980). The spherical, determinate nodules have flattened surface lenticulations, which, by restricting gaseous diffusion, probably contribute to the relative intolerance of cowpeas to short periods of anoxia (Minchin and Summerfield, 1976) compared with soybeans, for example, which have nodules with expanded lenticels and are correspondingly less sensitive to waterlogging (Wien et al., 1979a).

Research in controlled environments has shown that after 16 days of waterlogging, plants were 60 per cent smaller than controls (Minchin and

Summerfield, 1976) and that recovery of plants afterward was dominated by root and nodule growth, perhaps at the expense of initiation and growth of branches (Hong et al., 1977). Waterlogging during the vegetative stage of compact genotypes had the most severe effect on seed yield (reductions of about 48 per cent) (Minchin et al., 1978) and both nitrogen-fixing and non-nodulated plants suffered similar yield reductions. Work at IITA using unsterilized field soil showed even more drastic effects from flooding (Wien et al., 1979a). In one study, 4-day flooding at 3 weeks after emergence and again at flowering decreased total dry weight and reduced yields by 91 per cent. Yet no information on varietal differences to this constraint is available. In view of the probable interactions of root pathogens on the response of cowpea to flooding, screening for tolerance would need to be undertaken using unsterilized soil and should be conducted cooperatively with plant pathologists.

EDAPHIC FACTORS AND MINERAL NUTRITION

Cowpeas are grown on a wide variety of soils, from sands to heavy, expandable clays; they are said to grow best on well-drained, 'sandy substrates'. Whether grown in rotation with maize or cotton in the southern USA, or variously interplanted with millet, sorghum, cassava or maize in traditional farming systems of the tropics, cowpea is often subjected to tillage practices developed primarily for the companion crop (Stanton et al., 1966; Martin et al., 1976; Duke, 1980).

While zero tillage in conjunction with an appropriate herbicide can improve nodulation, vegetative growth, water-use efficiency and seed yield of cowpeas in the humid forest zone of Nigeria (Lal et al., 1978b), the effect varies even at nearby sites (Ezedinma, 1964a) and should be evaluated in semiarid environments.

Cowpeas share with most other green plants the ability to improve their nutrient uptake by mycorrhizal associations between their roots and soil fungi (IITA, 1977). As with other legumes, the potential for assimilation of nitrogen through symbiosis with Rhizobium creates special demands, notably for molybdenum and cobalt but also for boron, copper, phosphate and zinc (Munns and Mosse, 1980). However, neither mycorrhizae nor root nodules are essential for cowpea growth: both can be replaced by appropriate fertilizers. Thus, it is essential to evaluate and to quantify the changes in microbial dependence that are likely to result from applications of inorganic fertilizers, although this is seldom done. Not surprisingly, therefore, estimates of the nutrient requirements of cowpea differ widely, and a plethora of responses to various fertilizers has been reported (Summerfield et al., 1983). Recommendations for fertilization are often questionable. For example, it seems strange to recommend applications of 165 kg N, 67 kg P and 73 kg Mg/ha to achieve a seed yield of 807 kg/ha from non-nodulated cowpeas on the acid soils of Sierra Leone (Godfrey-Sam-Aggrey, 1975) when a judicious application of micronutrients and lime may allow nodule-dependent crops to yield well (Sellschop, 1962; Rhodes *et al.*, 1979). However, the role of lime in tropical agriculture has proved a most controversial subject (Dobereiner, 1974) and debates continue.

Clearly the discussions by Fox and Kang (1977), Munns (1977) and Munns and Mosse (1980) on current and potential fertility of tropical soils, and their collective recommendations for experiments to evaluate these problems are pertinent. Despite the dearth of reliable information on the mineral–nutrient requirements of cowpea, we can be confident that cultivars of subspecies *unguiculata* are potentially productive; that is, they can symbiotically fix quantities of nitrogen sufficient for large seed yields. Annual values for symbiotic fixation (kg N/ha) by cowpeas and, for comparison, soybeans in conducive field experiments averaged 198 and 103, respectively. Average values (4 years) for cowpeas and soybeans on farms in the USA were 25 kg/ha and 88 kg/ha, compared with a value for cowpea on farms in Egypt of 192 kg N/ha (Nutman, 1976).

The selection history of cowpeas and the edaphic conditions with which they have been forced to contend have proved indirect pressures for 'symbiotic potential', and little of this potential is expressed against the fertile backcloth of crop rotations in the USA (Summerfield, 1980). Cowpea breeders need to select for nodulation and nitrogen fixation (Gibson, 1980), which means they will need to restrict applications of inorganic fertilizers to their breeding plots (as they do at IITA). Otherwise, the symbiotic potential of cowpeas could decline in proportion to the breeding effort put into the crop.

MANIPULATION STUDIES

During the vegetative growth and reproductive development of branching, often indeterminate, plants such as cowpeas, the relations between 'sources' and 'sinks' for carbon, nitrogen and hormones are likely to change in complex ways. Nevertheless, an appreciation of these relations is necessary if the effects of environment and management practices on seed yield are to be fully understood. We have used three examples to illustrate the plasticity of cowpeas in response to various manipulative treatments involving defoliation or defruiting (Tables 4–6).

Clearly, the consequences of defoliation depend not only on the timing and the area of leaves removed but also on the relative age and portion of the leaves. Similarly, the outcome of defruiting depends not only on the number of fruits removed but also on their position within and between reproductive nodes and whether a given proportion of the expected number of fruits is taken off at the same time or over time. The examples illustrate that manipulative treatments alone can be difficult to interpret (e.g. in simulating

Table 4. Effects on seed yield (g/plant) of defoliating pot-grown, nodule dependent cowpea plants (determinate, compact cv. K2809) at first flower (Stewart, 1976)

Nominal reduction in leaf area (%)	Relative age bias of leaves removed*	Seed yield (and reduction in yield relative to undefoliated controls)
0 (control)	Not defoliated	32.5
80	Older (O)	11.0 (−66.2)
50	Younger (Y)	11.5 (−65.6)
50	Older	18.0 (−44.6)
50	(O + Y)	20.5 (−36.9)

* Achieved by removing trifoliate leaves acropetally (older) or basipetally (younger).

Table 5. Increments (cm^2) in the area of individual leaves of cowpea cv. K2809 plants during a 9-day period following various defoliation treatments (removal of entire leaflets or one-half or two-thirds of the area of individual leaflets to achieve the nominal defoliation (%) stated; from Huxley and Summerfield, 1976)

Nominal percentage leaf area removed		Leaf no. (1, the youngest to 6, the oldest)					
		1	2	3	4	5	6
Undefoliated							
(A)	0	495	282	93	9	15	15
Part of laminae of 'old' leaves removed							
(B)	16	557	300	94	−18	6	−15
(C)	33	510	287	−25	−25	−33	−19
(D)	50	374	174	60	22	−6	21
(E)	66	274	153	63	−8	4	−2
Part of laminae of 'young' leaves removed							
(B)	16	338	151	22	16	24	−8
(C)	33	385	210	67	38	45	−8
(D)	50	369	189	82	25	−1	−3
(E)	66	300	148	61	22	9	−1
Entire 'old' leaves removed							
(B)	16	502	372	102	1	17	—
(C)	33	538	314	68	80	—	—
(D)	50	488	275	18	—	—	—
(E)	66	484	327	—	—	—	—
Entire 'young' leaves removed							
(B)	16	—	405	105	62	3	11
(C)	33	—	—	124	23	−30	−25
(D)	50	—	—	—	65	30	9
(E)	66	—	—	—	—	−9	33

Table 6. Effects on seed yield (g/plant) of defruiting pot-grown nodule-dependent cowpea plants in various ways (Stewart, 1976)

Treatment	No. fruits/ plant	Seed yield		Actual seed yield relative to predicted (%)
		Actual	Predicted*	
Control	33	34.5	—	—
First 10 fruits left intact, all others removed	10	15.0	10.5	+42.9
First 25 fruits left intact, all others removed	25	34.0	26.1	+30.3
First 10 fruits left intact, next 15 removed, next 15 left intact, all others removed	25	37.5	26.1	+43.7

* Based on the ratio of fruit number relative to that of control plants.

depredations by leaf-chewing insects or those that attack flowers or fruits) because so many variables are involved. A comprehensive picture is only possible if a number of techniques are used to compile supporting evidence (e.g. defoliation, shading, carbon-14 tracer, and CO_2 enrichment techniques in studies on 'leafiness').

RESEARCH ON COWPEAS IN ARTIFICIAL CLIMATES

Researchers working on field plots are often sceptical about the usefulness of results generated from experiments on pot-grown plants in artificial environments—a traditional approach in physiologic research. Unfortunately, plants grown in controlled environments often fail to resemble, morphologically or phenologically, their counterparts in the field (i.e. they are artefacts reflecting the plant husbandry and culture techniques imposed on them). Compared with field-grown plants, grain legumes grown in pots seem especially prone to etiolation, abnormal branching, uncharacteristic leaf expansion, and untimely flowering (Summerfield, 1980). However, extensive research in artificial climates (cabinets, glasshouses and plastic houses) indicates that cowpea plants grown in these conditions can closely resemble those in the field, morphologically and phenologically in their nitrogen nutrition (Tables 7 and 8) and in biochemical traits too (Figure 10); but a prudent selection and precise

Table 7. Selected morphological attributes of cowpea cv. TVu 4552 grown in the field at Ibadan, Nigeria and in a simulated photothermal regimen of that location in growth cabinets (Summerfield et al., 1977)

Location	Days to first flower	Plant height (cm)	No. branches/ plant	No. main stem nodes	Terminal leaflet (cm)	
					Width	Length
Field, Nigeria	38	53	3.5	12.4	10.2	13.5
Cabinets, UK	37	52	3.3	10.3	8.5	12.0

Table 8. A comparison of the sources contributing to the total reduced N content (TRN) of mature seeds in more or less determinate cowpeas grown in controlled environments and in the field (from Eaglesham et al., 1977)*

Attribute	Plants in controlled environment (1973) cv. K 2809	Plants in the field at Ibadan (1976) cv. TVu 4552
TRN in seeds (%)[a]	63.2	63.7
Sources of seed N: Remobilized (%)[b] from		
Leaves	34.4	39.7
Stem + branches	13.0	8.6
Peduncles	2.9	4.3
Roots + nodules	2.9	ND
Total remobilized	53.2	52.6
Assimilated (%)[b] during reproductive period	46.8	47.4
TRN in non-seed components (%)[a]	36.8	36.3
Emergence to first flower (days)	43	35
Emergence to final harvest (days)	85	56
TRN in plant (mg)	4318	1321

* Both sets of plants depended primarily on dinitrogen fixation for their N supply.
[a] and [b] denote percentage TRN in plant and seeds, respectively.
ND, not determined.

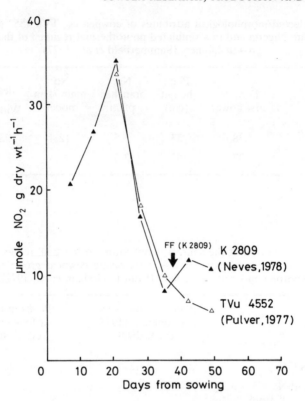

Figure 10. Temporal trends in nitrate reductase activity in determinate, compact cowpeas in the field at Ibadan, Nigeria and in a photothermal regimen simulating that environment in a plastics house in the UK (From Pulver, 1977 and Neves, 1978.)

regulation of pot size, growing media, nutrient supply and method of application, spectral quality of irradiance, diurnal temperature regimen, vapour pressure deficit and CO_2 are essential if, for example, plants are to be grown under artificial lighting in cabinets. Even then, plants grown in this way never experience the modest diurnal stresses (environmental 'noise') to which field crops are exposed and acclimated. Thus, extrapolations from research conducted in artificial climates must always be made with caution; predictability of field performance should be the measuring stick for the confidence of such extrapolations.

CONCLUDING REMARKS

In concluding his analysis of 'yield component compensation in crop plants', Adams (1967) stressed that a useful understanding of yield required detailed investigation and analysis—genetic, physiologic and phenologic—of the initiation and interrelations of characteristics of yield components and of mineral nutrition and the assimilation of organic compounds by the plant. He cautioned that such understanding, already then long sought, would come neither soon nor easily. Unfortunately, many later workers have failed to explore the dynamics of components of yield: they have measured relatively few of them and then only when the crop is mature.

For grain legumes in general, and cowpeas in particular, economic yield is principally a function of the number of seeds that mature (Sinha, 1977). For cowpeas, variations in yield caused by cultivar (Doku, 1970), environment (Stewart *et al.*, 1980), crop mixture (Wien and Smithson, 1979), stress including depredation by pests (Adams and Pipoly, 1980), and nitrogen source (Summerfield *et al.*, 1977a) are strongly correlated with the number of fruits that reach maturity. In general, significant negative correlations between either the number of seeds per fruit or the dry weight of individual seeds and the number of fruits per plant are much less common.

It is perhaps surprising then that traditional components-of-yield analyses equate seed yield in cowpeas with the product of just three components (the number of fruits that reach maturity, the average number of seeds in them and the mean weight of individual seeds), two of which are relatively homeostatic and seldom contribute substantially to variations in seed yield of any particular genotype.

Then again, as we have argued before (Summerfield *et al.*, 1978), these aggregated data fail to provide any direct understanding of why yield or any particular component of it varies; they do not provide any knowledge about the processes by which components of yield are determined nor how these processes are related to plant growth or to the environment. Furthermore, since biological yields are recorded only infrequently (and when they are, the methods of harvesting are different and often unreported), the relative investment of dry matter into fruits cannot be quantified. To classify as equal two genotypes that yield 100 units of seed without knowing their respective biomass is unacceptable (Bunting, 1971); to neglect root (and nodule) weight, both notoriously difficult and time-consuming to estimate in the field, could be serious since the ratio between dry-matter production above- and below-ground changes with time, with the environment, and with the genotype (Holliday, 1976); and to test progeny materials at only the current 'standard' density presupposes that large-yielding cultivars at that particular density will outyield all other combinations of plant type and density—which is patently invalid (IITA, 1975; Donald and Hamblin, 1976; Nangju *et al.*, 1976).

We have used a more detailed components-of-yield analysis (Summerfield *et al.*, 1983), and others have advocated a similar dynamic approach (Monteith, 1975), not only to identify those plant attributes and responses that seem most likely to influence adaptability to the aerial environment but also to predict the optimal environment for maximal yield, and to identify characters that best suit specific genotypes for particular situations (Summerfield, 1975). Greater attention should be devoted to factors responsible for variations in the number of fruiting nodes (Wien, 1973) and to the spatial and temporal relations between carbon metabolism and nitrogen nutrition with respect to not only fruit load per se but also the rate and synchrony of fruit growth (IITA, 1977; Summerfield, 1980).

Closer study is needed of the factors regulating leaf senescence after flowering, to separate, if possible, hormonal from nutritional effects. The role of endogenous hormones, particularly ethylene, in fruit set needs further attention. Comprehensive studies of development rates and durations of individual fruits in relation to whole plant growth and senescence are also needed. Detailed comparative growth experiments at sites of contrasting potential productivity (Table 1) should be conducted to pinpoint the environmental conditions leading to large seed yields. With the extensive data now available on cowpea responses to environment, modelling of crop growth should be attempted. This would allow prediction of growth response in as yet untested environments and would point out gaps in existing knowledge. Considerable comparative work would also be necessary to determine the field environments equivalent to those tested in controlled conditions. Indeed, the screening technique for photothermal effects on flowering advocated from research in artificial climates (Hadley et al., 1983), and the genetic analysis of these responses, should be tested and undertaken without delay.

Cowpea breeders are to be congratulated on their achievements in developing genotypes with multiple resistance to diseases (Allen, 1983) and to some pests (Steele et al., in press). Nevertheless, the new genotypes released to date have not always been well adapted to their intended environments (Figure 11). Thus, the improved yield potential of genotypes with multiple disease- and pest-resistance will not always be well expressed in farmers' fields because breeders selecting for resistance may have selected against 'environmental adaptation'.

With the enormous gap between yields under experimental conditions and in tropical farmers' fields, a major contribution to be made by physiologists is to improve current understanding of the responses of cowpeas to stresses. Close liaison will be necessary with plant pathologists, entomologists and plant breeders, to devise simple selection techniques that can be applied to large populations of breeding materials in screening for combined resistance to diseases, pests and environmental stresses (Goldsworthy, 1982). Such collaboration will not only extend and improve the scale and pace of the research effort but also the quality of the work done (Lawani, 1980) and the cultivars released.

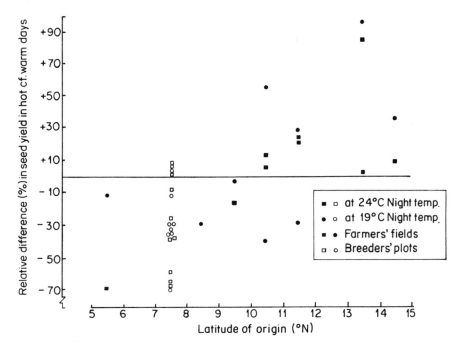

Figure 11. Relative seed yields (percentages) of cowpeas in (a) hot (33 °C) compared with (b) warm (27 °C) days during the reproductive period. Solid symbols denote genotypes collected from farmers' fields at various latitudes in Nigeria; open symbols represent 'improved' disease- and pest-resistant lines bred during the early 1970s (pre-1978) at IITA, Ibadan. Relative yields were calculated from the expression $[(a) - (b)/(b)] \times 100$

This review was prepared during tenure of a research award from the UK Overseas Development Administration (to R.J.S. and E.H.R.). We are pleased to acknowledge the work of Drs O. O. Ojehomon and W. M. Steele, whose combined imagination and research efforts provided the basis of the evolution of our understanding of the physiology of this crop; of Drs K. O. Rachie and P. R. Goldsworthy as successive leaders of an international research programme devoted to the improvement of all aspects of cowpea production and utilization; and of all the scientists who have been associated with the IITA grain legume programme whose collective research efforts have contributed so substantially not only to current knowledge but also to the likelihood of greater productivity from future cowpea crops. Professor A. H. Bunting has stimulated and inspired successive generations of researchers on this species. The technical assistance of Miss C. Chadwick and Mr. M. Craig in preparing this review is gratefully acknowledged.

PART 4

Breeding

5

Development of Improved Cowpea Varieties in Africa

B. B. SINGH and B. R. NTARE
International Institute of Tropical Agriculture, Ibadan, Nigeria

Before establishment of IITA, cowpea-improvement programmes were in progress in Nigeria, Niger, Senegal, Uganda, Kenya and Tanzania. In all these countries cowpea research received major attention from 1960 onward.

In Nigeria, before 1960, research on cowpeas concentrated on describing the crop's position within rotation systems. Subsequently, varieties were introduced from other countries and tested against land races. In all cases, yields were low, although the exotic varieties gave better yields than land races. Unsustained efforts were made to combine local and exotic genotypes but with no recorded successes, primarily because of damage from pests and photoperiod effects on the crop. The crop was, thus, considered difficult to work with (Ojomo, 1973). From the early 1960s the Federal Department of Agricultural Research, the Institute of Agricultural Research and Training at Ibadan, the University of Ife and the Institute for Agricultural Research at Samaru incorporated cowpeas in their research plans. Some of the advanced populations obtained at Ibadan from exotic and local crosses were tested in 1966 and were the sources for Westbred and Prima. Subsequent planned crosses and selections led to the development of the variety Ife Brown (Franckowiak *et al.*, 1973), which is a semideterminate plant type with upright podding habit. During the same period a vegetable cowpea variety, Dinner, was developed and released for cultivation (Ebong, 1971a). Considerable work was done on basic genetics, correlations, harvest index and mutation breeding at Ibadan (Ojomo, 1975).

Cowpea research at Samaru also progressed well. Crosses between locally adapted collections and variety New Era from the United States resulted in a large number of progenies that had desired seed quality and high-yield potential (Leleji, 1975). Further crosses between New Era progenies and Ife Brown gave rise to even better lines such as IAR 339-1, IAR 341, IAR 345 and IAR 355 (Leleji, 1979). Progress was also made in identifying and incorporating resistance to certain viruses (Bliss and Robertson, 1971). During this period nutritional factors such as protein and methionine content were considered, and progress was made in selections for a combination of high protein content and suitable grain type for Nigeria (Bliss, 1975).

In Senegal the first crosses with cowpea at Centre national de recherches agronomiques (CNRA) de Bambey were made in 1961 (Sene and N'Daiye, 1973) and the major emphasis thereafter was to develop high-yielding varieties. The major objectives were to develop varieties that were early maturing (70–75 days) and insensitive to daylength. Erect, determinate growth, lodging resistance and large creamy seeds with coloured eyes were also desired characters. However, after 1968, erect growth habit and lodging resistance received major emphasis. Advanced breeding lines were developed, from which varieties Ndambour and Mougne as indeterminate and Bambey 21, 23, etc. as determinate erect varieties came. From a study designed to evaluate breeding methods, Sene and N'Daiye (1973) observed the conventional bulk method to be more efficient than the pedigree or bulk-progeny test methods.

In Uganda a cowpea-improvement programme was initiated at Makerere University in 1965 (Rubaihayo et al., 1973). The programme started with collection of both local and exotic accessions (Mehta, 1970), which were screened for yield and other characters. From these collections five superior parents: Iron Grey and Pale Green from South Africa, SVs-47 from Tanzania, and Lolita-2 and Emma from Uganda were used in hybridizations.

In Tanzania, work on cowpea started in 1959 (Rai and Utkhede, 1973). In 1964 a trial of 25 best performers was conducted and SVs-3, SVs-66, as well as others, were found promising. A number of lines were also introduced from different countries and evaluated at Ilonga and Ukiriguru until 1973. In 1974, cowpea improvement was incorporated into the Tanzania Crop Research Project for which IITA provided technical assistance. The project lasted until 1982, and two disease-resistant varieties, TK-1 and TK-5, were released in 1980 (Singh and Mligo, 1981a). Also, two 60-day cowpea varieties, TKx 133-16D-2 and Cross 1-6E-2, were developed to ensure a successful crop during the short rainy season (December–January) (IITA, 1982). A great deal of work was done on resistance to different diseases (Patel, 1982b; Singh et al., 1982a,b) and on characterizing wild cowpeas (Singh and Mligo, 1981a).

In Burkina Faso, where cowpeas are mainly grown in association with sorghum and maize, progress in cowpea research dates from 1977, with the assistance of IDRC (International Development Research Centre) and IITA–SAFGRAD (Semi-Arid Food Grains Research and Development). In that programme two varieties, KN-1 (VITA 7), an introduction from IITA, and Gorom-Gorom (SUVITA-2), a local variety, are being widely grown. Recently, TVx 3236 (SUVITA-4), another selection from IITA, has been found to be adapted to the different zones and is being considered for release to farmers by the national programme. Breeding is under way and comprises germ-plasm screening, hybridization and selection for resistance to diseases and insects. Breeding for resistance to *Striga* and tolerance to drought has recently received added impetus.

In Mali the cowpea programme is supported by IDRC (International Development Research Centre) and has two components, breeding and agronomy. The breeding programme was initiated in 1980 with introduced varieties from IITA–SAFGRAD. Hybridization between these and local germ plasm has started.

In Togo one of the improved lines from IITA (VITA 5) was released for large-scale cultivation in 1981. In Botswana cowpea improvement is still in its infancy. Earlier, a number of improved materials from IITA were evaluated, and, from these lines, ER-7 has recently been released for large-scale production in the country.

Other countries that have started cowpea-improvement programmes include Kenya (Katumani, Coastal Province), Malawi (Bunda College of Agriculture), Rwanda, Zaire, Central African Republic, Niger, Benin, Ghana, Ethiopia and Zimbabwe. Most other national agricultural programmes in Africa collaborate with IITA so that cowpea varieties formulated by IITA can be evaluated, and the promising lines multiplied and released for general cultivation.

IITA's PROGRAMME

Cowpea-improvement programmes in Africa received a great deal of attention from 1970 onward, with IITA serving as a catalyst and as a centre for training and germ-plasm collection and improvement. IITA's research in the initial phases concentrated primarily on germ-plasm collection, evaluation and maintenance, and breeding for disease resistance. Subsequently, major emphasis was laid on breeding for insect resistance, early maturity, improved plant types and desired seed quality.

Progress in the 1970s

By 1975 a total of 7300 lines had been collected. These were evaluated and characterized for 46 characters (Rawal, 1975a). By 1980 the total cowpea germ-plasm lines at IITA consisted of 10,471 lines derived from various agroecological zones of the world. Today, there are more than 12,000 lines.

An international cowpea disease nursery (ICDN) programme was initiated by IITA in 1974 to identify broad-spectrum, stable resistance to isolates of pathogens from many different environmental conditions (R. Williams, 1975d). ICDN consisted of 100 resistant lines. These lines were sent to cowpea pathologists in several countries for evaluation over a diverse range of agroecological conditions. This programme led to the identification of several cowpea lines that possessed resistance to multiple diseases (Williams, 1977). These lines were further screened for resistance to savanna diseases (Allen *et al.*, 1981b) and to *Phytophthora* stem rot in Tanzania (Singh *et al.*, 1982b) (Table 1). A detailed review of pathological research on cowpea has been made by Allen (1983) and Emechebe and Shoyinka, chapter 11 in this volume.

Some of the lines combine high-yield potential with multiple disease resistance. Four lines—TVu 201(S), TVu 1190, TVu 1977 and TVu 4557—were found to be the best performers. These lines were subsequently described as VITA 1 (TVu 201, Rachie *et al.*, 1975b), VITA 3 (TVu 1190, Singh *et al.*, 1975), VITA 4 (TVu 1977, Rachie *et al.*, 1976), and VITA 5 (TVu 4557, Singh *et al.*, 1976). By systematic screening of 6000 lines entomologists at IITA also found sources of resistance to leafhoppers, aphids, thrips, Maruca, pod borers, and bruchids (Singh, 1977a) (Table 2). The levels

Table 1. Sources of resistance to different diseases in cowpea (Singh *et al.*, 1983)

Disease	Resistance sources
Anthracnose, rust, Cercospora, bacterial pustule, cowpea yellow mosaic virus	TVu 310, 345, 347, 410, 645
Cowpea yellow mosaic, cowpea mottle, cowpea aphid-borne mosaic, southern bean mosaic, golden mosaic	TVu 393, 493, 1185, 2755
Fusarium wilt	TVu 109-2, 347, 984, 1000
Bacterial blight	TVu 347, 410, 483-2, VITA 3, VITA 4
Scab	TVu 853, 1404, 1433, VITA 4
Septoria	TVu 456, 483-2, 486, 1433, VITA 4
Brown blotch	VITA 1
Root knot	VITA 1, VITA 4
Phytophthora stem rot	Ku 235

Table 2.   Sources of resistance to different insect pests in cowpea (Singh *et al.*, 1983)

Insect pest	Sources of resistance
Leafhoppers	TVu 59, TVu 123, TVu 662, TVu 662, VITA 3
Aphids	TVu 36, TVu 62, TVu 408, TVu 410, TVu 801, TVu 2896, TVu 3000
Thrips	TVu 1509
Bruchids	TVu 2027, TVu 11952, TVu 11953

of resistance for leafhoppers, aphids and bruchids were fairly high but those for thrips and Maruca were only moderate (Singh and Allen, 1980).

While the germ-plasm lines were being screened, a systematic hybridization programme was also initiated in 1971, and the elite breeding stocks from the University of Ife and some from Senegal were incorporated into the programme. A rapid hand-crossing technique was developed (Rachie *et al.*, 1975a), and a male sterile line was discovered (Rachie *et al.*, 1975b), prompting initiation of a population-improvement programme. By 1975 several breeding lines had been developed from planned crosses; these had higher yield potentials than earlier-described VITA lines and exhibited resistance to multiple diseases. A formal cowpea international testing (CIT) programme was initiated in 1975, and a trial consisting of 19 entries was sent to 55 locations in the tropical belt. Since then, CIT has been a regular part of IITA's programme and has been the main channel for interaction and dissemination of improved materials to various national programmes. Based on superior performance and multiple disease resistance, three breeding lines derived from crosses were described as VITA 6 (TVx 1193-7D), VITA 7 (TVx 289-46) and VITA 8 (TVx 66-2H) in 1978 (IITA, 1978). These three lines have brown seeds with smooth testae. Subsequently, a number of other breeding lines were found to be as good as, or better than, those described earlier. The most promising ones were TVx 1948-01F (VITA 9), TVx 1836-013J (VITA 10), 4R-0267-01F (VITA 11), and TVx 1843-1C (VITA 12). All have smooth seed coats and brown seeds, with the exception of TVx 1843-1C, which has red seeds.

Even though the VITA lines had high-yield potential and multiple disease resistance, no line produced seeds acceptable in Nigeria or in several other West African countries. The preferred types in West Africa produce large white or large brown seeds with rough testae. When the rough seeds are soaked in water they swell rapidly, with easy removal of the seed coat for food preparation. Although none of these lines became popular in West Africa, they provided an excellent base for further improvement with respect to insect resistance, seed quality, plant type, etc.

Breeding efforts in the 1980s

In view of the strict seed-type preferences in different regions and the damage caused by insect pests, the cowpea-breeding programme at IITA since 1980 has focused on these two characters along with multiple disease resistance. Considerable emphasis was also placed on developing extra-early cowpea varieties for areas with short rainy seasons and for areas where a catch crop after rice or wheat is possible. Because of their short duration, such varieties require fewer insecticide applications than do later-maturing varieties. Efforts were also made to develop bush-type varieties for production of tender pods for use as vegetables. This type of structure has promise as an alternative to climbing varieties that require trellis support.

Breeding for resistance to diseases is still a focus, but the aim of breeders now is to produce single lines with resistance to all the major cowpea diseases in the humid and subhumid tropics. Through a combination of screening under natural and supplemented infestation in the field and artificial screening in the glasshouse, all segregating populations are evaluated for resistance to several diseases from F_2 to F_6 generations. Screening for diseases prevalent in the humid tropics is done at IITA headquarters in Ibadan and for savanna diseases at Samaru. IT 82E-18, IT 83S-580-1, IT 83S-325-1, IT 83S-687-15, IT 83D-328-1, IT 83S-689-4 and IT 83S-951 all showed resistance to the 11 major diseases, and IT 82E-16, IT 82E-18, IT 82D-807, IT 82D-828, IT 82D-889 and IT 82D-950 have combined resistance to six major viral diseases (Table 3).

Systematic genetic studies have shown that resistance to anthracnose, Septoria leaf spot, brown blotch, scab, cowpea aphid-borne mosaic virus and cowpea yellow mosaic is rather simply inherited and, for each disease, is governed by one or two gene pairs. The exact gene symbols and detailed data are being published elsewhere. As apparent from these studies, all of the resistant characters are rather easy to manipulate in a breeding programme.

Good progress has also been made in breeding for resistance to pests, especially in exploiting the resistant sources found in the 1970s.

Aphids are a major problem in the dry regions, reducing yield not only directly but also indirectly by transmitting viral diseases. Resistant varieties have a high level of antibiosis, causing the death of the pests, and genetic studies have revealed that a single dominant gene controls resistance. A large number of breeding lines have been developed that combine aphid resistance with high-yield potential and multiple disease resistance. These are being evaluated in multilocational trials. Also, a backcrossing programme has been undertaken to incorporate aphid resistance into IITA's most promising varieties. An early-maturing variety IT 82E-60 and a medium-maturing variety TVx 3236, which were initially susceptible to aphids, now have resistance because of repeated backcrosses.

Resistance to thrips—insects that can cause yield losses of 100 per cent—has been found in TVu 1509 and TVu 2870 but is only moderate. TVu 1509 was crossed with Ife Brown in 1976 and one of the lines derived from this cross, TVx 3236-01G, showed moderate resistance and good seed quality (IITA, 1980). This variety has done extremely well throughout West Africa and is being cultivated on a large scale in Nigeria. Using TVx 3236 as a base, breeders have developed some advanced breeding lines that possess levels of resistance as good as TVx 3236; these are being evaluated in multilocational trials, and IT 82D-716 and IT 82D-952 are most promising.

Genetic studies elucidating the inheritance pattern have shown that resistance to thrips is a recessive trait, controlled by two recessive gene pairs.

The cowpea seed beetle, *Callosobruchus maculatus*, causes substantial losses of cowpea in storage. In fact, if only 5 per cent of production were lost because of bruchids, Nigeria alone would lose more than 40,000 t of cowpeas every year, amounting to about $100 million. Therefore, concerted efforts have been made during the past few years to exploit the resistance shown by TVu 2027. Several advanced breeding lines have been developed, and they combine bruchid resistance with high-yield potential, disease resistance and good seed quality. A number of these lines have already been evaluated in multilocational trials and have been sent to more than 55 countries for testing through CIT. All of the lines have produced high yields and are as resistant to bruchids as TVu 2027. Genetic studies have revealed that bruchid resistance is controlled by two recessive gene pairs.

A few selected breeding lines—TVx 3236, IT 81D-1137, IT 83S-742-11, IT 82D-716 and IT 81D-1020—have been evaluated for combined resistance to

Table 3.   Cowpea breeding lines with resistance to multiple viruses

Breeding line	CYMV	CAbMV	CuMV	CMeV	SBMV	CGM
IT 82E-16	1	1	1	2	1	1
IT 82E-18	2	1	2	3	2	1
IT 82D-807	1	2	2	2	1	1
IT 82D-828	1	2	2	1	1	1
IT 82D-889	1	1	2	1	1	1
IT 82D-950	1	2	2	1	1	1
Ife Brown (check)	4	4	2	4	3	1

Score* for different viruses

* 1 = resistant; 5 = susceptible.

diseases and pests. Of these, IT 82D-716 has performed well at many locations and holds great promise. It combines resistance to 10 diseases with thrips and bruchid resistance, along with acceptable seed quality and excellent plant type.

Several F_5 progenies from a cross between IT 81D-1020 and IT 82D-716 have been selected, and they combine resistance to major diseases and thrips, aphids and bruchids with good seed quality and plant type. These will be evaluated for yield performance in 1985.

A systematic programme was initiated to develop extra-early cowpea varieties that fit into multiple-cropping systems, and considerable progress has been made (Table 4). Most of the varieties are as good as, or better than, the current standard variety (Ife Brown) in yield and mature about 2 weeks earlier. The white-seeded, 60-day cowpea variety IT 82E-60 has already become popular in northern Kano and Bida regions of Nigeria where the rainy season is short. This variety does extremely well compared with local varieties. In Bida it is being grown on residual moisture in fields where paddy has been harvested. It has ensured double cropping where only one crop of rice used to be grown. All the early-maturing varieties have been distributed to the 55 countries of CIT.

Three bushy varieties have been developed that show great promise as alternatives to climbing varieties, which require staking for maximal yields. The improved varieties have yielded between 15 t and 18 t of green pods per hectare, with the first picking at 55 days after planting. Six additional pickings (once every 3–4 days) were done (Table 5). The pod characteristics and protein content were close to climbing varieties; therefore they were distributed in 1983 to several countries through CIT.

Table 4. Mean performance (kg/ha) of extra-early (60-day) cowpea varieties in 1982

	Seed colour	Yield (kg/ha)		
		IITA	Mokwa	Samaru
IT 82E-32	Red	1961	2109	1809
IT 82E-9	Black	2003	2078	1614
IT 82E-56	White	1810	1875	1049
IT 82E-5	Brown	1420	1845	2018
IT 82E-60	White	1374	1563	1127
Ife Brown (75-day variety as check)	Brown	2151	1484	1264
LSD 5%		428	307	444

Table 5. Performance of vegetable cowpea varieties in 1982 and 1983 at IITA

Variety	Plant type	Pod yield (kg/ha)		Pod length (cm)	Moisture content (%)	Percentage protein		
						Pods		
		1982	1983			Fresh weight	Dry weight	Matured seeds
IT 81D-1228-13	Bush	18913	14799	25	89.5	2.94	28	24
IT 81D-1228-14	Bush	16778	15131	29	89.3	2.78	26	25
IT 81D-1228-15	Bush	15468	15285	28	89.6	2.91	28	26
TVx 3442-27E	Climbing	14767	10151	32	89.3	2.78	26	25
Dinner	Climbing	12504	8725	31	89.1	2.73	25	26

General strategy

The main objective of IITA's cowpea breeding programme is to develop improved cowpea cultivars and to make these available to national programmes so that they may select or develop varieties well adapted to their specific growing conditions.

Every year four generations of breeding lines are advanced and, thus, within 2 years F_6/F_7 lines combining major attributes become available.

To ensure stability of performance and wide adaptability, IITA first evaluates materials at five locations in Nigeria: Onne, Ibadan, Mokwa, Samaru and Kano. These sites are on a north–south axis and represent different agroclimatic zones: Onne, forest zone with high rainfall; Ibadan, intermediate zone with bimodal rainfall; Mokwa, guinea savanna; Samaru, sudan savanna; and Kano, interphase between sudan and sahel savannas. Recently an additional test site has been established at Niamey, Niger, to represent the sahel. Preliminary trials are conducted the first year, and the varieties that perform best are tested in advanced trials the second year.

The top-performing breeding lines from the advanced trials are multiplied, and new trials are designed for international testing. The number of trials depends on the type of materials. For example, in 1984 four different cowpea international trials were constituted: CIT-1 had extra-early maturity lines, CIT-2 had medium-maturity lines, CIT-3 had bruchid-resistant lines and CIT-4 had vegetable cowpea varieties. Within each trial, efforts are made to include sufficient variability for seed type and quality to suit regional preferences.

After the trials have been formulated, a circular is sent to the cooperators in different countries informing them about the nature and composition of the

Table 6.　Improved cowpea lines from IITA released for seed multiplication in different countries

Country	Cowpea varieties
Nigeria	TVx 3236 released as Dan Knarda, which in Hausa means Son of Knarda; IT 82E-60 released as Ezorowo meaning new cowpea
Ghana	TVx 2724-01F, TVx 1843-1C, IT 81D-1069
Burkina Faso	VITA 7 released in 1980 as KN-1; TVx 3626
Cameroon	TVx 3236 multiplied by National Seed Projects for farmers' use, 1984
Benin	VITA 4, VITA 5 in 1981
Togo	VITA 5 released in 1981; TVx 3236
Liberia	VITA 4, VITA 7, TVx 3236, TVx 4577-02D
Tanzania	TK-1, TK-5 released in 1980 renamed Tumaini and Fahari, which in Swahili mean hope and pride
Zaire	VITA 5, VITA 7, TVx 3236
Central African Republic	VITA 1, VITA 4, VITA 5
Swaziland	IT 82E-18, IT 82E-27
Somalia	TVu 1502
Mauritius	TVx 1836-013J, TVx 3236, TVx 4654-44E
Zambia	TVx 456-01F, TVx 30-1G released as Shipipo and Muliana; TVx 3236, TVx 309-G
Botswana	ER-7
Zimbabwe	ER-7
Costa Rica	VITA 1, VITA 3, VITA 6
Brazil	VITA 7, TVx 1836-012J, VITA 6, VITA 3 named EPACE 1, EPACE 2, EMAPA-821, EMAPA-822; 4R-0267-01F (Manaus)
Venezuela	VITA 3
Nicaragua	VITA 3
Guatemala	VITA 3
Surinam	IT 82D-789, IT 82D-889
Jamaica	VITA 3, ER-7, TVx 2724-01F, TVx 1850-01F
Fiji Islands	VITA 1, VITA 3

continued

Country	Cowpea varieties
Guyana	ER-7, TVx 2907-020, TVx 66-2H, VITA 3 named as MINI-CA-I, II, III, IV
South Yemen	VITA 7
India	TVu 1977-01D (VITA 4), TVu 1502, TVx 1843-01C
Pakistan	VITA 4
Sri Lanka	TVx 930-01B, TVx 309-01G, VITA 4
Burma	VITA 4 as Yezin 1
South Korea	VITA 5

trials. Based on their requirements, individual cooperators request one or more sets of trials. Attempts are made to meet all requests, but it is often not possible because of large demands. For 1982–84, 358 requests were made, and 294 were met.

Each CIT set is provided with a computerized fieldbook comprising two copies of data sheets. One copy of data sheets is returned by the cooperators to IITA at the completion of the trials. The data are analysed, and a copy of the results is immediately sent to the cooperator. Later, the data from all the analyses are compiled in the form of a bulletin and sent to the cooperators.

The scientists in national programmes select the best lines and formulate their own multilocational trials in the following years. The most promising varieties are then either released to farmers or used as breeding stock. Several countries have released IITA's cowpea lines for general cultivation (Table 6).

Recently the international testing programme has been greatly strengthened by projects funded by IDRC, the European Economic Community (EEC), SAFGRAD and the United Nations Development Programme. These projects have stimulated considerable interest for cowpea research in Africa.

Cowpea Research, Production and Utilization
Edited by S. R. Singh and K. O. Rachie

6

Breeding Cowpeas to Suit Asian Cropping Systems and Consumer Tastes

S. N. MISHRA,* J. S. VERMA,* and S. J. B. A. JAYASEKARA†
* G. B. Pant University of Agriculture and Technology, Nainital, India; and †
Agricultural Research Station, Maha Illupallama, Sri Lanka

Cultivars from West Africa and the Indian subcontinent show similarity in leaf and flower shape but differ markedly in seed size and shape, because the human populations in the two areas have selected *Vigna unguiculata* for different purposes (Faris, 1965; Mehra *et al.*, 1974). In India, cowpea has been known since Vedic times, and India and China may be two secondary centres of origin, the former of the Catjang cowpea and the latter probably of the asparagus or yard-long beans. However, the crop could have been carried along the coastal and Indian trade routes (Sauer, 1952; Mehra, 1963; Singh *et al.*, 1971).

Cowpea is grown in Bangladesh, Burma, China, India, Indonesia, Korea, Nepal, Pakistan, Philippines, Thailand and Sri Lanka. Total area is estimated to be about 1.2 million hectares. On more than half the land, cowpea is grown for dry seeds as a monoculture and as a mixed crop with cereals.

In Sri Lanka, Indonesia and the Philippines, cowpea, more than any other species, is employed as groundcover in rubber plantations. In India and elsewhere it is grown as green manure in tea and sugarcane estates. Also, the ripe seeds are boiled and eaten, and, increasingly, they are being split and used as dhal. Cowpea flour or meal is also used to make sweets, cakes and porridge.

Improved varieties (Table 1) have evolved in many parts of Asia; however, most of the literature on plant improvement comes from India.

PLANT IMPROVEMENT

Basic studies on cytology, genetics and wide hybridization help breeders in hybridization and selection programmes. Sen and Bhowal (1960a) made a

Table 1. Important improved varieties of cowpea in Asia

Country	Promising variety	Use
Bangladesh	TVx 7-5H, VITA 7, Kalo, Sabuj, Sitsite	Grain, vegetable
Burma	TVx 1850-01E, TVx 7-5H, CP-2-3-1	Grain
China	Zhijiang 28-2	Vegetable
Hong Kong	GLB-19, Utah 52-70R	Vegetable
India	Co.1, Co.2, 24/8-2, C 152, Pusa Phalguni, Krishnamany	Grain, vegetable
	S 488, Branko, Conch, Cream pea, FS 68, EC 13060, P33-2-1, IC 8267, PLS 7, Culture 1, V-16, V-37, V-38, New Era, Rs 14, Gwalior K-11	Grain
	Russian Giant, Type 2, EC 4216, UPC 5286, K 397, Chinegra, HFC 42-1, FOS 1, Cowpea 74	Fodder, green manure, vegetable
	P85-2E9A, Red Seeded Sel, Brown Seeded Sel, P 460-1-1, Pusa Dofasli, Pusa Barsati, Philippines Early, EB 1, EB 2, EB 3	Vegetable
Indonesia	TVx 1836-19E, TVx 2894G, TVu 3629, BS 1, No. 2863	Grain, vegetable
Nepal	TVx 1850-01E, TVx 4661-07, IT 82E-16, IT 82E-32	Grain
Pakistan	62-157-00382, Swat, 62-151-00411	Grain, vegetable
Philippines	VITA 4, TVx 2939-09D, TVx 7-5H, All	Grain
	Season, Farve 13, IT 81D-1228-10, IT 81D-1228-13	Grain, vegetable
Sri Lanka	MI 35, Arlington, Bombay Cowpea, Sel 2664, Sel 238-2, TVx 33-1J, VITA 4, EG No. 2, EG No. 3	Grain
	TVx 930-01B, Lanka Kadala, BS 1, BS 2	Grain, vegetable
Thailand	TVx 7-5H, VITA 3, All Season, IT 82E-9	Grain

comprehensive study, indicating that the three cultivated species of cowpea had $2n = 22$ chromosomes. They are easily crossable among themselves, and the hybrids are fertile. Sen and Bhowal (1960a) asserted that the three cultivated types need not be considered as distinct species but should be regarded as members of a polymorphic species. A few crosses with the species

V. luteola have produced seeds but all crosses with other wild species have been unsuccessful (Sen and Bhowal, 1960a). The hybrid seeds of *V. luteola* × *V. catjang* were shrunken.

The large, showy flowers and untwisted keels make cowpeas one of the easiest legumes to manipulate, and, in India, detailed hybridization procedures were given as early as 1945 by Krishnaswamy *et al.* Recently, Kumar *et al.* (1976) updated the procedures, and Jatasra *et al.* (1980) compared the efficiency of two methods of crossing involving three varieties of *V. unguiculata*. The technique of Rachie *et al.* (1975a) substantially reduced the time taken for both emasculation and pollination and increased pod set from 18.6 per cent to 26.1 per cent. We (Mishra *et al.*, 1984) noted parental selectivity in hybridization, indicating that certain lines produce more pods and seeds/pod when used as female parents.

A total of 141 genes have been described for *V. unguiculata* for simply inherited traits, of which 60 genes have been described in India (Sen and Bhowal, 1961, 1962; Kolhe, 1970; Singh and Jindla, 1971; Premsekar and Raman, 1972; Apparao and Reddy, 1976; Singh and Patel, 1977; Venugopal and Goud, 1977).

Based on metroglyph analysis, 11 plant types have been recognized, and three occur in India. Indian material has been shown to be more variable than African materials in height, spread, position and degree of branches, as well as leafiness and fodder yield.

Economically important traits in reproductive and vegetative parts of cowpea are moderately heritable under most environmental conditions (Singh and Mehndiratta, 1969; Trehan *et al.*, 1970; Kohli *et al.*, 1971; Sohoo *et al.*, 1971; Bapna and Joshi, 1973; Bordia *et al.*, 1973; Veeraswamy *et al.*, 1973; Chandrappa *et al.*, 1974; Lakshmi and Goud, 1977; Tikka *et al.*, 1977). Broad-sense heritability for the two most frequently measured yield parameters, pod number and grain weight, ranges from 13.6 to 98.6 per cent and averages 46.2 per cent. From the diallel and line × tester analysis, several workers reported that both general and specific combining ability is important for grain yield (Tikka *et al.*, 1976; Kumar *et al.*, 1978; Prasad and Patel, 1978; Jain, 1982; Zaveri *et al.*, 1980, 1983). If so, then both additive and non-additive gene actions have a key role, although Singh and Mehndiratta (1969), Trehan *et al.* (1970), Veeraswamy *et al.* (1973), Tikka *et al.* (1977) and Kohli *et al.* (1971) suggested that grain, pod and fodder yields are all controlled largely by additive genes. According to Zaveri *et al.* (1983), the heterosis for seed yield can be attributed to heterosis in numbers of clusters and pods; Jain (1982) reported that green and dry-matter yields and seed yield showed considerable heterosis in a variety of crosses.

Correlation, multiple regression, discriminant function, coheritability and path coefficient analyses suggest that characters such as pod number, seeds/pod, and size of seed have considerable value in schemes of indirect selection for grain yield (Singh and Mehndiratta, 1969, 1970; Janoria and Ali, 1970; Mehndiratta and Singh, 1971; Bapna *et al.*, 1972; Premsekar and Raman, 1972; Bordia *et al.*, 1973; Chandrappa *et al.*, 1974; Kumar *et al.*, 1976, 1982, 1983; Lakshmi and Goud, 1977; Tikka *et al.*, 1977; Tikka and Asawa, 1978; Hanchinal *et al.*, 1979; Chauhan and Joshi, 1980; Dumbre *et al.*, 1982) and secondary characters such as cluster number, plant height, testa weight, pod length and peduncle length may also be of value. Leaf number, plant height, branch number, branch length, stem girth, protein content and digestibility are other characters that have potential in indirect selection for fodder yield (Dangi and Paroda, 1974; Chopra and Singh, 1977; Singh *et al.*, 1977; Tyagi *et al.*, 1978).

From studies of segregating generations, Veerupakshappa *et al.* (1980) reported that seed yield was positively correlated with pods/plant, seeds/pod and seed weight in F_2 and with pods/plant in backcrosses in a cross of C 152 × Lolita. In morphophysiolgic and sink parameter correlations, Pandey *et al.* (1981) observed that nodule weight was positively and significantly correlated with fresh plant weight, and shoot and root dry weights, indicating importance of symbiosis for photosynthetic productivity; pods/plant was positively correlated with lead area index (LAI), cumulative growth rate (CGR), and reltive growth rate (RGR); and net assimilation rate (NAR) influenced sink capacity. Pods-plant and seed weight had a positive relationship with yield and were the most important sink parameters.

The new varieties

The foremost need is to evolve new plant types that can be fitted in a series of multiple-cropping systems. To date, the main efforts have been devoted to reducing the time to maturity, inducing synchronized development in terms of pod formation, inducing photoperiod insensitivity, instilling an erect growth habit with peduncles above the leaf canopy and controlling the partitioning of dry matter. Mital *et al.* (1980) indicated that Rituraj, Red Seeded, 24/8-2, EC 42712, EC 43128 and EC 30040 offered a combination of three desirable characteristics—photoinsensitivity, long pods and earliness. The variety 24/8-2 also had field resistance to *Macrophomina* wilt. The rest of the stocks were desirable donor parents for agronomic characters.

New plant ideotypes have been developed after extensive hybridization programmes giving high seed yields. The new varieties are V-16, V-37 and V-38 in India (Jain, 1981) and MI 35 and Bombay Cowpea in Sri Lanka (Jayasekara, 1984). Some of the other promising materials of early and synchronous maturity (developed by IITA) in other parts of Southeast Asia are IT 82E-16, IT 82E-18 and IT 82E-32, which can produce 1.1–1.5 t/ha grain yield with only moderate management.

Combining a high harvest index and high dry matter production amounts to selection for yield, although a high harvest index alone does not necessarily mean high grain yields. In selecting for a high harvest index in cowpea, breeders increase the proportion of effective pods/plant and minimize flower and pod drops. Photoinsensitivity is desirable to fit the new varieties in multiple-cropping patterns. Varieties such as C 152, FS-68, Pusa Phalguni, Pusa Dofasli, Rituraj and Chirodi fit well into such systems and can be successfully cultivated during summer–spring as a catch crop (Jain, 1981; Parmar, 1981; Singh et al., 1981).

In addition to selection for increased yields, selection for resistance to diseases in Asia, as elsewhere, is a priority for cowpea breeders. Cowpea yellow mosaic virus, cowpea aphid-borne mosaic virus, bacterial pustule, bacterial blight, anthracnose, leaf spots, web blight and *Pythium* soft stem rot are the major diseases. Through systematic screening of germ plasm, breeders have identified various sources of resistance. Of 526 cultivars from different countries a few showed resistance to *Cercospora cruenta*, and many were resistant to cowpea mosaic virus (Ahmed et al., 1972). Several TVu and TVx lines have been identified (Singh and Reddy, 1982) as resistant to root-knot nematode, and lines IC 8, C 288, Patna Red Seeded and IC 7095B offer multiple virus resistance (Sheoraj and Patel, 1977, 1978). Other sources of resistance to multiple diseases have been identified by Jhooty et al. (1980), Karwasara et al. (1982) and Sheoraj and Patel (1977), including Iron, P 426, EC 107156, IC 290, IC 322, IC 464, IC 7093B, TVu 408P and TVu 410. Studies have indicated that most of the resistance to diseases and nematodes is governed by one or two genes.

Through planned hybridization (diallel selective mating, multiple crossing, backcrosses and convergent backcrosses) and selection from the segregating generations in epiphytotic conditions, breeding lines such as C 152, C 20, C 288, C 30, C 28, EC 42716, UPC 5286 and Pusa Barsati have been developed in India but none have high-yield potential.

Germ-plasm screening and breeding for insect resistance have not been as successful as efforts to instil varieties with resistance to diseases. The insects that cause damage to cowpea in Asia are aphids, beetles, leafhoppers, hairy caterpillars, pod-sucking bugs, pod borers and seed borers. Sri Ram et al. (1976) reported tolerance to leafhopper (*Empoasca* sp.), leaf beetle (*Pagria signata*) and semi-looper (*Plusia nigrisigna*). Chandola et al. (1970) screened exotic and indigenous lines of cowpea and noted that T_2 was the only variety resistant to *Bruchus* sp. in India. For resistance to *Aphis craccivora*, lines P 1473 and P 1476, which were completely free from aphids in studies by Dhanorkar and Daware (1980), are potential sources. Resistance to *Callosobruchus maculatus* was noted in varieties VIR 16, VIR 671, VIR 771, C 152 and 647-I (VIR 1982). Kolhe (1970) reported resistance to an unidentified beetle that was a storage pest.

The successes have been few, as isolating insect resistance and combining it with other desirable characters have eluded Asian breeders. We believe that hybridization programmes should be enhanced, with sequential and simultaneous use of resistant parents. Resistant sources may be obtained from local collections or from IITA's collection of exotic sources such as IT 81D-1007, TVx 3236, TVu 946, TVu 2027, IT 81D-985 (Singh et al., 1983).

Other characters

Another character that has been instrumental in making cowpea popular in many developing countries is drought-tolerance. For instance, it is more drought-resistant than mung bean and soybean (Pandey et al., 1983), and Yadava and Patil (1983a) reported that HFC 42-1, IGFRI 450 and IGFRI 457 varieties are highly drought-resistant.

Other characters that have received limited attention from cowpea breeders in Asia have to do with how well the legume fits into people's lives: whether it is attractive to consumers and suitable for local farming systems. Cowpeas' low digestibility and low amino-acid content are two quality factors that deserve attention; flatulence, cooking quality and fibre content are others. Also, livestock owners complain of the low palatability, digestibility and high crude fibre content in cowpea fodder.

In one study, cooked green pods were evaluated organoleptically and in terms of their days to flowering and yielding ability. Chinese Red, which is early-maturing and high-yielding, was judged as the best and was followed by PLS-46 with medium maturity and good cooking quality (Hanchinal et al., 1979).

In the Philippines, Rosario et al. (1981) indicated that varieties EGB 52, Acc 225 and Acc 22 are highly nutritious with considerable sulphur amino-acid content. In mature seeds, protein content ranged from 20 to 26 per cent and oil content from 0.3 to 1.7 per cent. Globulin was the predominant protein fraction (63–67 per cent of total N) and albumin was next.

Because of growth habit, prolonged periods of pod production of local varieties, and smallholdings, the cowpea is suited to subsistence rather than commercial farming in most parts of Asia. It has great value in multiple-cropping systems involving relay cropping, intercropping and mixed cropping. It also has great potential as a short-duration summer catch crop in several Asian countries.

To develop cowpea varieties suitable for rice-fallow after the second crop of rice (December–January) and parallel multiple cropping with sugarcane and other cash crops, breeders have selected cowpeas for such characters as earliness, compactness, seeds/pod, etc. and have derived some promising varieties from within the progenies of crosses between Virginia × C 152, Virginia × Iron Grey and Virginia × No. 2 (Mahadevappa et al., 1976). The selections gave higher yields than C 152 in parallel multiple cropping in sugarcane. In addition, these lines possessed bold seeds. Calicut 51, Calicut 78 and lines derived from crosses between Calicut 51 and Kalingi Payar as well as New Era and Kalingi Payar proved well suited for intercropping in cassava, banana and coconut gardens of peninsular India. When intercropped they provided increased grain, improved soil fertility and increased yield of the main crop. Mixed cropping of cowpea in groundnut in the monsoon areas of southern India is profitable, and in areas receiving early monsoon rains, cowpeas can be harvested before the main crop (such as cotton).

Cowpea Research, Production and Utilization
Edited by S. R. Singh and K. O. Rachie
© 1985 John Wiley & Sons Ltd

7

Achievements in Breeding Cowpeas in Latin America

EARL E. WATT,* ERIC A. KUENEMAN,* and JOAO PRATAGIL P. DE ARAÚJO†
* International Institute of Tropical Agriculture, Ibadan, Nigeria; and †
EMBRAPA/CNPAF (Empresa Brasileira de Pesquisa Agropecuaria/Centro
Nacional de Pesquisa—Arroz e Feijão), Goiânia, Goiás, Brazil

Cowpeas probably entered Latin America from West Africa during the seventeenth century with Spanish and Portuguese traders. In the tropical lowland ecology of Latin America, cowpeas have several advantages over dry bean (*Phaseolus vulgaris*), which is the preferred legume in many countries. Cowpeas have fewer diseases and insect pests, are more tolerant to drought and waterlogging, and are more tolerant to infertile soils and acid stress, than are dry beans. However, dry beans are better adapted to the ecology at high elevations than are cowpeas; consequently these two legume crops complement each other and only rarely compete within a region.

The largest, drought-prone region in Latin America is northeastern Brazil where cowpeas are frequently sown by farmers' practising shifting cultivation on the marginal lands. These cowpeas are frequently intercropped with maize and perennial cotton. In the extremely dry areas, cowpeas are intercropped with the spineless forage cactus, *Opuntia* spp. The preferred plant type for this dry ecology with erratic rainfall is one with medium duration (80–110 days) and prostrate growth habit; it covers the ground, reducing evaporation and controlling weed growth. Early-maturing, indeterminate varieties are also grown but to a lesser degree. Varieties with determinate growth habit fail when drought occurs during flowering and therefore are not recommended.

In the savanna ecologies with intermediate rainfall (600–1500 mm), cowpeas are frequently grown as a monocrop but occasionally with maize. Examples of this ecology include Bahía and Maranhão states of Brazil, the Llanos of Venezuela and Colombia, and some parts of Honduras and

Nicaragua. Erect, early and medium-maturing varieties are well suited to the savanna areas. Unfortunately, few such varieties have been made available to farmers in Latin America.

In areas of high rainfall, climbing cowpeas are frequently grown but on a small scale because staking costs prohibit large-scale production. Newly developed varieties of both grain and vegetable types that elevate the pods on peduncles can be grown without staking and should increase production. Examples of the high-rainfall areas in Latin America where cowpeas are grown include the Amazon River basin of Brazil and Peru, the coastal regions of Ecuador, Surinam and Guyana; and Trinidad.

The data on production are limited, but estimates are Brazil 600,000 t, Venezuela 10,000 t, Peru 5000 t, Panama 7000 t, El Salvador 6000 t, Haiti 2000 t, and others 30,000 t. Cowpea pods, consumed as a green vegetable, are important in the coastal regions of Ecuador, Guyana and Surinam; in the Amazon Basin of Brazil; and especially in Trinidad. In addition, in most countries, about 10 per cent of the crop is eaten as green immature seeds. In some parts of Brazil and Venezuela cowpea stover is highly valued by dairy farmers as dry-season pasture or hay. Unfortunately, no data are available on production of cowpea as vegetable pods, green seed or forage.

In contrast to consumers in West Africa, Latin Americans prefer smooth-seeded varieties. In Brazil, large, brown and cream-coloured varieties are generally preferred, although in some regions in the states of Rio Grande do Norte, Ceará, Piauí, Maranhão and Pará, white seeds with brown and black eyes are sought. In Venezuela, Peru and Surinam, brown, cream, and white seeds are common, and less emphasis is placed on seed size. In Central American and Caribbean countries, red-seeded varieties are preferred but other seed types including browns, blacks, speckled whites, and creams are also produced.

BREEDING TO OVERCOME PRODUCTION CONSTRAINTS

There are two common types of drought conditions that require different breeding strategies. Where the season is short, but water is available for the first 40 days, early-maturing varieties can escape the drought stress. In Latin America there are two farming systems where early-maturing cowpeas have obvious merit: in relay cropping in a rice- or maize-based system and in systems where cowpeas are sown on riverbanks as water recedes at the end of the rainy season.

Where rainfall is irregular, early maturing cowpea varieties are often not suitable. In Brazil some efforts have been made to select varieties for drought-prone ecologies (Guimarães et al., 1982), and varietal differences have been identified, although the mechanism of tolerance is not clear. Some cowpea lines showing promising levels of resistance include VITA 3, VITA 4 and TVx 1836-015J. Resistance, heritability and breeding methods are under study in Brazil.

Cowpea severe mosaic virus is widespread in the Americas and results in substantial yield losses. Agronomic practices that influence the presence of the beetle vectors can influence the prevalence of the disease, which is often high when maize is the predominant crop. Use of insecticides early in the growing season controls beetle populations and reduces the prevalence of disease. However, the most promising approach to the control of this disease is breeding for resistance. A resistant variety, Laura B, which was developed in Trinidad and released in Jamaica, carries the resistance gene or genes from line 10R3, developed in Puerto Rico (Thompson, 1977). In Brazil, CNC 0434 that carries genes conditioning immunity to severe mosaic virus was identified from an IITA disease subpopulation (Rios et al., 1982). CNC 0434 has white seeds and a high-yield potential and is currently being considered for varietal release in the state of Maranhão, Brazil. CNC 0434 is being used extensively as a parental source of resistance in EMBRAPA's breeding programme. It is also being used to convert IITA's high-yield potential varieties to resistant versions. Another useful source of resistance is Macaibo (Rios et al., 1980). Resistance to severe mosaic virus is genetically recessive; in some crosses one gene appears to control resistance, whereas in other crosses two-gene segregation is observed.

A group of viruses—the potyvirus complex—are primarily found in the dry ecology of northeastern Brazil, and they are all aphid-transmitted and generally seed borne. A useful source of field resistance is found in BR-1/Poty released in Brazil (EMBRAPA, 1984). TVu 410 was the donor source.

Cowpea golden mosaic virus, also found primarily in Brazil, is serious in the transition zone between the wet and dry regions in the north. There are many sources of resistance to this disease including BR-1/Poty and Pitiuba.

Powdery mildew (Erysiphe spp.) is frequently present in the state of Bahía in Brazil and has also been observed by one of us (E.A.K.) in Ecuador. There is considerable genetic variability in the disease organism and resistance to it may not be stable. However, some levels of field resistance have been noted in VITA 7 and in Manaus (Rios and Neves, 1982a).

Scab (Sphaceloma spp.), a fungal disease attacking stems, leaves and especially pods, occurs frequently in environments where days are hot and nights are cool, such as the high plateaus of northeastern Brazil. Sources of resistance include TVu 1888 and Kalkie.

Cowpea smut caused by Entyloma vignae is a sporadic disease in Brazil. VITA 4 offers useful levels of resistance to smut.

The cowpea curculio (Chalcodermus spp.) is found from North America to Brazil, boring into and depositing its egg in the seed of 10-day-old pods. The seed quality of infected pods is reduced drastically. Sources of resistance include CR 17-1-13 identified in South Carolina and CNCX 10-2E identified in Brazil.

Leafhopper burn caused by Empoasca spp. is common, especially in dry environments. VITA 3 is resistant, and breeders at IITA and Brazil have been able to transfer the resistance into other genetic backgrounds.

Table 1. Cowpea cultivars released in Latin America and the Caribbean

Name	Country	Year	Reference
Arauca, Orinoco	Venezuela	1956	Marcano and Linares, 1956
US 62	Colombia	1957	Anonymous, 1957
IPEAN-V69	Brazil	1970	Araújo, 1983
Bush Sitão	Trinidad	1973	
Tuy	Venezuela	1975	Barrios and Ortega, 1975
Pitiuba	Brazil	1976	Araújo, 1983
Romefa	Panama	1978	Rodriguez et al., 1978
VITA 3	Nicaragua	1979	
Laura B	Jamaica	1981	
Manaus	Brazil	1981	Araújo, 1983
VITA 3, ER-7, TVu 66-2H, TVx 2904-03E	Guyana	1982	Ross (personal communication)
IPA 201, IPA 202	Brazil	1982	Araújo, 1983
EPACE 1, EPACE 6	Brazil	1982	Araújo, 1983
Apure, Unare	Venezuela	1982	Boscan (personal communication)
IPA 203	Brazil	1983	Araújo, 1983
EMAPA-821, EMAPA-822	Brazil	1983	EMBRAPA, 1983
BR-1/Poty	Brazil	1984	Araújo, 1983

RELEASE OF COWPEA VARIETIES

Several countries have released cowpea varieties (Table 1) derived primarily from mass selection of introduced germ plasm and occasionally from local varieties that were not pure lines. A few varieties such as Laura B and BR-1/Poty were developed from crosses made in Latin America.

In collaboration with EMBRAPA in Brazil, IITA is expanding its activity in Latin America and is optimistic that progress will be rapid because of the ties with national programmes. Compared with other crops in Latin America, cowpea has received little research, although it has great potential for production.

Cowpea Research, Production and Utilization
Edited by S. R. Singh and K. O. Rachie
© 1985 John Wiley & Sons Ltd

8

Improved Cowpea Cultivars for the Horticultural Industry in the USA

R. L. FERY
United States Department of Agriculture, Agricultural Research Service, US Vegetable Laboratory, Charleston, South Carolina, USA

Cowpea was introduced into the New World in the latter part of the seventeenth century, and the crop has been cultivated in the southern parts of the United States since the early eighteenth century. Although once an important agronomic crop in the United States, cowpea is grown and consumed on a limited basis at present but is relatively important horticulturally still in the United States. An extensive industry exists to supply the processed and dry peas that are marketed nationwide. Cowpeas are sold to the American consumer under a variety of names, e.g. field pea, southern field pea, acre pea and crowder pea. Dry blackeye types are usually marketed as blackeye beans or blackeye peas. Most American horticulturists use the term southernpea when referring to cowpea being grown as a vegetable, and the term cowpea is seldom, if ever, used in the American marketplace.

Cowpea is grown commercially in all of the southern states, and it is a popular home garden item throughout this region. Dry-seed industries are extensive in both California and Texas. Both fresh and dry peas are processed; Georgia is the leading processor of fresh peas and Texas leads in processing of imbibed 'dry' peas. Accurate production figures are not available, but the sum of the best obtainable estimates by state indicates that about 80,000 ha of cowpeas are grown in the United States each year. The states of Georgia, California and Texas are the leading producers, and together they account for about 65 per cent of the total production. Cowpea culture in the United States is quite similar to that used for the other table legumes, e.g. the common bean (*Phaseolus vulgaris*) and the lima bean (*P.*

lunatus), and extensive use is made of the practices and equipment that have been developed for these crops.

Plant breeders have played an important role in the development of the horticultural cowpea industry as it exists in the United States today. This review presents an account of the current status of the cowpea-breeding effort in the United States.

CULTIVAR CLASSIFICATION

There is a broad range in the characteristics of cowpea cultivars that are popular in the United States. Although several researchers (Piper, 1912; Brittingham, 1946; Ligon, 1958) have proposed schemes to classify cowpea cultivars using various seed traits, most modern American horticultural cultivars can be classified as being of the blackeye, crowder or cream types. The peas produced by each type have a distinctive appearance and flavour. Each type appeals to a unique market. Thus, the horticultural cowpea is not just one product, but several, and the plant breeder must appreciate fully the attributes of these cultivar classes if he or she hopes to be successful in developing improved cultivars.

The blackeye cultivars are the most popular, with California Blackeye 5 (CB5) serving most of the dry-seed industry. A variant, the pinkeye, is more popular as a processing vegetable crop because it contains less water-soluble pigment in its seed coat and, when cooked, produces a lighter-coloured liquor. The shelled, fresh, pinkeye peas are attractive and have a pleasant mild flavour when cooked. Besides these qualities, the blackeye/pinkeye type had an early reputation for good yield. Much of the early breeding work in the United States was for disease resistance in the blackeye types.

The term 'crowder' refers to a pea type that produces peas closely crowded in the pod. The shape of the peas is often irregular, distorted or globular. Although there are a number of seed-coat colours and patterns in this group, the brown crowders are by far the most popular. The crowders as a group have the largest peas. Cooked crowder peas have a strong or astringent flavour, a granular texture and are dark in colour with a dark liquor; they are the least popular of the three types (Hoover, 1966). Crowder cultivars are often used for canning, and those with purple hulls are widely grown for the fresh market because of their inherently long shelf-life.

The cream- or white-seeded type cultivars have peas with little or no eye. Cream peas have a much milder, less starchy flavour, are succulent and have a better appearance than the blackeye or crowder peas. The flavour is often compared to that of the butter bean (*P. lunatus*). The liquor of the cooked product is light. The creams have long been popular with the American consumer. For example, Piper in 1912 noted that white-seeded cultivars sold at a higher price than other cultivars because of the popularity as a table vegetable. As a group the cream peas are much more diverse than blackeye or crowder types, and breeders have made much progress in recent years in developing cultivars with improved plant habit, disease resistance and pea and pod characteristics.

There are a number of cultivars popular with some American home gardeners and consumers that do not fit well into the simplified blackeye–crowder–cream classification. Many of these cultivars are derived from old agronomic types, e.g. Clay, Whippoorwill and Bluegoose, and are capable of producing good yields under a wide range of conditions. Most have coloured seeds, and the cooked product is dark, with a pronounced, somewhat strong flavour. These peas are often referred to as 'field peas' in the American marketplace. Additionally, cowpeas sometimes are grown for their tender, immature pods or 'snaps'; the cultivar Snapea was developed specifically for this purpose. Some packers include a small portion of 'snaps' in their processed product.

BREEDING METHODS

Most plant-breeding programmes in the United States have short-term goals, i.e. the quick development of cultivars to meet immediate contingencies and industry requirements. However, some programmes are working to develop improved breeding populations and to conserve genetic variability.

The pedigree system of breeding a self-pollinated crop is the predominant method used by cowpea breeders in the United States. This method has proved suitable for the short-term objective of developing cultivars with new combinations of horticultural characteristics and disease resistance. Also, the pedigree method is useful for genetic studies where the genotype of individual plants must be determined.

The backcross breeding method has proved efficient for transferring single-gene resistance to specific diseases into cowpea cultivars. For example, my colleagues and I (Fery et al., 1982) used this procedure to transfer the Cls-1 gene, which provides resistance to Cercospora leaf spot (C. cruenta), into the susceptible crowder cultivar Colossus.

A combination backcross–pedigree breeding method has been used in some programmes to transfer a desired trait from a relatively unadapted genetic background into a well-adapted commercial background. Only one, two or three backcrosses are made, and the remainder of the breeding is handled with conventional pedigree procedures. This approach allows the recovery of favourable gene combinations in the adapted, recurrent parent and still retains the benefits of transgressive segregation for horticultural traits such as yield and adaptation. It has been used to develop horticultural cowpea lines with resistance to the cowpea curculio (Chalcodermus aeneus Boheman).

COWPEA BREEDING AND EVALUATION PROGRAMMES

Cowpea breeding and evaluation programmes have existed in the United States since the latter part of the nineteenth century. The first artificial hybrid was made by C. L. Newman at the Arkansas Agricultural Research Station in 1893 (Piper, 1912). In 1904 an early USDA scientist, W. A. Orton, initiated a cowpea-breeding programme to develop cultivars resistant to *Fusarium* wilt (*F. oxysporum tracheiphilum*) and root-knot nematode (*Meloidogyne* spp.) (Piper, 1912). In 1912 Piper published descriptions of 220 cowpea cultivars, 50 catjang cultivars and 35 asparagus bean cultivars, and concluded that a considerable number of the American cowpea cultivars originated in the United States. In 1920 there were about 15 major cowpea cultivars grown in the United States, and an additional 45 were grown on a limited scale (Morse, 1920): three of the major cultivars, Victor, Arlington and Columbia, were products of USDA plant-breeding programmes.

Most of the current US breeding effort on cowpeas is by horticulturists working for public agencies. A recent search of the approximately 24,000 research resumés in the USDA's Current Research Information System database (coverage included projects from the 58 state agricultural experiment stations and the five USDA research agencies) identified 20 state institutions and one federal agency with active or recently completed cowpea research projects. Cowpea research is being conducted at 28 different locations in the US. Eight of the state institutions (Auburn University, Clemson University, Louisiana State University, Mississippi State University, Texas A&M University, University of Arkansas, University of California and University of Georgia) and the Agricultural Research Service of USDA have ongoing research programmes with clearly identifiable cowpea-breeding objectives. Three other state institutions (New Mexico State University, University of Florida and University of Minnesota) have active cowpea cultivar-evaluation projects. Other institutions have research projects to develop improved technologies and methods for cowpea breeders. For example, these projects are addressing such topics as the genetic and physiologic aspects of economically important cowpea traits, the identification of new sources of resistance to diseases and insects, the identification of new sources of tolerance to herbicides, air pollution and heat and drought stress, ways to improve protein quality and photosynthetic and nitrogen-fixation efficiencies and the development of improved evaluation procedures for various traits of economic interest. A total of 42 cowpea cultivars, virtually all horticultural types, has been released in the United States since 1960, and there are usually a number of suitable cultivars for environments in which cowpeas are grown (Ligon, 1958; Minges, 1972; Cowpea Varietal Nomenclature Study Group, 1978; Fery, 1981).

Regional testing of advanced breeding lines is coordinated by the Regional Southernpea Cooperative Trials, a part of the Southern Cooperative Vegetable Trials sponsored by the Southern Region of the American Society for Horticultural Science (Yarnell, 1948). The cooperative trials, in which lines are tested in many locations, provide breeders with as much information in a single season as they could obtain at single locations over many seasons. Only advanced breeding lines are accepted for testing; the Southernpea Trials are not intended to be a substitute for ordinary cultivar trials. The 1984 trials were grown at 15 sites in 8 different states, and there were 6 entries and 3 checks each in the replicated and observational portions of the trials. The trials are coordinated, planned and reported annually by a chairperson.

There are two other commodity-oriented organizations that play an important role in coordinating US cowpea-breeding programmes. The first, the Cowpea Improvement Conference, is open to all scientists with an interest in cowpea improvement. This organization, which is quite informal, meets annually, usually in conjunction with the annual meeting of the Southern Region of the American Society for Horticultural Science. The second organization, the *Vigna* Crop Advisory Committee, is a national working group of specialists providing analysis, data and advice about activities necessary for effective acquisition, evaluation, preservation and use of genetic resources within the genus *Vigna*. It reports to the Cowpea Improvement Conference, the National Coordinator for Plant Germplasm, the National Germplasm Committee and the National Plant Genetic Resources Board. The committee's activities are facilitated through the office of the chairperson, Plant Genetics and Germplasm Institute, USDA Agricultural Research Service, Beltsville, Maryland.

BREEDING OBJECTIVES

Many of the general objectives of all US cowpea-breeding programmes are the same. These include such traits as increased yield, plant habit, improved adaptation and increased tolerance to environmental stress. Resistance to fungal and bacterial pathogens, viruses, nematodes and insects, particularly those that significantly reduce yields, is universally important. Many of these traits are under a high degree of genetic control (Fery, 1980), and American cowpea breeders have been quite successful in developing improved cultivars for all segments of the industry.

Processing cultivars

Virtually all horticultural cowpeas grown for fresh processing are mechanically harvested and are selected accordingly. To be suitable for processing a cultivar must be even maturing, with 85–90 per cent of the pods approaching maturity simultaneously. The plant habit must be upright and compact, although recent improvements in harvesting equipment have lessened the importance of this trait. Earliness is important, and the breeder must be cognizant of consumer and industry preferences. Such traits as wholeness, consistency, freedom from defects, flavour, colour and texture must be carefully evaluated for both canned and frozen products. Recently there has been increased interest among both processors and plant breeders in the green testa (*gt*) gene. This gene conditions a green seed coat that persists to dryness, and the processed product has an improved consumer appeal because of the uniform colour (Chambliss, 1974). The cultivar Freezegreen is homozygous for the *gt* gene (Chambliss, 1979).

Fresh market and home garden cultivars

The development of improved cultivars for fresh market and home gardens is a major objective of the majority of the US breeding programmes. Such cultivars must produce an attractive pod that is easy to hand-shell, fills well, has a high number of peas and remains harvestable for an extended time. Cultivars with pods that dry quickly or have a short shelf-life are not suitable for fresh-market use.

Disease resistance

Breeding cowpeas for resistance to the most destructive pathogens has been an effective method of minimizing disease losses in the United States. Resistance has been identified for most diseases that limit yield (Fery, 1980; Meiners, 1981). Breeding for multiple resistance to fungal diseases is a major objective in most breeding programmes. Additionally, some programmes are addressing resistance to selected viral and bacterial pathogens.

Root-knot nematode causes economic losses to cowpeas throughout the United States. Although several US cultivars exhibit excellent resistance, many of the important horticultural cultivars are susceptible or exhibit only an intermediate level of resistance. Resistance is conditioned by one of two alleles at the *Rk* locus; the dominant allele *Rk* conditions the high level of resistance and the recessive allele rk^i conditions the intermediate level of resistance (Fery and Dukes, 1980, 1982). One of the objectives of the USDA breeding programme is to incorporate the *Rk* allele into leading horticultural cultivars.

Insect resistance

Although several insects are serious pests of cowpeas in the United States it has only been in recent years that breeders and entomologists have begun to search for resistant germ plasm. The Auburn University and USDA programmes are developing cultivars with resistance to the curculio, the major insect pest of the cowpea in the southeastern United States. USDA recently released the curculio-resistant cultivar Carolina Cream (Fery and Dukes, 1984). Cowpea lines with resistance to *Lygus* bugs, *L. hesparus*, have been identified at the University of California at Davis (UCD), and these lines have recently been entered into the UCD cowpea-breeding programme (Moshy *et al.*, 1983).

CONCLUSIONS

Although US cowpea breeders have been successful in developing cultivars suitable for highly specialized production and marketing systems, they still face formidable challenges. Cultivars are needed that exhibit increased resistance to a wide array of diseases and insects, better tolerance to environmental stress, increased nitrogen-fixing capacity, better seed qualities and improved efficiency in the utilization of limited soil nutrients. Compared with many other crops the cowpea has received little attention from plant breeders, and a large effort needs to be made to break yield barriers. The plant breeder has an important role to play in the development of improved production systems. For example, cultivars with modified plant growth habits might be better suited for use in modern, intensely managed farming systems than cultivars currently available. Weed problems could be solved by cultivars with tolerance to an existing herbicide. Finally, the cowpea is important in the diets of many people, and the plant breeder has an obligation to improve the nutritional, flavour and aesthetic aspects of the cowpea products consumed.

Cowpea Research, Production and Utilization
Edited by S. R. Singh and K. O. Rachie
© 1985 John Wiley & Sons Ltd

9

Breeding for Resistance to Drought and Heat

A. E. HALL and P. N. PATEL
University of California, Riverside, USA

Drought and high temperatures often occur together in the semiarid zones where cowpeas are grown, and they have interactive adverse effects on plant growth and development. However, in this analysis we consider them separately, emphasizing high night temperatures, the detrimental effects of which are apparent with or without drought (Warrag and Hall, 1984b). Resistant cultivars are defined as having the ability to produce relatively high yields of grain when subjected to stresses, such as drought and heat.

BREEDING FOR RESISTANCE TO DROUGHT

Drought is defined as the occurrence of a substantial water deficit in the soil or atmosphere. The first step in breeding for resistance to drought is to determine the type and frequency of droughts that occur in the production areas for which improved cultivars are being developed. Hydrologic budget analyses, based on information about rainfall, soil-water storage (as determined by soil physical characteristics and rooting) and crop water-use characteristics can be used to determine the type and frequency of droughts. Most mechanisms whereby crops resist drought will be mainly effective against specific types of drought. For example, short-cycle, synchronous-flowering varieties are needed where the average period of water supply is short but reasonably reliable. In contrast, more indeterminate, longer-cycle varieties with sequential flowering are needed where the average period of water supply is longer but less reliable.

Deeper and more extensive root systems may increase resistance to drought if, in an average year, present cultivars do not exploit most of the 'available' water in the soil profile. However, to be most effective, quantitative characters such as the extent of rooting should be expressed at intermediate levels (Hall, 1981). The extra water gained by increased rooting must more than compensate for the plant's additional investment of carbohydrates in roots (Passioura, 1982). The levels of rooting or earliness that are optimal depend upon the environment and genetic background (as they influence the hydrologic budget and plant responses to drought), and the extent of useful plasticity in character expression influences breadth of adaptation (Hall, 1981).

Developing shorter-cycle cowpea varieties

Hydrologic budget analyses based upon 44 years of rainfall data predicted that varieties of annual crops with cycle lengths of 75 days from sowing to maturity are needed for the northern, drier part of the semiarid zone in Senegal (Dancette and Hall, 1979). In this zone, average rainfall since 1968 has been substantially less than in earlier years, and it is likely that varieties with shorter cycles (*ca.* 65 days) may be more effective.

In developing shorter-cycle cowpeas we concentrated on selecting strains with a shorter vegetative stage, by selecting for earlier flowering, rather than selecting for a shorter reproductive stage because grain yield of cowpea is much more dependent upon the amount and activity of leaves present during the reproductive stage than during the vegetative stage (Turk and Hall, 1980b). Selecting for earlier flowering should not reduce yield potential as much as selecting for a shorter reproductive period.

Transgressive segregants with early synchronous flowering were selected in California from progeny from a cross between two cowpeas that flower at the same time (California Blackeye 5 and Bambey 23 flower within 50–60 days from sowing in California depending upon temperatures). The earliest selections flowered 4 days earlier than the parents at Riverside, California, but under the higher night temperatures in Senegal the parents and selections flower at the same time (36 days from sowing). Several of the selections have been more productive than local cultivars in dry years in Senegal and Sudan (Table 1) where they have a cycle length of 60 days with late season drought and more than 70 days under well-watered conditions. Short-cycle cowpeas have also been developed by IITA. The IT 82E-strains were early in Senegal, and IT 82E-18 and IT 82E-56 gave good yields under dry conditions in Senegal in 1983 (Cisse *et al.*, 1984). Some of these strains appear to be photoperiod-insensitive (IT 82E-13, IT 82E-16 and IT 82E-18), flowering early under long-day conditions in California, whereas others—IT 82E-3, IT 82E-10, IT 82E-32, IT 82E-56 and IT 82E-60—flowered late under these conditions (unpublished data of I. Dow El Madina, University of California, Riverside). California summer conditions appear to provide a useful environment for selecting for earliness that is expressed over a broad range of environments.

Table 1.   Comparison of early and local cowpeas

| | Senegal* | | | | Sudan† |
| | 1982 | | 1983 | | 1983 |
	Bambey	Louga	Bambey	Louga	El Obeid
Useful rainfall (mm)	452	181	315	135	230
Cowpea strains	Grain yields (kg/ha)				
CB5 × B23					
1-2-1	2324	663	1315	250	468
1-11-1	2290	949	1253	188	
1-12-3	2406	1091	1816	206	500
3-4-1	2033	919	1445	290	
3-4-13	2418	1026	1422	216	355
CB5	2354	922	1355	195	625
Local					
Bambey 21	2263	699	1303	51	—
Garnel Kabish	—	—	—	—	135
Gambaru	—	—	—	—	169
LSD 0.05	NS	213	NS	81	170

* Data reported by Cisse et al. (1984).
† Unpublished data provided by the Western Sudan Agricultural Research Project.

Developing varieties with improved rooting

In tropical semiarid zones, rainfall can exceed crop-water use during part of the rainy season (Hall and Dancette, 1978), and water is then available deep in the soil profile which is only fully accessible to varieties with deep, extensive root systems. Available methods for evaluating the root systems of plants are laborious and severely limit the number of genotypes that can be screened. A new method has been developed to evaluate rooting indirectly under field conditions (Robertson et al., submitted for publication). It is based upon the time at which symptoms appear in leaves as an indication of when roots reach a herbicide band placed deep in between rows of plants. Consistent differences among extreme genotypes have been observed over three seasons, and earliness of symptom appearance was associated with greater extraction of moisture deep in the soil (Robertson et al., submitted for publication). Diverse cowpea strains have been screened, and genotypes have been discovered (e.g. Grant, which is similar to California Blackeye 3, and Bambey 21), which may be useful as parents for conferring improved rooting (Table 2). We are currently attempting to combine the improved rooting of

Table 2.   Herbicide screening of cowpea strains for rooting in 1983

UCR No.	Cowpea strain	Origin	Number of days to first symptoms*
55	Bambey 21	Senegal	52.8 a
122	Grant	California	53.1 ab
237	Cross 1-6E-2	Tanzania	53.6 abc
248	Quarenta dias	Brazil	53.6 abc
262	Reata	Mexico	53.6 abcd
251	CNCX 27-2E	Brazil	53.7 abcd
244	Pitiuba	Brazil	54.0 abcde
216	TVx 3236	IITA	54.1 abcde
255	CNCX 105-5E	Brazil	54.2 abcdef
243	TKx 9-11D (TK-1)	Tanzania	54.5 abcdefg
170	CB5	California	54.6 abcdefg
39	PI 293579	Texas	54.6 abcdefg
72	VITA 3	IITA	54.7 abcdefg
175	58-57	Senegal	55.1 abcdefg
260	Gamusa	Mexico	55.1 abcdefg
250	CNCX 24-016E	Brazil	55.3 abcdefg
177	TVx 1841-0-1-E	IITA	55.4 abcdefg
247	Serido	Brazil	55.5 abcdefg
261	Malu	Mexico	55.6 abcdefg
74	VITA 5	IITA	55.7 abcdefg
217	VITA 7 (KN 1)	IITA	55.9 abcdefg
	7977	UCD	56.0 bcdefg
40	PI 302457	—	56.3 cdefg
73	VITA 4	IITA	56.4 cdefg
176	TVx 1193-FD	IITA	56.5 cdefg
52	88-63	Niger	56.7 cdefg
71	VITA 1	IITA	56.7 defg
180	TVx 1836-015J	IITA	56.9 efg
182	4R-0269-1F	IITA	57.3 fg
179	TVx 309-1G	IITA	57.3 fg
156	8006	UCD	57.4 g
239	TVx 133-16D-2	IITA	57.6 g

* Figures followed by different letters differ significantly at the 5 per cent level.

Grant and the superior canopy architecture of 8006, which is a breeding line developed at the University of California, Davis. Selection for improved rooting is being conducted at Riverside, under stored soil moisture without irrigation, in soil that has a high bulk density (1.5–1.8 g/cm^3) and offers considerable resistance to root development. A major part of the semiarid zone of Senegal has a 'Dior' soil, which offers considerable resistance to root development. Improved capacity for root growth could be particulariy useful in hard soils where the extent and intensity of rooting are usually poor and where improved root growth results in substantial increases in the amounts of water and nutrients extracted from the soil.

Selecting for increased accumulation of osmotica

Several scientists have proposed that drought resistance may be increased by selecting for increased osmotic adjustment (Turner and Jones, 1980). We have observed little osmotic adjustment in the leaves of cowpeas and little difference in leaf osmotic potential among 100 contrasting cowpea genotypes (Shackel and Hall, 1983). In future, screening for genotypic differences in osmotic potential in roots may prove useful because accumulation of osmotica in root cells could increase turgor and provide sufficient additional force to increase root growth in mechanically resistive soils. For shoot tissues, osmotic adjustment at night, as occurs in sorghum (Shackel et al., 1982), could be adaptive if reproductive processes that are sensitive to low turgor occur at night.

Developing varieties with improved efficiency for using water and partitioning carbohydrates

Varieties with improved seasonal water-use efficiency would produce more biomass in dry environments, if they also have the same total seasonal water use. Unfortunately, methods for evaluating water-use efficiency based upon simultaneous measurements of photosynthesis and transpiration are laborious and severely limit the number of genotypes that can be screened. Theoretical analysis by Farquhar et al. (1982) demonstrated that changes in carbon-13 in dried leaf tissue provide a rapid and sensitive measure of relative, integrated, intrinsic water-use efficiency under field conditions. With this technique, researchers have discovered genotypic differences in water-use efficiency in wheat (Farquhar and Richards, 1984), and we have begun to use Farquhar's method to screen cowpea genotypes under drought.

Because increases in grain yield of determinate cereal crops have been strongly associated with increases in harvest index, it is possible that the productivity of the more determinate types of cowpea could be improved by selecting for increased harvest index. Cowpea strains selected visually for early production of mature pods had increased proportions of shoot dry matter in grains and grain yield under stored soil moisture (Hall and Grantz, 1981). However, high harvest index would not be appropriate for cowpea varieties intended for the production of both grain and forage, nor for the more indeterminate varieties needed where rainy seasons are long but unreliable with a high probability of droughts in the middle of the season. Synchronous flowering, as is present in Bambey 23, may be used to enhance physiologic determinancy and increase harvest index while retaining greater morphologic plasticity than is present in anatomically determinate genotypes. Synchronous flowering types will exhibit determinancy only if a substantial number of the early flowers set pods; they will be able to continue to produce floral nodes, flowers and leaves late in the season if early pod set is low due to midseason stress. In contrast, anatomic determinancy is fixed, providing no opportunity for regrowth late in the season.

Evaluating the extent of drought resistance in advanced lines

Drought resistance depends upon many interacting plant characteristics, and progress in breeding should be monitored. Multilocational yield trials are a necessary part of breeding programmes, and trials in dry sites can provide relative measures of differences in drought resistance among advanced lines. Systems for providing controlled levels of available water can be used to evaluate genotypic resistance to a range of drought intensities in 1 year at one location. The line-source sprinkler has been proposed as a technique for evaluating resistance over a range of drought intensities (Hall *et al.*, 1977). Unfortunately, the efficiency of the line-source approach is severely limited by methodological and statistical problems that constrain data analysis and separation of genotypic differences, and the water regimens and droughts imposed are usually unnatural.

The method of Fischer and Maurer (1978) has distinct advantages for evaluating the adaptation of advanced lines to semiarid environments. With this approach, yields of individual varieties must be determined under drought (Y_D) and well-watered (Y_W) conditions. Data on the average yield of all varieties under drought (X_D) and well-watered conditions (X_W) are used to calculate drought intensity (D):

$$D = 1 - X_D/X_W$$

Then the drought susceptibility (S) of individual varieties can be calculated:

$$Y_D = Y_W(1 - SD)$$

Varieties with average susceptibility or resistance to drought have an S value of 1.0. Values of S less than 1.0 indicate less susceptibility and greater resistance to drought, with a value of $S = 0.0$ indicating maximum possible drought resistance (no effects of drought on yield). In evaluating drought resistance with this method it is most effective to impose paired wet and dry treatments on varietal trials at key field sites. This can be achieved in many soil conditions, where horizontal transfer of water is not substantial, by use of surface irrigation for the wet treatment. For example, plants could be grown on single-row beds, with adjacent, paired four-row, wet and dry plots. The central furrow of the wet plot would be irrigated with sufficient frequency to maintain an adequate water supply to the two centre rows, which would be harvested to obtain a measure of yield potential (Y_W). All of the other furrows would receive the dry treatment, which would be natural rainfall or limited irrigation. The two centre rows of the dry plot would be harvested to obtain an estimate of yield under drought (Y_D). Varieties or advanced lines would be randomized, as paired wet–dry treatments, within a block composed of a set of beds and furrows, and four to six randomized blocks could be used. Conventional statistical analyses could be used to evaluate genotypic differences in Y_W, Y_D and S for individual wet–dry varietal trials, or for several multilocation wet–dry trials. Genotypic differences in S would be most readily detected in experiments where the intensity of drought is intermediate (e.g. D values of 0.5). The S values of different types of cowpea varieties may vary when they are subjected to droughts of different intensities, or at different stages of plant growth, but this variation simply indicates that they are better adapted to certain types of drought and environmental conditions.

A case for varietal intercrops

In tropical semiarid zones, rainfall during different years varies, favouring short-cycle, synchronous-flowering, erect varieties one year and longer-cycle, more indeterminate varieties the next. In addition, early cowpea varieties have an advantage over long-cycle varieties because they provide food during the period just before the main cereal harvest when food is often in short supply, whereas longer-cycle, more indeterminate varieties provide more forage than do early, erect varieties. It may not be possible to develop extremely plastic varieties that can produce early harvests of cowpea grain while still retaining the ability to respond to later rains, producing additional grain and forage. Consequently, to increase stability of yields, farmers in tropical, semiarid zones may have to sow both types of cowpea varieties.

Sowing both types of cowpea in alternating rows as a varietal intercrop has advantages over planting sole crops in separate fields, and over intercrops with different species, such as cereals. With the first pulse of rain the early, erect variety would produce grain and then senesce. With subsequent rains the later, spreading variety would spread over the adjacent rows occupied by the senescing early variety and produce additional harvests of grain and then a forage harvest. However, mechanized weeding may be more difficult with a varietal intercrop than with sole crops of erect varieties. Varietal intercrops of cowpeas rotated with cereal crops, such as pearl millet, have advantages over cowpea–millet intercrops. Rotating constrasting species makes possible more efficient use of manure or fertilizer (it should be applied to the most responsive crop, which usually is the cereal), and can slow the build-up of pests and diseases for which only one of the species acts as a host.

BREEDING FOR RESISTANCE TO HEAT

Resistance to the stress caused by high temperatures requires that limiting plant processes are not irreversibly damaged. All plant processes are irreversibly damaged if high enough temperatures are imposed for sufficient time. Consequently, the key questions for breeders are:

O What aspects of high temperatures (considering temperature levels and durations at different times during the season and day) cause significant reductions in productivity in different climates?

O What plant processes and stages of development are most sensitive to high temperatures and are responsible for reductions in productivity?

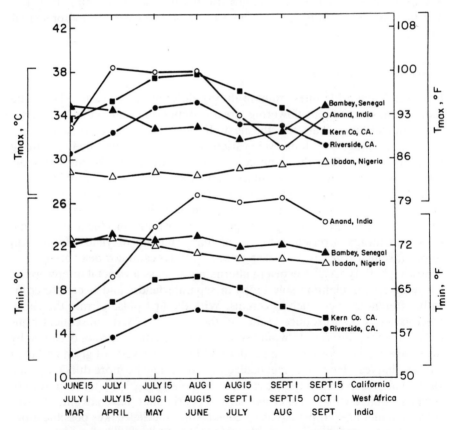

Figure 1.   Early planting and late planting period of flowering

For cowpeas, considering the natural variations in temperature that occur in the tropics and subtropics (Figure 1) and studies on cowpea response to temperature (Warrag and Hall, 1984a,b), we have concluded that high temperatures at night can be much more damaging to grain yield of cowpeas than high temperatures during the day. Growth chamber and field studies demonstrated that the temperatures that commonly occur at night in the tropics can cause male sterility (Warrag and Hall, 1984b) and substantially reduce grain yield by increasing floral abscission and decreasing the number of pods/m^2 (Figures 2 and 3). Male sterility, as induced by high night

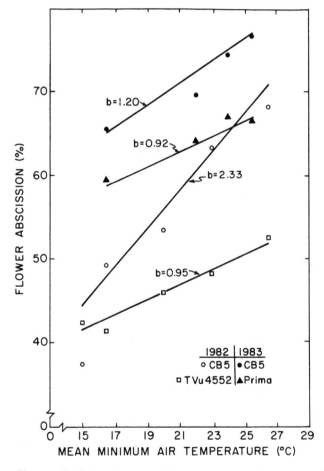

Figure 2. Flower abscission in contrasting cowpea genotypes at various levels of minimum night air temperature in field conditions during flowering. (Nielsen and Hall, 1985.) Reproduced by permission of Elsevier Science Publishers

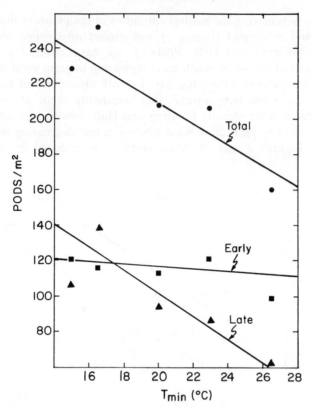

Figure 3.   Pod density of early flowers ■ (anthesis occurred during the first 6 days of higher temperatures), late flowers ▲ (anthesis occurred 7 or more days after temperatures were raised), and all flowers ● of cowpea cultivar CB5. (Nielsen and Hall, 1985.) Reproduced by permission of Elsevier Science Publishers

temperatures, is mainly due to lack of anther dehiscence, which results from incomplete pollen development (Warrag and Hall, 1984b). Experiments involving the transfer of plants between growth chambers with moderate and high night temperatures (Figures 4 and 5) established that the stage of pollen development most sensitive to high night temperatures occurs 5–7 days before anthesis. High temperatures are also detrimental to other aspects of floral development in certain cowpea varieties. For example, they delay or inhibit the development of floral buds in CB5, Bambey 21, PI 218123, TVx 12-01E and IT 81D-1020 (P. N. Patel and I. Dow El Madina, unpublished data) and induce seed-coat browning in TVu 4552 (Nielsen and Hall, 1985) and this character is inherited as a single dominant gene. We also observed seed-coat browning in TVx 3236-01G plants growing in Senegal. High day

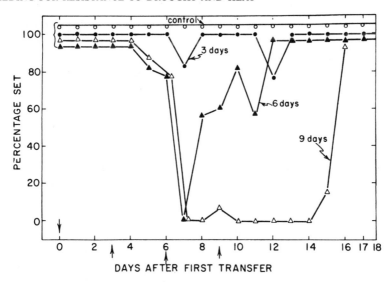

Figure 4. Percentage pod set of plants transferred from a growth chamber at 33/22°C to a chamber at 33/30°C day/night temperature for periods of 3, 6 and 9 days, and then returned to the 33/22°C chamber. (Warrag and Hall, 1984b.) Reproduced by permission of Elsevier Science Publishers

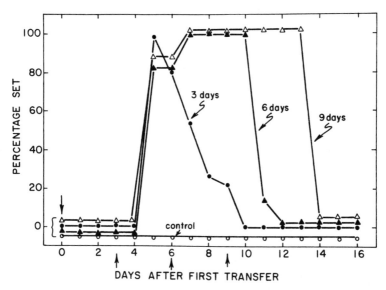

Figure 5. Percentage pod set of plants transferred from a growth chamber at 33/30°C to a chamber at 33/22°C day/night temperature for periods of 3, 6 and 9 days, and then returned to the 33/30°C chamber. (Warrag and Hall, 1984b.) Reproduced by permission of Elsevier Science Publishers

temperatures can result in seeds with asymmetric cotyledons under growth chamber conditions (Warrag and Hall, 1984a) and in hot field conditions in California (Mackie, 1946).

In our programme for breeding heat-resistant cowpeas, we developed a technique for screening for tolerance to heat at flowering (Warrag and Hall, 1983). This technique consists of sowing diverse cowpea strains in fields in Imperial Valley, California, in early June. In an average year most effective screening can be done only with genotypes that begin flowering during late July and early August when temperatures are extremely hot (Figure 6). In these conditions a few heat-tolerant strains had abundant pod set, whereas the majority of strains exhibited little or no pod set (Figure 6 and Table 3). Unfortunately, this technique is not useful for screening late-flowering strains, especially those that are sensitive to photoperiod, because the day lengths are longer than 14 hours during July in Imperial Valley.

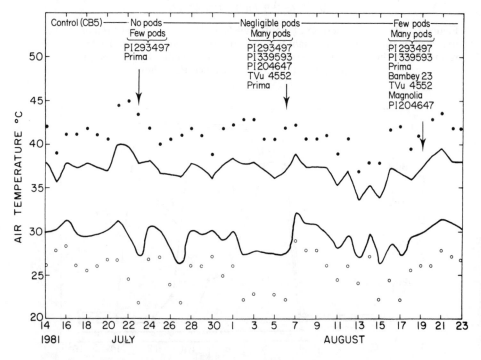

Figure 6.   Observations concerning pod set of cowpea strains in Imperial Valley, California, Shelter air temperatures, including daily maximum (●), daily minimum (○), and the effective day and night temperatures, calculated from the mean of the daily mean, and daily maximum and minimum, respectively. (Warrag and Hall, 1983.) Reproduced from *Crop Science,* **23**, 1088–1092, by permission of the Crop Science Society of America

Table 3.   Visual evaluation of heat tolerance at flowering of 58 cowpea genotypes in Imperial Valley, California, 1981, based upon ability to set pods (Warrag and Hall, 1983)

	Country of origin
High	
Prima	Nigeria
TVu 4552*	Nigeria
PI 204647	Turkey
Moderate to high	
PI 339593	South Africa
Moderate	
PI 293497	USA
Bambey 23	Senegal
Magnolia	USA
PI 307558	USA
Low to moderate	
PI 292899	Hungary
PI 293550	USA
PI 339604	South Africa
Bambey 31	Senegal
Bambey 25	Senegal
Bambey 24	Senegal
TVu 354	Nigeria
TVu 3046	India
TVx 12-01E	Nigeria
CB7	USA
CB	USA
Early Pinkeye	USA
TVu 161	USA
PI 292892	South Africa
CPI 30783	Burma
CPI 11900	Argentina
CB5	USA
Low	
PI 165486	India
PI 177579	Turkey
PI 218123	Pakistan
PI 220210	Afghanistan
PI 220850	Afghanistan
PI 221731	South Africa
PI 293522	USA
PI 293570	USA
PI 302457	USA

continued

Table 3. cont.

	Country of origin
Low	
N 70	Burkina Faso
Bambey 12	Senegal
Bambey 13	Senegal
Bambey 14	Senegal
Bambey 15	Senegal
Bambey 21	Senegal
Bambey 22	Senegal
Bambey 26	Senegal
Bambey 27	Senegal
Bambey 28	Senegal
Bambey 29	Senegal
Bambey 30	Senegal
Bambey 32	Senegal
VITA 4	Nigeria
VITA 5	Nigeria
TVu 401	USA
TVu 984	Nigeria
TVu 1016-1	USA
C-152	India
CPI 77123	Soviet Union
CPI 77122	Soviet Union
PI 170859	Turkey
CPI 30780	Burma
PI 124609	India

* Identification codes are: TVu, TVx and VITA, IITA Nigeria; PI, Plant Introduction No. USA; Bambey, ISRA Senegal; CPI, Plant Introduction No. Australia. Reproduced from *Crop Science,* **23,** 1088–1092 by permission of the Crop Science Society of America.

Our studies indicate that Prima and TVu 4552 have substantial heat tolerance, which is conferred mainly by the same recessive gene. Growth chamber studies indicated that TVu 4552 has greater heat tolerance than Prima (Warrag and Hall, 1983), but Prima appears to be a better parent. TVu 4552 has at least two detrimental characters: heat-induced browning of the seed coat and a tendency for embryo abortion at the end of the pod where it is attached to the inflorescence. We have made crosses to attempt to transfer heat tolerance from Prima and TVu 4552 to CB5, 7977, 58–57, Bambey 21, Mougne, Ndiambour, and IT 82E-18. Segregating F_2 progeny are screened for ability to set pods under high temperatures in Imperial Valley. Selections are screened again in Imperial Valley at the F_3 or F_4 level. In Imperial Valley in the summer the temperatures are so high that the expression of many agronomic characters is distorted, so selection for characters influencing growth habit, canopy architecture and seed quality is conducted in more normal production environments. The high levels of pod set in Imperial Valley by selections, compared with heat-sensitive parents, indicate that progress is being made in incorporating heat tolerance. Preliminary yield tests of heat-tolerant selections (F_5 generations) were conducted in California in 1984, and evaluations of their agronomic characters have been initiated in Senegal.

Cowpea strains with tolerance to high night temperatures should be extremely useful where minimum night temperatures exceed 22°C during early flowering stages, e.g. in tropical regions of West Africa and parts of India (Figure 1). We have observed that several cowpea strains developed by empirical breeding procedures in one of the hottest cowpea-production regions of the world (New Delhi, India) have some tolerance to high night temperatures. In contrast, many cowpea strains developed in West Africa have low tolerance to high night temperatures (Table 3). We have not yet evaluated the heat-tolerance of traditional, late-flowering cowpea varieties from West Africa. For the future it is necessary to determine whether selection for tolerance to high temperatures adversely affects cowpea performance at lower temperatures, because this would necessitate the development of different varietal types for tropical and subtropical zones.

This research was supported by grant MSU/AID/DSAN-XII-G-0261 from the United States Agency for International Development to the Bean/Cowpea Collaborative Research Support Program. We appreciate the collaboration with our colleagues in the Institut Senegalais de Recherches Agricoles.

PART 5

Pathology

Cowpea Research, Production and Utilization
Edited by S. R. Singh and K. O. Rachie
© 1985 John Wiley & Sons Ltd

10

Worldwide Occurrence and Distribution of Virus Diseases

G. THOTTAPPILLY and H. W. ROSSEL
International Institute of Tropical Agriculture, Ibadan, Nigeria

Among the numerous pathogens affecting cowpea, viruses are known to infect cowpeas either at one stage or throughout the life of the plant. The effects of virus diseases can be devastating, and they are a major constraint to large-scale production.

This survey covers the major viruses attacking cowpeas and complements other recent and relevant reviews by R. Williams (1975a), Singh and Allen (1979, 1980), Iwaki (1979), COPR (1981), Beniwal and Nene (1983), Boswell and Gibbs (1983), Allen (1983), and Rossell and Thottappilly (in press). More than 20 viruses have been reported on cowpea from different cowpea-production areas (Table 1). Many are of local importance, and others occasionally occur as natural infections of minor significance, e.g. bean common mosaic virus (Sachchidananda *et al.*, 1973), bean yellow mosaic virus (Brierly and Smith, 1962; Harrison and Gudauskas, 1968), peanut mottle (Demski *et al.*, 1983), cowpea rugose mosaic virus (Santos *et al.*, 1981), cowpea green vein-banding virus (Lin *et al.*, 1981b), cowpea severe mottle virus (M. T. Lin, personal communication), alfalfa mosaic virus (Kaiser *et al.*, 1968; Kaiser, 1979), tobacco streak virus (Kaiser *et al.*, 1982), pea leaf roll virus (Kaiser and Danesh, 1971; Kaiser, 1972; Kaiser and Schalk, 1975), broad bean wilt virus (Tsuchizaki *et al.*, 1970; Selassie-Gebre *et al.*, 1976; Xi *et al.*, 1982), tobacco ring spot virus (Ganacharya and Mali, 1981; Patil and Mali, 1982), cowpea little leaf disease (Beningo and Favali-Hedayat, 1977; Talens, 1977), tomato spotted wilt virus (Boswell and Gibbs, 1983), urd bean leaf crinkle virus (Kolte and Nene, 1972, 1975; Boswell and Gibbs, 1983), cowpea stunt virus (Iwaki, 1979) as well as diseases probably caused by

Table 1. Viruses reported from cowpea and their properties

Virus	Distribution	Transmission			Virus particles		
		Sap	Vector	Seed (%)	Shape	Size (nm)	Virus group
Cowpea mosaic	Africa, Surinam, Cuba, USA	+	Beetle	1–15	Isometric	24, diameter	Comovirus
Cowpea severe mosaic	Americas	+	Beetle	3–10	Isometric	25, diameter	Comovirus
Cowpea aphid-borne mosaic	Africa, Europe, Asia, Australia, South America	+	Aphid	0–40	Filamentous	750, long	Potyvirus
Blackeye cowpea mosaic	USA, Brazil, India, Taiwan	+	Aphid	3–55	Filamentous	750, long	Potyvirus
Bean common mosaic, cowpea strain	India, Australia	+	Aphid	25–40	Filamentous	750–925, long	Potyvirus
Bean yellow mosaic, cowpea strain	USA	+	Aphid	?	Filamentous	750, long	Potyvirus
Peanut mottle	USA	+	Aphid	Low	Filamentous	750, long	Potyvirus
Cowpea rugose mosaic	Brazil	+	Aphid	?	Filamentous	730, long	Potyvirus
Cowpea green vein-banding	Brazil	+	Aphid	Low	Filamentous	750, long	Potyvirus
Cowpea severe mottle	Brazil	+	Aphid	?	Filamentous	750, long	Potyvirus
Cowpea chlorotic mottle	North, Central America	+	Beetle	0	Isometric	25, diameter	Bromovirus
Cowpea mottle	Nigeria	+	Beetle	0–10	Isometric	30, diameter	Ungrouped
Cucumber mosaic	Probably worldwide	+	Aphid	4–26	Isometric	28, diameter	Cucumovirus
Cowpea ring spot	Iran	+	?	15–20	Isometric	26, diameter	Cucumovirus

Disease	Distribution		Vector		Particle shape	Particle size	Group
Cowpea mild mottle	Ghana, Kenya, Nigeria, Ivory Coast, probably worldwide	+	Whitefly	0–90	Flexuous rod	650, long	Carlavirus
Sunnhemp mosaic including cowpea chlorotic spot	India, Nigeria, USA	+	?	4–20	Rigid rod	300, long	Tobamovirus
Southern bean mosaic, cowpea strain	America, West Africa, India	+	Beetle	3–4	Isometric	28, diameter	Sobemovirus
Cowpea golden mosaic	Africa, India, Pakistan, Brazil	–	Whitefly	0	?	?	?
Alfalfa mosaic	Iran	+	Aphid	?	Bacilliform	Model lengths of 58, 49, 38 and 29, long with 18 wide	Alfalfa mosaic virus 'group'
Tobacco streak	USA	+	Thrips	?	Isometric	28, diameter	Ilarvirus
Pea leaf roll	Iran	–	Aphid	0	Isometric	25, diameter	Luteovirus
Broad bean wilt	Europe, Japan, China	+	Aphid	?	Isometric	25, diameter	Broad bean wilt virus 'group'
Tobacco ring spot	India	+	Thrips	0	Isometric	28, diameter	Nepovirus
Cowpea little leaf	Philippines, probably India	–	Aphid	Low rate	Isometric	25, 30 diameter	?
Cowpea stunt	Indonesia	–	Aphid	?	?	?	?
Urd bean leaf crinkle	India	+	Beetle	?	Isometric	30, diameter	Ungrouped
Tomato spotted wilt	Worldwide but restricted on cowpea	+	Thrips	?	Isometric	85, diameter	Tomato spotted wilt group
Cowpea witches' broom and cowpea phyllody	India	–	?	?	Mycoplasma		

mycoplasma-like organisms, cowpea witches' broom and cowpea phyllody (Ramaiah and Narayanaswamy, 1974; Varma et al., 1978). We describe here the major cowpea viruses—diseases that are of economic importance.

COWPEA MOSAIC VIRUS

In 1924 Smith described a mosaic disease of cowpea in the southern USA that was transmitted by the bean leaf beetle Cerotoma trifurcata. At that time no further characterization work was done and, hence, present comparisons are difficult. From Trinidad, Dale (1949) reported a seed-borne cowpea mosaic virus, which was transmissible by the beetle Cerotoma ruficornis. Chant (1959) reported from Nigeria a mosaic disease of cowpea and described the causal agent as cowpea yellow mosaic virus (CYMV). Shepherd and Fulton (1962) reported a cowpea mosaic virus disease from Arkansas that was serologically related to the Trinidad cowpea mosaic virus. By comparing three isolates from Surinam with that of Trinidad's cowpea mosaic and CYMV from Nigeria, Agrawal (1964) classified the Trinidad cowpea mosaic virus and two isolates from Surinam as the severe strain of cowpea mosaic virus, and the Nigerian cowpea yellow mosaic virus and another isolate from Surinam (SB) as the yellow strain of cowpea mosaic virus. According to Van Kammen (1971), all isolates of this virus can be separated into two subgroups, a severe and a yellow group.

Later, Swaans and Van Kammen (1973) made a detailed comparison of these isolates and concluded that the viruses should be considered distinct because of their differences in host range, symptomatology and antigenic properties; absence of homology in nucleotide sequence; and lack of complementation between heterologous nucleoprotein components. As a result, the name 'cowpea mosaic virus' (CPMV) has been retained for the yellow strain (Van Kammen and De Jager, 1978) while cowpea severe mosaic virus has been chosen for the severe strain (De Jager, 1979).

However, it would have been better to retain the name cowpea yellow mosaic virus (CYMV) instead of cowpea mosaic virus because cowpea yellow mosaic virus is the original name given by Chant (1959) when he described the virus from Nigeria and many 'diseases' described as 'cowpea mosaic' are in fact not caused by the cowpea mosaic virus. The name 'cowpea mosaic' is used loosely in the literature to describe symptoms caused by various cowpea viruses. Hence, much confusion exists in association with the name 'cowpea mosaic'.

Cowpea mosaic virus (CPMV) has been reported from Nigeria (Chant, 1959), Surinam (Van Hoof, 1963), Kenya (Bock, 1971), Cuba (Kvičala et al., 1970), Tanzania (Patel and Kuwite, 1982), Togo (Singh and Allen, 1979), and the USA (McLaughlin et al., 1977). In Nigeria the virus has not been detected thus far in ecological zones having a northern guinea savanna climate. Yield reductions for CPMV of from 60 to 100 per cent have been reported from Nigeria (Chant, 1960; Wells and Deba, 1961; Shoyinka, 1973; Gilmer et al., 1974).

The many varied symptoms produced by CPMV differ with cowpea variety and isolates. Systemic symptoms in susceptible varieties range from an inconspicuous light-green mottle to a distinct yellow mosaic, leaf distortion with significantly reduced growth, and premature death of the plant (Bliss and Robertson, 1971). CPMV is easily transmissible by sap and by various beetles. In Africa the beetle *Ootheca mutabilis* has been reported to be an efficient vector (Chant, 1959; Bock, 1971), but *Paraluperodes quaternus* (Chrysomelidae) and *Nematocerus acerbus* (Curculionidae) have also been found to transmit the virus (Whitney and Gilmer, 1974).

In Surinam, *Cerotoma variegata* and in Cuba *C. ruficornis* are the vectors (Van Hoof, 1963; Kvičala et al., 1973). Jansen and Staples (1971) have listed *C. trifurcata, Diabrotica balteata, D. undecimpunctata howardii, D. virgifera* and *Acalymma vittatum* (all chrysomelid beetles) as vectors of the virus. Beetle vectors remain viruliferous for 1–2 days to more than 8 days (Chant, 1959; Jansen and Staples, 1971). Whitney and Gilmer (1974) have also reported transmission by two species of thrips (*Sericothrips occipitalis* and *Taeniothrips sjostedti*) and by two species of grasshopper (*Cantatops spissus* and *Zonocerus variegatus*), but their report needs to be confirmed. CPMV has also been reported to be seed-borne in cowpea but in general at low levels (Gilmer et al., 1974).

CPMV is characterized by isometric particles averaging 24 nm in diameter, having two kinds of nucleoprotein particles that are similar in size but differ in their single-stranded RNA content. In addition, particles void of RNA are also produced by most isolates. RNA species enclosed within different types of particles represent separate parts of the viral genome (Van Kammen and De Jager, 1978).

Owing to its common occurrence, epidemic potential and general severity, CPMV is considered one of the most important cowpea virus diseases in Africa. Surveys within Nigeria have revealed that CPMV most commonly occurs in the middle belt, which has a typical savanna climate. CPMV has yet to be found in northern latitudes where most of Nigeria's cowpea crop is grown. The majority of locally grown cowpea varieties (large white, rough-seeded) have been shown to be highly susceptible to this virus. CPMV has also been found to occur in soybean within Nigeria's soybean-growing areas, though at low levels of incidence. Possible control methods include cultural practices, vector control with insecticides and the use of resistant cultivars. The roguing of infected plants at the onset of the first symptoms also reduces disease spread (Williams, 1975a). CPMV incidence can further be reduced by intercropping (Shoyinka, 1975). However, the use of insecticides as a control measure has not been found to be economically feasible in Nigeria. The best and most practical method of control may be resistant cultivars. (CPMV-resistant cowpea lines have been reported from Nigeria—Wells and Deba, 1961; Robertson, 1965; Williams, 1975e.)

At IITA, all promising cowpea germ-plasm accessions as well as intermediate or advanced breeding materials are being routinely screened for resistance to CPMV. As such, most of IITA's improved cowpea materials are resistant to CPMV.

COWPEA SEVERE MOSAIC VIRUS

Cowpea severe mosaic virus (CSMV) (De Jager, 1979), which previously was known as a severe strain of cowpea mosaic virus, has been reported from the USA, Trinidad, Puerto Rico, El Salvador, Costa Rica, Venezuela, Surinam, Brazil and Peru (Smith, 1924; Dale, 1949; Shepherd and Fulton, 1962; Shepherd, 1964; Agrawal, 1964; Debrot and De Rojas, 1967; Caner et al., 1969; Lin, 1979; Pio-Ribeiro and Paguio, 1980; Gabriel et al., 1981; Lin et al., 1981a, 1982a; Valverde et al., 1982a,b). Inoculated leaves of cowpea develop chlorotic or necrotic lesions. Portions of the major veins turn red and become necrotic. Trifoliate leaves show severe mosaic with blistering, distortion and necrosis. Plants inoculated as young seedlings show necrosis of the epicotyl just below the primary leaves and then collapse. However, isolates of the Arkansas-type do not induce systemic necrosis, mosaic or distortion, and are less severe. They induce pronounced vein-clearing on lower leaves (Agrawal, 1964; Shepherd, 1964).

In Trinidad and Surinam, prevalences up to 100 per cent in cowpea crops have been reported (Dale, 1949; Van Hoof, 1963), and virus infection has been reported to reduce about 50 per cent of the plant's fresh weight, as well as the number and weight of cowpea pods (Debrot and De Rojas, 1967). In Costa Rica, reductions of 84.8, 82.1 and 55.6 per cent in yield were noted when the virus infected cowpea before, during and after flowering, respectively (Valverde et al., 1982b).

Cowpea severe mosaic is transmitted by beetles mainly of the family Chrysomelidae. About 10 species have been listed as vectors, of which C. ruficornis and C. trifurcata are among the most important (Smith, 1924; Dale, 1949, 1953; Walters and Barnett, 1964; Debrot and De Rojas, 1967; Jansen and Staples, 1971; Perez and Cortes-Monllor, 1971; Slack and Fulton, 1971; Sanderlin, 1973; Gonzalez et al., 1975). The beetles acquire the virus after being on the source plant for only 5 min (Dale, 1953) and can remain infective from 1 to 2 weeks (Dale, 1953; Walters and Barnett, 1964). Transmission efficiency and persistence in the vector increase with increasing length of acquisition and inoculation access (Dale, 1953; Jansen and Staples, 1971). The common weed Phaseolus lathyroides is also susceptible to the virus and may act as a reservoir (Alconero and Santeago, 1973). An 8 per cent seed transmission has been reported for the Trinidad isolate in asparagus bean (Dale, 1949), and the Arkansas isolate can be transmitted to 10 per cent of cowpea seed (Shepherd, 1964). Haque and Persad (1975) have recorded 3.3–5.8 per cent seed transmission in four of seven cowpea selections in Trinidad. The virus is characterized by isometric particles of approximately 25 nm in diameter. It has two kinds of nucleoprotein particles that are morphologically identical but contain different, single-stranded RNA molecules. Both particles are necessary for infection to occur (De Jager, 1979).

Control measures include regular weeding to remove the virus and vector reservoirs, and use of resistant cowpea varieties. Some varieties developed at IITA have proved to be resistant in various locations within the tropics (Allen, 1977; Fulton and Allen, 1982). Rios and Neves (1982b) reported cowpea lines Macaibo and FP7732-2 to be immune to CSMV.

COWPEA APHID-BORNE MOSAIC VIRUS

Cowpea aphid-borne mosaic virus (CAbMV) has been reported to occur in Italy (Vidano and Conti, 1965; Lovisolo and Conti, 1966), Kenya (Bock, 1973), Nigeria (Ladipo, 1976; Rossel, 1977), Morocco (Fischer and Lockhart, 1976a), Tanzania (Patel and Kuwite, 1982), Uganda (COPR, 1981), Zambia (COPR, 1981), Iran (Kaiser and Mossahebi, 1975), Japan (Tsuchizaki et al., 1970), India (Mali et al., 1981), the Philippines (Beningo and Favali-Hedayat, 1977), Australia (Behncken and Maleevsky, 1977), Brazil (Lima et al., 1981; Lin et al., 1981b), Egypt (COPR, 1981), Cyprus (Taiwo et al., 1982), Iraq (Fegla et al., 1981), People's Republic of China (COPR, 1981) and Indonesia (Iwaki et al., 1975; Iwaki, 1979).

Symptoms due to CAbMV infection include a severe mosaic, the severity being dependent on the interactions between host cultivar and virus strain. Infected plants show dark-green vein-banding, leaf distortion, blistering and stunting (Bock and Conti, 1974). The European (type) strain causes a severe distorting mosaic in cowpea (Lovisolo and Conti, 1966). The African (neotype) strain induces irregular, angular broken mosaic, and the African vein-banding strain induces a broad dark-green vein-banding (Bock, 1973). The African mild strain induces a very mild mottle, with little or no effect on plant growth.

Yield losses of 15–87 per cent due to infection by CAbMV have been reported from Iran (Kaiser and Mossahebi, 1975), and complete loss of an irrigated cowpea crop in northern Nigeria was tentatively attributed to an aphid-borne virus disease (Raheja and Leleji, 1974).

CAbMV is readily transmitted by sap inoculation and by the aphids in a non-persistent manner (Atiri, 1982). *Aphis craccivora, A. gossypii, A. medicagenis, Macrosiphum euphorbiae* and *Myzus persicae* are reported to be aphid vectors (Vidano and Conti, 1965; Bock, 1973; Atiri et al., 1984).

CAbMV is seed-borne in cowpea, but seed transmission is dependent upon the virus strain and cowpea cultivar (Aboul-Ata *et al.*, 1982). Seed transmission is usually low (0.3 per cent) but has been reported to be 2.5–19 per cent for an Indian isolate (Phatak, 1974; Mali *et al.*, 1983) and 1.1–39.8 per cent for an Iranian isolate (Kaiser and Mossahebi, 1975). Some cowpea lines possessing resistance to seed transmission have been identified (Ladipo, 1977). Although susceptible to CAbMV, cowpea lines K-39, L-868, L-1552 and P 1476 in India do not sustain seed transmission (Mali *et al.*, 1983), and selection for absence of seed transmission may prove to be a useful approach for virus control.

CAbMV is characterized by flexuous, filamentous particles about 750 nm long. Within the African strain the neotype and mild strains are serologically identical; the vein-banding strain, although related, is distinguishable (Bock, 1973).

Similarly, cowpea aphid-borne mosaic and blackeye cowpea mosaic viruses, although serologically related, are not identical (Taiwo *et al.*, 1982) nor are the cowpea varietal reactions to them.

Virus isolates that have some properties similar to those of CAbMV but have not been fully characterized include the cowpea mosaic viruses of McLean (1941), Yu (1946), Van Velsen (1962), Abeygunawardena and Perera (1964) and Shankar *et al.* (1973) and the asparagus bean mosaic virus of Snyder (1942).

In surveys throughout Nigeria during the past several years, CAbMV has been found to occur in all ecologic zones and is considered one of the most widespread and important viral diseases of cowpea.

CAbMV is difficult to detect in cowpeas by means of the SDS-agar gel diffusion test, although positive reactions have been obtained by scientists using infected *Nicotiana benthamiana* plants as test samples. To provide a reliable and quick method for identifying CAbMV, IITA researchers developed a test using enzyme-linked immunosorbent assay (ELISA) and an antiserum with a titre of 1/1024 in microprecipitin tests. When an extraction buffer consisting of 0.25 M potassium phosphate (pH 7.5) with 0.1 M EDTA and 0.25 per cent sodium sulphite was used, positive reactions were obtained up to a dilution of 1/625 with infected young cowpea (TVu 76) plants (10 days after inoculation).

At IITA where all potential breeding parents are routinely screened for resistance to CAbMV, resistance has been found in several germ-plasm introductions (Ladipo and Allen, 1979a; H. W. Rossel, unpublished data). Some varieties have combined resistance to cowpea (yellow) mosaic (CYMV), southern bean mosaic (SBMV), and cowpea mottle (CMeV) viruses (Williams, 1977; Ladipo and Allen, 1979b).

BLACKEYE COWPEA MOSAIC VIRUS

Blackeye cowpea mosaic virus (BlCMV) is reported from the USA (Anderson, 1955a; Zettler and Evans, 1972; Lima *et al.*, 1979), India (Mali and Kulthe, 1980a,b; Sekar and Sulochana, 1983), Taiwan (Chang, 1983), and Brazil (Lin *et al.*, 1981b,c). Also, Taiwo *et al.* (1982) identified Kenyan, Nigerian and Japanese isolates of CAbMV as BlCMV on the basis of the serological reactions and host response. Isolates from Morocco, Cyprus and Iran, however, were identified as belonging to the CAbMV group.

BlCMV produces both local and systemic symptoms on blackeye cowpea. Local symptoms include large reddish lesions that spread along the veins. Systemic symptoms are severe mottling, distortion, yellowing, mosaic and vein necrosis.

BlCMV is readily transmitted mechanically and in a non-persistent manner by *Macrosiphum solanifolii*, *A. gossypii* and *M. persicae* (Anderson, 1955a; Mali and Kulthe, 1980a,b; Pio-Ribeiro *et al.*, 1978).

BlCMV is seed-transmitted; the percentage of transmissions is from 3.5 to 55, and is dependent upon the cowpea cultivar and virus isolate interaction (Anderson, 1957; Gay and Windstead, 1970; Zettler and Evans, 1972; Mali *et al.*, 1983).

BlCMV is characterized by flexuous filamentous particles of about 750 nm (Lima *et al.*, 1979; Pio-Ribeiro *et al.*, 1978; Taiwo *et al.*, 1982).

Resistance to BlCMV in cowpea has been reported (Lima *et al.*, 1979; Taiwo *et al.*, 1981) and has been found to be conditioned by a single recessive allele (Walker and Chambliss, 1981).

COWPEA CHLOROTIC MOTTLE VIRUS

Cowpea chlorotic mottle virus (CCMV) was reported for the first time from Georgia, USA (Kuhn, 1964), and is now endemic in the southern USA (Kuhn, 1964; Harrison and Gudauskas, 1968) and Central America (Gamez, 1976). Field-infected cowpea plants usually develop bright yellow mottling, which can be distinct chlorotic rings, mosaic-like areas or complete and uniform yellowing of leaflets.

CCMV is readily sap-transmissible. The bean leaf beetle, *Ceratoma trifurcata*, and the spotted cucumber beetle, *Diabrotica undecimpunctata*, transmit the virus (Hobbs and Fulton, 1979; Walters and Dodd, 1969). According to Gay (1969), the virus is present in seed coats but not in cotyledons or embryos of immature undried cowpea seeds and is not seed-transmitted.

Mixed infections with CCMV and SBMV cause a synergistic reduction in cowpea growth and yield (Kuhn and Dawson, 1973).

CCMV particles are isometric and about 25 nm in diameter (Bancroft *et al.*, 1968). The strains include the type cowpea strain, a soybean strain (Kuhn, 1968), a *Desmodium* strain (Walters and Dodd, 1969), and the yellow stipple strain of beans from Costa Rica (Fulton *et al.*, 1975). The strains specific for soybean and *Desmodium* produce less intense mottling than does the type strain. All the strains are serologically identical (Bancroft, 1971).

Since CCMV is not an economically important disease, there have been no major control efforts. Although all commercial varieties of cowpea are susceptible to CCMV, cowpea cultivars PI 147652, PI 186458, PI 186465, PI 194203, PI 255811, PI 297561 and PI 297562 have been identified as resistant (Sowell *et al.*, 1965).

COWPEA MOTTLE VIRUS

Robertson (1963) reported a cowpea virus from western and eastern Nigeria, which he called cowpea mottle virus (CMeV). This virus causes mottling or bright yellow mosaic in cowpea. It also induces distortion and reduction in leaf size and a witches'-broom syndrome. Cowpea yield reductions of more than 75 per cent have been reported (Robertson, 1963; Shoyinka *et al.*, 1978).

CMeV is readily transmitted by sap inoculation. It also naturally infects Bambarra groundnut (*Vigna subterranea*) and induces systemic symptoms such as green mottle, stunting, and shortening of petioles (Robertson, 1963; Rossel, 1977).

The vector is the galerucid beetle, *Ootheca mutabilis*, which acquires the virus within 10 min, transmits it within 1 h, and remains infective for 5 days (Shoyinka *et al.*, 1978). *Medythia quaterna* also transmits CMeV to cowpea but less efficiently than *O. mutabilis* (Allen *et al.*, 1981a). Although Robertson (1966) failed to detect seed transmission of CMeV through many thousands of seeds harvested from infected cowpea plants, Shoyinka *et al.* (1978) recorded up to 10 per cent seed transmission from naturally infected, and up to 7 per cent from artificially inoculated plants. Subsequent studies with five cowpea cultivars failed to detect an incidence higher than 0.4 per cent seed transmission (Allen *et al.*, 1982). This virus has not been reported outside Nigeria, although it is thought to occur in other regions in Africa having similar ecologic conditions. Extensive surveys in Nigeria revealed that the virus typically occurs in cowpea when grown in association with Bambarra groundnut. Geographically, within Nigeria, the virus almost exclusively occurs in the riverine area of the middle belt, which has a southern guinea savanna climate. This belt is where most of the Bambarra groundnut is also grown.

CMeV particles are isometric and measure about 30 nm in diameter. They contain single-stranded RNA, which sediments as a single component. CMeV is a distinct virus not related to other beetle-transmitted cowpea viruses, such as cowpea chlorotic mottle, cowpea (yellow) mosaic, cowpca severe mosaic, and southern bean mosaic virus (Shoyinka *et al.*, 1978; Bozarth and Shoyinka, 1979). Screening revealed that, although not readily available in many cowpea germ-plasm introductions, acceptable levels of resistance to CMeV do exist. Neither immunity nor totally symptomless infections were encountered among 562 introductions tested (Allen, 1980), but experiments on the usefulness of the available sources of high tolerance revealed that under field conditions virtually no natural infection occurs in such genotypes (Allen *et al.*, 1982).

CUCUMBER MOSAIC VIRUS

Cucumber mosaic virus (CuMV) causing cowpea diseases has been described in different countries. Among the diseases are the cowpea strain of cucumber mosaic virus (CuMV-CS) reported from the USA (Anderson, 1955b; Harrison and Gudauskas, 1968; Kuhn, 1968), Morocco (Fischer and Lockhart, 1976b), Spain (Díaz-Ruíz, 1976), the Philippines (Talens, 1979), Brazil (Lin *et al.*, 1981c), Taiwan (Chang, 1983), Nigeria (Robertson, 1966), Egypt (El-Din *et al.*, 1980) and Iraq (Fegla *et al.*, 1981) as well as cowpea banding mosaic (CpBMV) and cowpea necrosis (CpNV) from India (Sharma and Varma, 1975a). CuMV is one of the most common and most variable viruses known.

CuMV-CS has been reported to cause yield reductions of 14.2 per cent in the USA (Pio-Ribeiro *et al.*, 1978). Local symptoms on inoculated leaves include poorly developed chlorotic areas or reddish necrotic rings; systemic symptoms are mild mottle and distortion. A few varieties develop severe mottle, distortion and considerable reddish vein necrosis. Symptoms in most varieties are mild or partial (Anderson, 1955b).

CuMV-CS is seed-borne to the extent of 4–26 per cent in cowpea (Anderson, 1957; Fischer and Lockhart, 1976b), and seed transmission varies with cultivar, ranging from 3.5 to 31.5 per cent (Mali *et al.*, 1983). All the cucumoviruses infecting cowpea are transmitted by aphid vectors—*Aphis craccivora*, *A. euonymi*, *A. fabae*, *A. gossypii* and *Myzus persicae*—in a non-persistent manner (Chenulu *et al.*, 1968; Sharma and Varma, 1975b, 1982a; Fischer and Lockhart, 1976b).

CuMV has isometric particles about 28 nm in diameter containing single-stranded RNA with a tripartite genome (Francki *et al.*, 1979). CpBMV and CpNV also have isometric particles, ranging from 25 to 30 nm in diameter, and belong to the cucumovirus group (Sharma and Varma, 1975a).

CpBMV and CpNV are serologically distinct, and CpBMV is serologically related to a cowpea mosaic virus reported by Chenulu *et al.* (1968) and relabelled CpBMV by Sharma and Varma (1975a) to distinguish it from the beetle-transmitted CPMV.

At IITA, further host range and serological studies of the cucumovirus occurring in cowpeas and lima beans in Nigeria have revealed that this virus should be considered a strain of cucumber mosaic virus (CuMV). Although host-range differences exist between this and other CuMV strains, they are no larger than differences among other compared strains. In all studies the virus has produced systemic symptoms in *Nicotiana glutinosa* and *Cucumis sativus*. At IITA a close serological relationship was found to exist between a cowpea isolate and various CuMV isolates obtained from crops in temperate regions. CuMV is widespread in cowpeas within Nigeria and is difficult to identify solely on the basis of symptoms. Agar-gel diffusion tests with crude sap from cowpeas have also given inconsistent results. To produce a more reliable diagnostic method we at IITA developed a direct (double-antibody sandwich) ELISA method, which is now used routinely.

Antiserum against CuMV was produced from cowpeas after purification with phosphate buffer, PEG precipitation, differential centrifugation, and sucrose density-gradient centrifugation. The antiserum had a titre of 1/128 in agar-gel diffusion tests with purified virus as an antigen. A gamma globulin coating 1 μg protein/mℓ and a conjugate dilution of 1/250 provided the best results with infected cowpea and *N. glutinosa*. We were able to detect CuMV in frozen tissues as well as in cowpea leaves, which were dried over calcium chloride. Using crude juice from infected cowpeas we detected the virus to dilutions between 1/625 and 1/3125.

Since cucumoviruses are non-persistent, treatment with insecticide does not protect cowpea from infection (Sharma and Varma, 1982b). Seemingly, the best method of control is to use virus-free seed and resistant cultivars.

At IITA, elite cowpea breeding lines and potential breeding parents are routinely evaluated for resistance to CuMV. A few germ-plasm accessions were identified as resistant to this virus. Generally, however, symptoms due to infection with CuMV are rather mild. In view of the high epidemic potential of a commonly seed-borne virus and quarantine implications, resistance to seed transmission is another character that is being selected for at IITA. Seed lots of elite materials, multiplied for international testing, are evaluated for seed-transmission incidence. Introductions showing a seed transmission rate of greater than 2 per cent are eliminated. This has shown to be sufficient to prevent any noticeable virus spread in the following year's multiplication plots.

Cowpea ring spot virus (CpRSV), which belongs to the cucumovirus group, has been described from Iran (Phatak *et al.*, 1976). It causes chlorotic and necrotic local lesions on primary leaves of cowpea, whereas systemic symptoms vary according to cultivar and are temperature-dependent. Systemic symptoms include vein clearing and circular chlorotic spots that coalesce to a chlorotic mottle on trifoliate leaves. Some cultivars develop systemic necrotic streaking or ring spots, often causing leaf drop. In addition, the lamina becomes distorted, and infected plants are stunted.

The virus is readily sap-transmissible and can also be transmitted by seed (15–20 per cent). No vector has yet been found to be associated with this virus, however, and *M. persicae* has been shown to be unable to transmit the virus. CpRSV has isometric particles of about 26 nm in diameter. Though not related serologically to CuMV, CpRSV has been classified as a probable member of the cucumovirus group, based on its sedimentation coefficient, instability at high salt concentrations, and coat-protein molecular weight (Phatak *et al.*, 1976; Matthews, 1979).

COWPEA MILD MOTTLE VIRUS

Cowpea mild mottle virus (CMMV), a member of the carlavirus group, occurs in Ghana, Kenya, Nigeria and Ivory Coast in cowpeas (Brunt and Kenten, 1973, 1974; Fauquet and Thouvenel, 1980; Anno-Nyako, 1983). CMMV is also reported in groundnut in India (Iizuka *et al.*, 1984) as well as in soybean in Thailand and Ivory Coast (Iwaki *et al.*, 1982; Thouvenel *et al.*, 1982). CMMV is found in, and is known to have spread to, Israel and Fiji (Brunt, 1983). Recent studies indicate that CMMV is serologically unrelated to 12 recognized members of the carlavirus group (Brunt, 1983).

Naturally infected cowpea plants exhibit either a mild systemic mottle or are symptomless, but inoculated plants of susceptible cultivars develop necrotic lesions on primary leaves and severe systemic chlorosis and necrosis on trifoliate leaves. Differences in symptom severity on cowpea plants make economic assessment of the disease difficult, but it is nonetheless considered a definite threat to the successful production of other leguminous crops, such as peanut, pigeon pea and jack beans, all of which are at times intercropped with cowpeas in Ghana (Brunt and Kenten, 1973).

CMMV isolates are transmitted by the whitefly vector, *Bemisia tabaci*, either in a semipersistent (soybean isolate) (Iwaki *et al.*, 1982) or non-persistent (peanut isolate) manner (Muniyappa and Reddy, 1983). The Nigerian isolate is transmitted by *B. tabaci*, in a semipersistent manner (Anno-Nyako, 1983). CMMV is seed-borne in cowpea, soybean and French bean but not in *Nicotiana clevelandii*. The level of seed transmission is higher in cowpea and soybean (90 per cent) than in French bean (15 per cent) (Brunt and Kenten, 1973).

CMMV has straight to slightly flexuous, fragile filamentous particles (650 × 13 nm in diameter).

SUNNHEMP MOSAIC VIRUS

The cowpea diseases caused by the tobamovirus group have been variously described in different countries. These include cowpea chlorotic spot (Sharma and Varma, 1975a), catjang mosaic (Capoor *et al.*, 1947; Capoor and Varma, 1956; Nariani and Kandaswamy, 1961), nylon cowpea seed-borne mosaic

(Kulthe and Mali, 1979) from India, cowpea mosaic from the USA (Toler, 1964), and legume or cowpea strain of tobacco mosaic (TMV-CS) (Lister and Thresh, 1955; Chant, 1959) from Nigeria. All have been reported to be caused by, or to be related to, sunnhemp mosaic virus (SHMV) (Bawden, 1956; Kassanis and Varma, 1975), which is a definitive member of the tobamovirus group (Gibbs, 1977).

Prevalence of cowpea chlorotic spot virus (CpCSV) under field conditions has been reported to range from 2 to 20 per cent and to reduce yield from 5 to 25 per cent in India (Sharma and Varma, 1975b). However, TMV-CS has not been reported to cause significant cowpea yield reductions in Nigeria (Chant, 1960).

Symptom expression on cowpea varies among isolates. TMV-CS causes only a mild green mottle on trifoliate leaves (Chant, 1959), whereas CpCSV causes chlorotic spots that later coalesce on primary and trifoliate leaves, fine vein-clearing in young leaves, reduction in lamina, dwarfing, early defoliation and poor pod formation (Sharma and Varma, 1975a).

SHMV has rigid rod-shaped particles 300 nm long and contains single-stranded RNA (Kassanis and Varma, 1975). CpCSV also has rigid, rod-shaped particles 300 × 17 nm in diameter that are serologically related to SHMV and common tobacco mosaic viruses (Sharma and Varma, 1975a). Other cowpea seed-borne isolates from India have similar particle morphology and are serologically related to common tobacco mosaic virus (Kulthe and Mali, 1979; Nariani et al., 1980).

No potential vector has been reported. Isolates of CpCSV and catjang mosaic as well as the nylon cowpea isolate have been reported to be 4–20, 4 and 18 per cent seed-borne, respectively, in cowpea.

The best method of control appears to be the use of virus-free seeds and resistant cultivars. Cowpea cultivars Early Sugar-crowder, Suwanne and Taylor are tolerant to catjang mosaic (Capoor and Varma, 1956) and CM-11 and V-16, to the nylon cowpea isolate (Kulthe and Mali, 1979).

SOUTHERN BEAN MOSAIC VIRUS

Field occurrence of southern bean mosaic virus (SBMV) has been reported from the USA (Kuhn, 1963; Gabriel et al., 1981), Ghana (Lamptey and Hamilton, 1974), Nigeria (Shoyinka, 1974; Ladipo, 1975; Shoyinka et al., 1979), Ivory Coast (Fauquet and Thouvenel, 1980; Givord, 1981), and India (Singh and Singh, 1974; A. Varma, personal communication).

Inoculated primary leaves are often symptomless; occasionally chlorotic lesions appear and ring spots may form. Systemic symptoms include vein-clearing followed by a mild to severe mottling or coarse mosaic pattern on leaves. SBMV has isometric particles 28–30 nm in diameter and contains a single-stranded RNA (Shepherd, 1971). It is readily transmitted by sap inoculation, but infectivity of the virus is confined to the Leguminosae family. The cowpea strain of SBMV (SBMV-CS) infects peas in addition to cowpeas but not beans, whereas Ghana strain (SBMV-GH) infects both cowpea and beans but not peas. Systemic symptoms are not produced in bean. Only SBMV-CS has been reported to be seed-borne in cowpea, the level of seed transmission being 3 to 4 per cent (Shepherd and Fulton, 1962). SBMV-GH is also seed-borne but generally at a low level (1.3 per cent). The bean leaf beetle, *C. trifurcata*, has been reported to be an efficient vector of SBMV-CS (Walters and Henry, 1970). The maximum amount of transmission to cowpea occurs in 5–8 days following acquisition feeding by the vector. In laboratory studies Allen *et al.* (1981) transmitted SBMV by *O. mutabilis*, which retained the virus for 13 days following 24 h acquisition feeding.

From glasshouse screening of cowpea lines, Ladipo and Allen (1979b) reported 23 lines that were resistant to three isolates as well as CPMV. One cowpea line (TVu 1948) remained symptomless in all tests.

COWPEA GOLDEN MOSAIC DISEASE

In 1977 cowpea golden mosaic disease was first discovered at Onne near Port Harcourt in eastern Nigeria. Infected cowpea plants have large bright yellow patches on their leaves, and, under severe infection, the entire leaf surface turns bright yellow. The virus is transmitted by the whitefly in a persistent manner but is not seed- or sap-transmitted (Anno-Nyako *et al.*, 1983).

There are similar descriptions of diseases in cowpea that are not sap-transmissible but only whitefly-transmitted. These include cowpea yellow fleck from India (Sharma and Varma, 1976) and cowpea bright yellow mosaic from Pakistan (Ahmad, 1978). According to a recent publication, a similar disease occurs in Kenya, Tanzania and Niger (COPR, 1981). Also, a whitefly-transmitted cowpea disease is reported from Brazil (Lin and Rios, chapter 13 in this volume). These diseases may all be caused by a similar or identical virus. So far no information is available on virus particles of any of these diseases.

Most of the germ plasm used in IITA's breeding programme has been screened for resistance to cowpea golden mosaic, and sources of resistance have been identified. The procedure was simple: plants were evaluated in the fields near Port Harcourt where epiphytotic conditions are so extensive that normally no plant is free from infection.

COWPEA STUNT DISEASE

Cowpea stunt, reported from the USA, is caused by a synergistic interaction of BlCMV and CuMV (Pio-Ribeiro *et al.*, 1978). The disease is characterized by severe stunting of field-infected plants: leaves are small, mottled, blistered and malformed (Pio-Ribeiro *et al.*, 1978).

California Blackeye seed is reduced by 14.2 per cent by CuMV and 2.5 per cent by BlCMV, whereas yield of doubly infected plants is reduced by 86.4 per cent. Double infection also reduces leaf weight, stem weight and root weight by 94.3, 89.3 and 87.3 per cent, respectively.

Chang (1983) reported from Taiwan that CuMV and BlCMV were isolated from cowpea plants with severe symptoms. When plants were inoculated with CuMV and BlCMV, simultaneously or in sequence, leaves above those inoculated expressed rugose mosaic symptoms, and pods developed abnormally and became necrotic.

Cowpea stunt is readily transmitted mechanically and, upon sap inoculation, some cowpea cultivars exhibit a striking initial shock (acute phase), whereas others exhibit a chronic disease. *Cassia obtusifolia* and cucumber (cv. Marketer) are recommended to separate the mixture of viruses (BlCMV and CuMV) from stunted cowpea plants (Pio-Ribeiro *et al.*, 1978). In California Blackeye and Knuckle Purple Hull, both CuMV and BlCMV are seed-transmitted from either single or double infection. The aphid vector *M. persicae* transmits both CuMV and BlCMV from single and mixed infections in a non-persistent manner. The level of aphid transmission from single infections is 17.1 per cent (BlCMV) to 22.8 per cent (CuMV) and from double infections is 13.8 per cent (CuMV) to 15.8 per cent (BlCMV) (Pio-Ribeiro *et al.*, 1983). The inheritance pattern of resistance to BlCMV and consequently to cowpea stunt has been determined by Pio-Ribeiro *et al.* (1980). According to these scientists the necrotic reaction is controlled by an allele at a single locus, which exhibits incomplete dominance.

Cowpea stunt disease reported from Indonesia is distinct from the cowpea stunt reported from either the USA or Morocco: it is transmitted by aphids (*A. craccivora*) in a persistent manner but not by sap inoculation, and the particles have not been observed (Iwaki, 1979).

CONCLUDING REMARKS

Numerous virus diseases have been reported to occur in cowpea, but the available literature does not detail the economic importance of the diseases. We expect that the increase in research in the tropics will be accompanied by much more information on such aspects.

The infection of the same plant by two or more viruses in the field has led to confusion in disease identification. Accurate identification requires a combination of various diagnostic methods, including host-range and vector-transmission studies as well as electron microscopy and serological examination.

Since no direct control method such as viricides exist, most efforts involve one or more evasive measures intended to reduce sources of infection from within and outside the crop: using virus-free seed that has been produced in isolated areas where no viruses occur and where vectors that transmit the viruses do not exist or can be controlled; choosing the planting time to avoid vectors; and roguing diseased plants where practical. Crop rotation, weeding to remove alternative hosts of viruses and vectors, and the use of barrier crops to prevent movement of vectors are also worthwhile. The use of insecticides to prevent virus transmission by vectors has been only partially successful because the insects normally transmit the virus before they die from the effects of the insecticides. The best single control method is the use of resistant varieties, and the first step toward developing such varieties is large-scale screening and selection. In general, however, no single measure can prevent virus spread. For profitable cultivation all measures that can be feasibly integrated should be considered.

Cowpea Research, Production and Utilization
Edited by S. R. Singh and K. O. Rachie
© 1985 John Wiley & Sons Ltd

11

Fungal and Bacterial Diseases of Cowpeas in Africa

A. M. EMECHEBE* and S. A. SHOYINKA†
*Faculty of Agriculture/Institute for Agricultural Research; and †CEC/IITA Project, Institute for Agricultural Research, Ahmadu Bello University, Zaria, Nigeria

Diseases are induced by viruses, bacteria, fungi, nematodes, parasitic flowering plants and adverse environmental factors. This paper considers the diseases induced by fungi and bacteria in Africa.

AGROECOLOGICAL ZONES

For the sake of convenience we have recognized four agroecological zones of cowpea production in Africa, namely: the rain forest; the guinea savanna; the sudan savanna and the sahel; and the areas with cool, humid climates typified by medium altitudes (500–1200 m) and high rainfall in central, including northwestern Cameroon, eastern, and southern Africa.

In the low-altitude rain forest the greatest losses occur because of seed decay and seedling damping-off. Nine diseases (in order of importance: brown blotch, anthracnose, web blight, *Fusarium* wilt, 'Cercospora' leaf spot, bacterial pustule, soft rot, bacterial blight and brown rust) also cause serious economic losses. These diseases, when considered along with viruses and nematodes, cause damage surpassing that done by insects.

In the guinea savanna the major diseases are fewer (*Sphaceloma* scab, brown blotch, *Septoria* leaf spot, bacterial blight and web blight), but scab and brown blotch are particularly destructive (Emechebe, 1980, 1981a), and finding resistance to them has been difficult (Singh *et al.*, 1983).

The number of cowpea diseases encountered in the sudan savanna is considerably smaller than that of either the forest belt or the guinea savanna. This is perhaps expected, given the harsher environment, limited rainfall, relatively low relative humidity, high temperatures during the growing season, and the long dry season (8–9 months) that begins with a cold spell occasioned by the harmattan winds in West Africa.

Nevertheless, *Septoria* leaf spot (induced by *Septoria vignae*), brown blotch (*Colletotrichum capsici*) and bacterial blight (*Xanthomonas campestris* pv. *vignicola*) are important. In addition, a widespread epiphytotic of ashy stem blight and stem canker induced by *Macrophomina phaseolina* occurred in most of the West African sudan savanna in 1983 (V. Parkinson, personal communication). The severe drought that occurred in the same region in 1983 probably contributed to the epidemic, as did infection of cowpea plants by *Striga gesnerioides*. Of minor importance in this zone is basal stem rot induced by *Fusarium solani*.

A good proportion of the land for agricultural production in central, eastern and southern Africa lies in areas with elevations of 500–1200 m above sea level (ASL); the temperatures are lower than those of the lowland areas, while annual rainfall is usually more than 750 mm. Of the major diseases, blight, induced by *Ascochyta phaseolorum*, attains epiphytotic proportions only under cool humid conditions found at elevations above 1000 m (Allen, 1979, 1983). Singh *et al.* (1982) reported a localized epiphytotic of *Phytophthora* stem rot that was first observed in 1979 in Ilonga (Tanzania), which at about 600 m ASL, just fits into the elevation range. Before 1979 the disease, which is devastating in Australia (Purss, 1957), was unknown in Africa (Williams, 1976).

Some of the 16 major diseases afflicting cowpea are important in two or more agroecological zones (Table 1). What we consider as minor cowpea diseases at both the low and the intermediate altitudes of the rain forest and guinea savanna are listed in Table 2.

SYMPTOMS AND EPIDEMIOLOGY

The symptoms of major diseases (Table 1) are noteworthy.

Anthracnose

Individual lesions of anthracnose (Onesirosan and Barker, 1971; Williams, 1975b) are lenticular, sunken and tan to brown. Cowpea varieties that are highly susceptible develop spreading lesions that rapidly coalesce to girdle stems, branches, peduncles and petioles. Symptoms on pods are less prominent. Resistant lines exhibit a hypersensitive response by developing tiny necrotic flecks or lenticular, shiny, reddish-brown lesions (up to 5 mm long) on which no sporulation occurs.

The pathogen is seed-borne (Esuruoso, 1975; Emechebe and McDonald, 1979) and is also dispersed by rain splash, air currents and contact (COPR,

1981). It survives the dry season in infected seeds (in which it can survive for at least 2 years) and on diseased stem tissues either on the soil surface or buried (Onesirosan and Sagay, 1975; COPR, 1981).

Ascochyta blight

Lesions of *Ascochyta* blight (COPR, 1981) first appear as small, irregular leaf spots with grey to light brown centres surrounded by a light greenish-yellow border. They develop rapidly and become large, light-brown lesions with concentric, target-like markings. Lesions often crack in the centre when the dead tissue finally drops. Similar lesions may develop on stems and pods.

The pathogen is seed-borne, and spread is favoured by wet conditions (COPR, 1981). It survives in diseased crop residues and is destructive only when the crop has been predisposed to infection by unfavourable growing conditions, by wind, by insect damage or by prior infection with another pathogen (Sutton and Waterston, 1966; Pegg and Alcorn, 1967; Singh *et al.*, 1978). Nevertheless, complete defoliation and loss of yields can result from early infection.

Bacterial blight

Tiny, water-soaked dots on the under surface of the leaves constitute the primary symptoms of bacterial blight (Williams, 1975c). These remain small, but the surrounding tissue becomes necrotic and is tan to orange with a yellow halo. Necrotic areas on heavily infected leaves coalesce to form larger lesions that retain the individual dark spots of the initial infection. The pathogen also infects the stem and peduncle and causes water soaking of pods from which it enters the seed.

In India, seed transmission of bacterial blight causes seedling mortality and bushy, stunted plant growth; secondary spread causes only leaf blight (Shekhawat and Patel, 1977). The pathogen is seed-borne (Shekhawat *et al.*, 1977) and is also spread by wind-driven rain and soil, insects and infected plant debris (Kaiser and Vakili, 1978; COPR, 1981).

Bacterial pustule

Lesions of bacterial pustule (Williams, 1975c) begin as tiny, dark water-soaked dots on the underside of the leaves. The dots enlarge to become roughly circular spots (1–3 mm in diameter) and, when young, appear as water-soaked pustules on the underside of the leaf and as dark brown necrotic spots on the upper surface. Older, large pustules become dry, sunken in the centre, and water-soaked around the margin (Williams, 1975c). Heavily

Table 1.　Major fungal and bacterial disease of cowpea in Africa

Common name	Pathogen	Estimate of crop loss/ importance on susceptible varieties	Ecological zone in which major damage is done	Distribution	Selected reference
Seed decay and seedling damping-off	*Pythium aphanidermatum, Rhizoctonia solani* (teliomorph = *Thanatephorus cucumeris*), *Colletotrichum capsici, Macrophomina phaseolina* (sclerotial stage = *Rhizoctonia bataticola*)	*P. aphanidermatum* and *R. solani* induced up to 75% seedling mortality during cool, wet weather in Ibadan, Nigeria. In Zaria, Nigeria, *C. capsici* induced up to 50% pre- and post-emergence damping-off	Rain forest for *P. aphanidermatum* and *R. solani*; rain forest and guinea savanna for *C. capsici*; sudan savanna for *M. phaseolina*	Widespread	Alabi, 1981; Emechebe 1981a; Oladiran and Okusanya, 1980; Onuorah, 1973; Williams, 1973a
Brown blotch	*Colletotrichum* spp., especially *C. capsici* but occasionally *C. truncatum*	Up to 75% crop loss under protracted wet field conditions. Second most important disease in the rain forest, guinea savanna and sudan savanna	Rain forest, guinea savanna, sudan savanna	Widespread	Allen *et al.*, 1981b Emechebe, 1981a; Okpala, 1981; Singh and Allen, 1979, 1980
Anthracnose	*Colletotrichum lindemuthianum*	Reduced yields by 38–50% in Ibadan, Nigeria, in 1972	Rain forest, especially during cool, wet weather	Eastern, southern and western Africa	Onesirosan and Barker, 1971; Williams, 1973b, 1975b,c
Web blight	*Rhizoctonia solani*	Caused 24–40% crop loss in Ibadan, Nigeria	Rain forest, southern guinea savanna	Widespread	Oyekan, 1979

Disease	Pathogen	Notes	Zone	Distribution	References
Wilts	*Fusarium oxysporum* f. sp. *tracheiphilum*, *Sclerotium rolfsii* (teliomorph = *Corticium rolfsii*)	*Fusarium* wilt incidence in three varieties ranged from 15% to 55% under natural epiphytotic conditions at Ibadan, Nigeria, during the second cropping season in 1974; 20% mortality due to *Sclerotium* wilt reported elsewhere (India)	Rain forest	Nigeria, Uganda, Tanzania for *F. oxysporum* f. sp. *tracheiphilum*; *S. rolfsii* has widespread distribution	Booth, 1971; Holliday, 1970; Oyekan 1975; Ramaiah *et al.*, 1976; Teri, 1984
'Cercospora' leaf spot	*Pseudocercospora cruenta* (teliomorph = *Mycosphaerella cruenta*), *Cercospora canescens*	Up to 20% and 40% yield reduction induced by *P. cruenta* and *C. canescens*, respectively	Rain forest	Widespread	Schneider *et al.*, 1973; Williams, 1975b
Bacterial pustule	*Xanthomonas* sp. (probably a strain of *X. campestris* pv. *vignicola*)	Yield losses to 27% recorded at Ibadan, Nigeria	Rain forest	Nigeria, Kenya, Tanzania	Allen, 1983; Ekpo, 1978
Pythium soft stem rot	*Pythium aphanidermatum*	Up to 11% mature plant mortality in Ibadan, Nigeria	Rain forest, southern guinea savanna	Nigeria, Tanzania	Onuorah, 1973
Sphaceloma scab (first description of scab named it as Cladosporium spot)	*Sphaceloma* sp. (the presumed teliomorph, *Elsinoe phaseoli*, has not been detected on cowpea)	Up to 60% crop losses in both experimental and farmers' fields around Zaria, Nigeria. Total crop losses can occur during epiphytotics at flowering and early pod filling of determinate, photoperiod-insensitive cultivars	Guinea savanna	Kenya, Ethiopia, Burkina Faso, Uganda, Zambia, Zimbabwe, Tanzania, Rwanda, Nigeria	Anon, 1955; Bates, 1955; Mukiibi, 1979; Emechebe, 1980, Allen, 1983

continued

Table 1 cont.

Common name	Pathogen	Estimate of crop loss/importance on susceptible varieties	Ecological zone in which major damage is done	Distribution	Selected reference
Septoria leaf spot	*Septoria vignae, S. vignicola, S. kozopolzanskii*	Yield losses up to 30% induced by *S. vignae* observed in Samaru, Zaria, Nigeria; elsewhere (in India) up to 50% yield depression was induced by *S. vignicola*; severe attack by *S. vignae* reported in Kenya and Zimbabwe	Guinea and sudan savanna	*S. vignae* in East Africa, Zimbabwe, Nigeria; *S. vignicola* in East Africa; *S. kozopolzanskii* in Zambia	Wilkinson, 1927; Wickens, 1936; Hopkins, 1950; Allen, 1979, 1983; Emechebe, 1981c; Rawal and Sohi, 1983
Bacterial blight and canker	*Xanthomonas campestris* pv. *vignicola*	Yield depression in moderately susceptible Ife Brown in 1975 and 1976 was 26 and 18%, respectively at Ibadan, Nigeria, where damage by bacterial blight is much less than that in the guinea savanna	Guinea savanna, much lower severity in rain forest	Widespread	Sabet, 1959; Bellington, 1971; Ekpo, 1978; Kaiser and Ramos, 1979; Allen, 1981
Macrophomina charcoal rot, ashy stem blight, stem canker, root rot	*Macrophomina phaseolina* (sclerotial stage = *Rhizoctonia bataticola*)	Important in cowpea subjected to extreme moisture levels. The 1983 drought in most of West African savanna resulted in severe, pandemic ashy stem blight that caused total crop loss in most areas	Sudan savanna and the sahel	Widespread	Tompkins and Gardner, 1935; Luttrell and Weimer, 1952; Esuruoso, 1975; Williams, 1975c; Emechebe and McDonald, 1979; Oladiran and Okusanya, 1980; COPR, 1981

Disease	Pathogen	Remarks	Ecology	Distribution	References
Yellow blister (false rust)	*Synchytrium dolichi*	Usually minor, but localized severe epiphytotics frequently occur in Uganda	Areas of medium elevation (>500 m) with high rainfall and cool temperatures	Widespread	Mukiibi, 1969; Emechebe, 1975; Singh and Allen, 1979, 1980; COPR, 1981; Allen, 1983
Brown rust	*Uromyces appendiculatus*	Highly susceptible varieties can be almost completely defoliated by midflowering time resulting in severe yield loss	Rain forest, southern guinea savanna, medium elevation (500–1200 m) areas of East Africa	Widespread	Fromme, 1924; Snowdon, 1927; Singh and Allen, 1979, 1980; Allen, 1983
Ascochyta blight	*Ascochyta phaseolorum*	Severe epiphytotics can result in complete defoliation and destruction of shoots and pods, with complete destruction of the crop	Medium elevation (500–1200 m) in Southern, East and Central Africa including north-western Cameroon	Widespread, especially in East and Southern Africa	Hansford, 1932; Anon, 1955; Bates, 1957; Staples, 1958; Emechebe, 1975; Singh and Allen, 1980; Allen, 1983
Phytophthora stem rot	*Phytophthora* sp.	A restricted, severe epiphytotic in Ilonga (Tanzania) in 1979 killed all cowpea varieties in the affected plot. Out of 1781 cowpea lines screened for resistance in 1980 and 1981 only two were resistant; the rest were very susceptible and died 20–25 days after infection	Medium altitude sites in savanna	Restricted to Ilonga (Tanzania)	Singh *et al.*, 1982

Table 2.   Minor fungal and bacterial diseases of cowpea in three ecological zones of
Africa

Common name	Pathogen
Low-altitude, rain forest	
Sclerotium wilt	*Sclerotium rolfsii* Sacc. (*Corticium rolfsii* Curzi)
Fusarium collar and dry root rot	*Fusarium solani* (Mart.) Sacc. (*Nectria haematococca* Berk. & Br.)
Ascochyta blight	*Ascochyta phaseolorum* Sacc.
Corynespora target leaf spot	*Corynespora cassiicola* (Berk. & Curt.) Wei
Leaf smut	*Protomycopsis phaseoli* Ramak. & Subram.
Lamb's tail pod rot (*Choanephora* pod rot)	*Choanephora cucurbitarum* (Berk. & Rav.) Thaxt.; *C. infudibulifera* (Currey) Sacc.
Pink rust	*Phakospora pachyrhizi* Syd.
Basal stem rust	*Aecidium sp.* (probably *A. caulicola* P. Henn.) (?? aecidial stage of *Uromyces appendiculatus*)
Dactuliophora zonate leaf spot	*Dactuliophora tarrii* Leakey
Powdery mildew	*Erysiphe polygoni* DC
Aristatoma leaf spot	*Aristatoma guttulosum* Sutton; *A. oeconomicum* (Ellis & Tracy) Tehon
Purple stain	*Cercospora kikuchii* (Mat. & Tom.) Gardner
Low-altitude, guinea savanna	
Cercospora leaf spot	*Pseudocercospora cruenta* (Sacc.) Deighton
Charcoal rot, ashy stem blight, stem canker, root rot	*Macrophomina phaseolina* (Tassi) Goid. [sclerotial stage = *Rhizoctonia bataticola* (Taub.) Butler]
Brown rust	*Uromyces appendiculatus* (Pers.) Unger
Fusarium wilt	*Fusarium oxysporum* Schl. f. sp. *tracheiphilum* (E. F. Smith) Snyder & Hansen
Basal stem rot	*Schlerotium rolfsii* Sacc. (Corticium rolfsii Curzi)
Fusarium collar and dry root rot	*Fusarium solani* (Mart.) Sacc. (*Nectria haematococca* Berk. & Br.)
Powdery mildew	*Erysiphe polygoni* DC
Dactuliophora zonate leaf spot	*Dactuliophora tarrii* Leakey
Lamb's tail pod rot	*Choanephora cucurbitarum* (Berk. & Rav.) Thaxt; *C. infundibulifera* (Currey) Sacc.

continued

Common name	Pathogen
Medium altitude areas of rain forest and guinea savanna	
Cercospora leaf spot	*Pseudocercospora cruenta* Sacc. (Deighton); *Cercospora canescens* Ell. & Mart.
Septoria leaf spot	*Septoria vignae* P. Henn.; *S. vignicola* Vasat Rao; *S.* kozopolzansii Nikolajeva
Powdery mildew	*Erysiphe polygoni* DC
Bacterial blight	*Xanthomonas campestris* pv. *vignicola* (Burkholder) Dye
Leaf smut	*Protomycopsis phaseoli* Ramak. & Subram.
Dactuliophora zonate leaf spot	*Dactuliophora tarrii* Leakey

infected leaves turn yellow and fall. Complete defoliation of susceptible plants can occur.

The pathogen is seed-borne (COPR, 1981), and the disease spreads rapidly in rainy conditions.

Brown blotch

Hypocotyl lesions that originate from seed-borne infection are signs of brown blotch (Emechebe, 1981a); they are purplish-brown and occur at and above soil level. Those that do not cause seedling death persist on the growing and mature plant. Some of these subsequently girdle the stem base, causing wilting and death of the plant. Secondary stem lesions are initially dark brown; eventually a straw-coloured, sporulating central portion of the lesion is surrounded by a dark-brown border. Early infection of the flowering axis produces brown necrosis that causes either complete floral abortion or distortion and shrivelling of immature pods. Pods at all stages of maturity can be infected. Pod lesions are various shades of brown. Sporulation of the fungus on dry pods under humid conditions produces alternating brown and black bands, the latter being due to acervuli. Lesions on petioles and veins are dark brown, those on the veins being elongate. Occasionally, discrete dark-brown lesions occur in the interveinal areas of the lamina.

The disease is seed-borne (Emechebe and McDonald, 1979; Emechebe, 1981a), and all parts of the seed, including the plumule and radicle, can be infected (Alabi, 1981). Secondary spread in the field is mostly by rain splash and wind-blown rain. The pathogen also survives the dry season on infected debris on the soil surface (Okpala, 1981).

Brown rust

Lesions of brown rust (COPR, 1981) start as minute, almost white, slightly raised pustules on the leaves. In young plants the pustules, containing light-brown uredospores, cover the leaves, which wilt quickly in periods of sporadic rainfall. Pustules may be surrounded by a yellow halo. Severely affected leaves shrivel and fall. In older plants, leaves that have not been completely destroyed have blackened masses of teliospores.

The pathogen—*Uromyces appendiculatus*—builds up rapidly during the sporadic rainfall at the beginning and end of the rainy season (Williams, 1975c). It is favoured by cloudy, humid weather with temperatures between 22 and 28°C. The spores are spread by wind and insects and through contact with animals (including humans) and farm implements (COPR, 1981). The teliospores survive the dry season in the crop residues.

'*Cercospora*' leaf spot

Until recently, the pathogens of two diseases were regarded as species of *Cercospora, C. cruenta* and *C. canescens* (Williams, 1975c; Singh and Allen, 1979), but the former is now regarded as *Pseudocercospora cruenta* (Allen, 1983). Some data on the differences between the two diseases and their pathogens are provided by Vakili (1977).

Spots induced by *C. canescens* are roughly circular, cherry-red, and up to 10 mm diameter; and when numerous, they cause the leaves to become chlorotic and fall. Leaf spots induced by *P. cruenta* begin as a chlorosis on the adaxial surface, which becomes dotted with necrotic spots that, in turn, enlarge until the whole area is necrotic and coloured brown. On the abaxial surface, *C. canescens* lesions are red, whereas lesions caused by *P. cruenta* exhibit profuse sporulation in which masses of conidiophores appear as downy grey–black mats. Symptoms are not usually seen until flowering time, but in susceptible varieties the disease builds up rapidly and causes premature defoliation (Schneider *et al.*, 1976; Vakili, 1977). Spindle-shaped lesions on petiole, peduncle and stem have been reported by Vakili (1977).

Both pathogens are seed-borne (Williams, 1975c) and survive the dry season on diseased leaves as well as in seeds. Formation and release of spores are favoured by humid weather. Spores are spread by wind and water splash. A dense plant population and moderately warm temperatures aid the spread (Amin *et al.*, 1976; COPR, 1981).

Macrophomina charcoal rot and ashy stem blight

The seedling phase of disease caused by *Rhizoctonia bataticola* is commonly known as charcoal rot (Tompkins and Gardner, 1935; Luttrell and Weimer,

1952; COPR, 1981). It is characterized by cotyledonary lesions that appear as circular, dark-brown or black spots of various sizes. These coalesce, and frequently the entire cotyledon is blackened and contains numerous sclerotia. The hypocotyl lesions are linear and black. In the mature plant the most common symptom is dry or wet, greyish-black sunken lesions on the lower stem and roots associated with a crust-like accumulation of black sclerotia. The lesions enlarge to form elongated cankers with grey centres and brown margins, which quickly girdle the stem. The surface of the canker is slightly sunken and eventually splits and frays. Wilting then occurs. Premature senescence is associated with the sclerotial stage.

The pathogen is seed-borne (Sinha and Khare, 1977a). The disease in the mature plant results from secondary infection by air-borne conidia and sclerotia, which are also spread by irrigation water, and farming implements. Disease development is favoured by extremes of soil-moisture levels.

Phytophthora stem rot

Plants at all stages of growth are susceptible to infection by *Phytophthora vignae* (Purss, 1957; Singh *et al.*, 1982a). The characteristic lesions are brown and commence at ground level, extending upward, often reaching the top of the stem. The lesion is sunken and completely girdles the stem.

The margin is water-soaked and, in moist weather, is covered by a light fungal growth bearing sporangia. Internally, a brown discoloration extends into healthy tissue. The root system rapidly decays and is usually covered by a mat of white and pink fungal spores. Leaves of affected plants turn yellow and, in dry weather, the plant wilts and dies. Wilting of affected seedlings is rapid (Purss, 1957). Susceptible plants inoculated with a suspension of sporangia at Ilonga (Tanzania) developed large, water-soaked greyish leaf spots (Singh *et al.*, 1982a).

Phytophthora vignae is soil-borne and can survive in crop debris for long periods. Plants are severely affected during wet weather when water stands in the field for a few days. High humidity exacerbates the disease.

Pythium soft stem rot

Pythium soft stem rot (Onuorah, 1973; Williams, 1975c) is characterized by a grey–green, water-soaked girdle of the stem that extends from soil level to, and sometimes including, the lower branches. The infected area is slimy to the touch, and the stem cortex, which becomes packed with oospores, is easily stripped off. During periods of high humidity, copious growth of white cottony mycelia occurs at the stem base. Infected plants quickly wilt and die. The pathogen is soil-borne and at Ibadan, Nigeria, is favoured by warm, wet overcast weather (Williams, 1975c).

Seed decay and seedling damping-off complex

Seedling damping-off complex occurs in seedlings before and after emergence; it is induced by at least four pathogens (Table 1) (Onuorah, 1973; Williams, 1975c; Emechebe, 1981a). The postemergence seedling mortality induced by *R. solani* is characterized by reddish-brown lesions that are usually limited to the collar regions of the hypocotyl at which point the diseased seedling collapses. The lesions induced by *Pythium aphanidermatum* move rapidly up the hypocotyl, appear grey–green and wet, and the seedling undergoes a watery collapse. Seeds very severely affected by *C. capsici* fail to germinate. Seedlings collapse because this pathogen induces purplish-brown lesions that girdle the stem at soil level, or at the cotyledonary node. Lesions may spread from the petiole of an affected unifoliate leaf to the shoot growing point, which then dies.

Septoria leaf spot

Leaf spot caused by *Septoria* (Williams, 1975c) is characterized by bright red to dark red, roughly circular to irregular spots, 2–4 mm wide, which appear almost identical on both surfaces of the leaf. The disease can be distinguished from leaf spot induced by *C. canescens* by its smaller and more concentrated spots, which give the leaf a freckled appearance. Heavily spotted leaves turn yellow and fall.

The pathogen is seed-borne and transmitted, and is also spread by rain splash.

Sphaceloma scab

All above-ground parts of plants are affected by *Sphaceloma* scab (Emechebe, 1980). The first symptom on an infected young leaf is puckering of the lamina. Heavily puckered portions of the lamina become pale green or yellowish-green and develop white spots. Individual lesions on leaf veins are usually white, 1–15 mm long, while interveinal lesions on leaflets are circular (1–5 mm in diameter) and also white. The central portion of old lesions sometimes falls out to produce shot holes. Petiole lesions are oval to oblong–elliptical and initially dark brown, buff or white. Mature lesions on the stem are usually oblong–elongate and, under favourable conditions for the disease, coalesce and entirely cover parts of the stem. Lesions on the pods (a few to about 200/pod) are mostly ovoid, initially rusty brown; some are almost black when chlamydospores are formed. Lesions similar to those on the pods occur on the peduncle. Heavy scabbing of the flowering axis either causes flower and pod abortion or completely prevents flower formation. The pathogen is dispersed by rain splash, and the disease is aggravated by protracted periods (3–4 days) of wet weather (Emechebe, 1980); it survives the dry season on infected plant residues (Donli, 1983).

Web blight

The pathogen of web blight (Williams, 1975c; Singh and Allen, 1979; Allen, 1983) infects leaves and young stem tissues and can totally destroy the leaf canopy of the crop during periods of heavy rain with continuous overcast skies. The initial symptoms are small, circular brown spots. These enlarge, often showing concentric banding, and become surrounded by irregular shaped, water-soaked areas. Under humid conditions the lesions develop rapidly and coalesce, leading to extensive blighting and defoliation. The affected aerial parts of the plant become covered with sclerotia, which resemble a coarse sand dust. Affected leaves are often bound together by a web of brown hyphae of the fungus.

Web blight is induced by an aerial type of *R. solani*, which Onesirosan (1977) suggested has only cursory association with the soil. It survives in the absence of the host on infected debris (Onesirosan and Sagay, 1975), weed hosts and seed-borne inoculum (Onesirosan, 1975; Emechebe and McDonald, 1979).

Wilts

Fusarium wilt symptoms (Oyekan, 1975; COPR, 1981) include stunting of the affected plant and chlorosis, drooping, premature shedding and withering of leaves. The vascular tissues of the affected plant are discoloured brownish-purple; the discoloration often extends throughout the plant. The lower stem may become swollen before any chlorosis occurs. Veinal discoloration also occurs, and the affected plant finally wilts.

Apart from inducing damping-off disease of cowpea seedlings, *Sclerotium rolfsii* induces wilt of older plants. The initial symptoms include small, brown water-soaked dots that expand into irregular necrotic spots (2–3 mm in diameter). Later, shot holes appear on the necrotic tissue. Affected parts of the stem and root rot at soil level with dark-brown lesions appearing adjacent to or below the soil level. The lesions may girdle the stem. They become covered with white mycelia in which white–brown sclerotia are embedded. The sclerotia turn dark brown, and eventually the plant wilts and dies.

Sclerotium rolfsii survives in crop residues and weed hosts, which are sources of primary infection at the beginning of the new season. It is most harmful in light sandy soil, and shading as well as crowding increases disease prevalence. Sclerotia are dispersed during farming operations and by wind and runoff water.

Fusarium oxysporum f. sp. *tracheiphilum* survives in infested soil (as chlamydospores) for many years. Dry conditions tend to trigger irreversible wilting. The fungus is seed- and soil-transmitted, as is *S. rolfsii* (COPR, 1981).

Yellow blister (false rust)

The first symptom of yellow blister is the appearance on the leaf of small yellow blisters, greasy in appearance. These increase in number and may eventually cover the whole leaf surface. Petioles, peduncles, pods and stems are usually affected as well. Later, the blisters burst to release orange–brown spores. Eventually, raised warty orange–brown scabs are formed, and the plants are often distorted. The spores of the pathogen are dispersed by wind.

ECONOMIC IMPORTANCE

In reviewing the economic importance of the various diseases we found it convenient to classify them into the five groups based on the plant parts or growth stages that are most adversely affected by the diseases or their phases: seedling diseases, stem diseases, wilts, pod and flowering cushion diseases, and foliar diseases. These groups overlap somewhat as some pathogens (for example, *P. aphanidermatum, R. solani* and *C. capsici*) induce economically important diseases in more than one of the five classes (Table 1).

Seedling diseases

Williams (1973a) reported that seedling mortality induced by *P. aphanidermatum* and *R. solani* resulted in as low as 25 per cent seedling stand in cowpea sown during cool, wet weather in Ibadan, Nigeria, in 1972. In contrast, Onuorah (1973) observed only 1 per cent seed decay induced by *P. aphanidermatum* but a high percentage of losses from the pathogen in soft stem rot. Sowing of untreated cowpea seeds heavily infected with *C. capsici* in Zaria gave a seedling establishment of 24 per cent compared with 88 per cent establishment for healthy seeds of the same variety (Emechebe, 1981a).

Stem diseases

The major stem diseases of cowpea in Africa comprise anthracnose, *Pythium* stem rot, ashy stem blight, and *Phytophthora* stem rot. Anthracnose can cause up to 35–50 per cent yield reduction in susceptible varieties in the rain forest areas of Nigeria (Williams, 1973b). Onuorah (1973) reported 11 per cent deaths of established plants caused by *Pythium* soft rot. He noted that the plants died too late in the life of the crop for the healthy plants to take advantage of reduced competition from the dead plants. Onuorah (1973) then concluded that the potential yield loss was about 11 per cent.

We could find no published data on the damage caused by ashy stem blight, but our unpublished observations, as well as those by V. Parkinson (1983, personal communication), are that complete crop failures can be caused by the disease in the sudan savanna in several countries of West Africa.

Phytophthora stem rot is a relatively new disease in Africa; it is hoped that it is still confined to Ilonga (Tanzania) where it was first observed in 1979 (Singh *et al.*, 1982a). It is one of the most important diseases of cowpea in Queensland (Australia); for example, Purss (1957) reported total destruction of crops on farms that varied in size from 1 to 12 ha. In Ilonga, out of 1781 lines screened for resistance to the disease, only two were found to be resistant; the rest were destroyed by disease within 20–25 days of inoculation (Singh *et al.*, 1982a). A particularly worrying aspect of the disease found in Ilonga is that the varieties resistant to one or more of the four races of pathogen in Australia were all destroyed by the disease in Tanzania. This suggests that either a new race of *P. vignae* or an entirely different species of *Phytophthora* occurs in Tanzania.

Wilts

As far as we could determine, *Fusarium* wilt has only been reported from three African countries, namely Nigeria, Tanzania and Uganda (Holliday, 1970; Booth, 1971; Oyekan, 1975; Teri, 1984).

Studies in Ibadan, Nigeria, revealed a *Fusarium* wilt incidence of 21, 15 and 55 per cent respectively in Ife Brown, TVu 4557 and Prima (Oyekan, 1975).

We could find no published references for damage caused by *Sclerotium* wilt in Africa; however, workers in India have reported 20 per cent mortality caused by *Sclerotium* wilt and leaf blight in 5–10-day-old cowpea seedlings (Ramaiah *et al.*, 1976).

Foliar diseases

The major foliar diseases of cowpea include bacterial blight, bacterial pustule, *Ascochyta* blight, '*Cercospora*' leaf spots, brown rust, false rust, *Septoria* leaf spot and web blight. Bacterial blight caused 26.4 and 18.1 per cent yield loss in Ife Brown in Ibadan in 1975 and 1976, respectively (Ekpo, 1978). The disease is more destructive in the savanna zone where we have observed almost complete crop loss in susceptible varieties. Also, Ife Brown is considered to be moderately resistant to bacterial blight (Allen *et al.*, 1981b) so that higher losses would be expected to occur on more susceptible varieties in the rain forest.

Ekpo was quoted by Allen (1983) as having reported 1.8 and 26.6 per cent crop losses due to bacterial pustule in resistant and susceptible varieties, respectively.

· Crop losses caused by 'Cercospora' leaf spots have received far more attention than those caused by any other foliar disease. Studies by several workers have shown that leaf spot induced by P. cruenta is more important than that induced by C. canescens (Schneider, 1973; Schneider et al., 1976; Fery and Dukes, 1977b; Fery et al., 1977; Vakili, 1977). For example, yield reduction induced by P. cruenta was 42 per cent while that induced by C. canescens was 18 per cent (Schneider, 1973). Williams (1975a) reported 20 and 40 per cent yield loss due to leaf spots induced by P. cruenta and C. canescens in Ibadan, Nigeria. Oyekan (1979) has reported a yield depression of 28–40 per cent caused by web blight in southwestern Nigeria. Other data (Emechebe, 1981c) indicated a yield loss of about 30 per cent caused by Septoria leaf spot in northern guinea savanna. In India, yield losses up to 50 per cent have been reported for leaf spot induced by S. vignicola. We found no quantitative data for yield losses caused by Ascochyta blight or brown rust, but severe epiphytotics occur for each of the two diseases and cause premature defoliation and heavy crop losses or total destruction of the crop (Emechebe, 1975; Singh and Allen, 1979, 1980; Allen, 1983). Finally, localized epiphytotics of yellow blister (false rust) occur in Uganda (Mukiibi, 1969; Singh and Allen, 1979), although the disease is regarded as minor (Allen, 1983), at least in comparison with any of the other foliar diseases.

Pod diseases

Sphaceloma scab and brown blotch are the only two important pod diseases of cowpea in Africa. A crop loss of 75 per cent under protracted wet weather in Zaria, Nigeria, has been reportedly caused by brown blotch (Emechebe, 1981a). Sphaceloma scab has been observed to cause complete crop failure in susceptible varieties that pod during conditions conducive for disease development in the guinea savanna, where up to 60 per cent crop loss has been recorded both in farmers' fields and in experimental plots in the field.

CONTROL MEASURES

The major diseases can be controlled by the adoption of appropriate cultural practices; by judicious use of pesticides, especially fungicides; by use of resistant cowpea varieties and by integrated disease management that implies the complementary use of all these control methods.

Cultural practices

The cultural practices proposed for control of diseases are based on pathogen survival, inoculum spread for secondary infection and the environmental conditions that enhance disease development.

All the major diseases survive the dry season on infected crop residues in the form of vegetative bacterial cells, fungal mycelia or resistant structures such as chlamydospores (e.g., *Sphaceloma* sp., *F. oxysporum* f. sp. *tracheiphilum*), oospores (e.g. *P. aphanidermatum, Phytophthora* sp.), sclerotia (e.g. *S. rolfsii, R. solani*), and teliospores (*U. appendiculatus*). These primary sources of inoculum could be reduced if plant residues were collected and burned at the end of the crop season. The usual practice of igniting heaps of collected stubble when winds are favourable often leaves parts of the infected tissues untouched and is, therefore, of doubtful value (Palti and Rotem, 1983).

Anthracnose, *Ascochyta* blight, bacterial blight, bacterial pustule, brown blotch, '*Cercospora*' leaf spots, charcoal rot/ashy stem blight, *Septoria* leaf spot, web blight, *Fusarium* wilt and *Sclerotium* wilt are all induced by seed-borne pathogens. Their temporal and spatial dispersal could be reduced or prevented if farmers sowed only pathogen-free seeds, which, in most cases, can be produced from crops grown under conditions not conducive for disease development or treated with fungicidal sprays for control of foliar, stem and pod diseases. Such a regimen is followed by IITA in producing seeds for multilocational trials (Singh and Allen, 1980). Recent data from the Institute for Agricultural Research (IAR, 1984) indicate that well-timed fungicidal sprays even during the main crop season can result in a susceptible crop's virtual freedom from the three main fungal diseases of the guinea savanna (namely scab, brown blotch and *Septoria* leaf spot). Seeds produced in such plots can then be sold to farmers. The farmers, for their part, can obtain relatively healthy seeds from pods that are entirely free from disease signs or symptoms. The pods used for seeds must be threshed separately from those for consumption to avoid contamination.

Appropriate crop rotation helps control all the major diseases, but the choice of crops is difficult for soil-borne pathogens that form perennating structures or have a wide host range. For example, *M. phaseolina* affects so many crops that only a few species such as field pea and rice are resistant (COPR, 1981).

Four of the diseases are enhanced by high plant populations—'*Cercospora*' leaf spots, *Pythium* stem rot, *Sclerotium* wilt and web blight; their infection rates can be reduced in moderate plant populations.

Where possible to synchronize planting date with rainfall, losses from pod and flower cushion phases of brown blotch and scab can be reduced. Planting should be at dates that enable podding to commence just toward the end of the rains (IAR, 1982, 1983).

In areas where *Phytophthora* stem rot is prevalent, farmers should not plant cowpeas in poorly drained soils (COPR, 1981). In contrast, light sandy soil and shading exacerbate *Sclerotium* wilt, which can be controlled if one rogues out infected plants and ploughs contaminated topsoil. Avoiding mechanical damage to stem bases and heaping soil around the bases during cultivation can help. In areas with two growing seasons per year, farmers have to skip a season to avoid carryover of inoculum of all the pathogens. This 'closed season' is particularly important where only a few weeks separate the two growing seasons, as in southern Nigeria.

Chemical control

Chemical control of some of the major fungal diseases can be effected through application of fungicides to the seeds or through foliar fungicidal sprays. Soil drenches have also received some attention (for example, Oladiran and Okusanya, 1980). Williams (1975a,b), Singh and Allen (1980), and Allen (1983) have concluded that disease control through fungicidal sprays is not feasible at the farmers' level and that this technology is only relevant in experimental plots. In Nigeria, however, some cowpea is now being produced by large-scale farmers who have the money to purchase fungicides and spraying equipment. In these farms the indications to date are that fungicidal sprays are not uneconomic.

Dry fungicidal treatment of seeds can be used by all classes of farmers. Unfortunately, data are available for only a few diseases, even though all except three of the major diseases (brown rust, *Phytophthora* stem rot and false rust) are definitely seed-transmitted. A fourth (*Sphaceloma* scab) is at least carried with the seed. Seed treatment has been shown to be effective against brown blotch (controlled with benomyl or carbendazim, 1 g/kg seed), seedling mortality (controlled with chloroneb, 2 g/kg seed), and charcoal rot (controlled with carbendazim or benomyl, 1 g/kg seed, or thiram/captan at 2.5 g a.i./kg seed) (Williams, 1975a; Sinha and Khare, 1977b; Alabi, 1981; COPR, 1981).

Various workers have reported excellent control of foliar diseases with fungicidal sprays. One of us (Emechebe, 1981c) observed effective control also of *Sphaceloma* scab, brown blotch and *Septoria* leaf spot in northern Nigeria with a mixture of benomyl and mancozeb. Similarly, good control of brown rust in Kenya with triadimefon (Bayleton®) has been reported (Singh and Musymi, 1979), and Oyekan (1979) working in southwestern Nigeria controlled both web blight and '*Cercospora*' leaf spots with captafol spray. Schneider *et al.* (1973) obtained effective control of '*Cercospora*' leaf spot with one or two postflowering applications of benomyl. Thiabendazole and zineb have been found effective against *Septoria* leaf spot and *Ascochyta* blight, respectively (COPR, 1981). Where *Pythium* stem rot is a problem, however, its severity can be increased by the use of fungicides, such as benomyl (Williams and Ayanaba, 1975).

Resistant varieties

Until recently there was virtually no popular cowpea variety with acceptable levels of resistance to the major fungal and bacterial diseases prevalent in cowpea-production areas in Africa. Thus, one of the most popular, high-yielding varieties of cowpea in Nigeria, Ife Brown, is susceptible to diseases in both the rain forest and the savanna zones. Also, some IAR varieties (e.g. IAR 48, 176B, 353, 355), which are also high-yielding in Nigeria, are susceptible to at least one major disease in the savanna zone for which they were developed. This vulnerability of high-yielding varieties to disease has been receiving attention from plant breeders at national research institutes and at IITA.

A considerable achievement was the development by IITA of two varieties that combine high yield with resistance to many diseases and a few insect pests. The first to be developed was TVx 3236, which is resistant to anthracnose, bacterial blight, brown blotch and web blight, although it is susceptible to leaf spot induced by *P. cruenta* and highly susceptible to *Septoria* leaf spot (IITA, 1984). Screening we undertook at Samaru in 1983 (IAR, 1984) showed that TVx 3236 is also resistant to scab; this finding was confirmed by recent genetic studies (Abadassi, 1984). TVx 3236 is being superseded by another variety, IT 82D-716, which in addition to having all the desirable characteristics of TVx 3236, is resistant to leaf spot induced by *P. cruenta*. The only handicap of the two varieties with respect to bacterial and fungal diseases in the savanna is their susceptibility to *Septoria* leaf spot. Abadassi (1984) has worked out the inheritance of resistance to *Septoria* leaf spot and has shown that resistance is a dominant trait involving two loci. It should therefore be relatively easy to incorporate resistance to it into IT 82D-716.

Whether the resistance will be durable is uncertain, but four recent reviews (Fery, 1980; Hamblin, 1980; Meiners, 1981; Allen, 1983) have discussed the subject at various depths. There is some concern because resistance to each of the 10 diseases discussed in the reviews is controlled by one or two genes; in all cases at least one dominant gene is involved. Abadassi (1984) has just completed studies on three other diseases and has shown that resistance to brown blotch is monogenic with partial dominance of susceptibility over resistance. He further showed that resistance to scab is monogenic recessive.

Plant pathologists generally agree that resistance inherited as a dominant trait is fragile, and that the pathogen may evolve virulent races in its own selection process (Meiners, 1981). While further discussion of this topic is considered to be outside the scope of this paper, we advocate studies of the variability in each of the pathogens against which monogenic, major gene resistance has been incorporated into cowpea varieties. We also emphasize that breeders, in selecting for vertical resistance, should be careful not to erode the horizontal resistance in cowpea, which is, presumably, comparatively large but is often underestimated in breeders' plots (Singh and Allen, 1980).

Integrated disease management

Only complementary use of the three methods will ensure effective control of diseases, especially as several diseases attack cowpea in each of the agroecological zones. For example, all the cultural practices recommended for the control of anthracnose and other major diseases in the rain forest should be adhered to, whether or not a resistant variety like TVx 3236 or IT 82D-716 is being grown; apart from providing some measure of control of the other diseases to which the two varieties are susceptible, the cultural practices could delay the emergence of large populations of fit, novel races of *C. lindemuthianum* with matching virulence genes against the resistance genes of the varieties. Also, the two improved varieties may be protected against several other diseases, to which they are susceptible, if seeds are treated with appropriate fungicides; where economic, fungicidal sprays should also be used to control diseases.

Cowpea Research, Production and Utilization
Edited by S. R. Singh and K. O. Rachie
© 1985 John Wiley & Sons Ltd

12

Cowpea Diseases in Tropical Asia and Control in Rice-based Cropping Systems

T. W. MEW,* F. A. ELAZEGUI* and Y. P. S. RATHI†
*International Rice Research Institute, College, Laguna, Philippines; and †G. B. Pant University, Pantnagar, India

As elsewhere, cowpea in Asia suffers from many viral, fungal, bacterial and nematode diseases. The prevalence rates vary as the ecosystem changes, but some diseases are fairly widely distributed. The major viral diseases include bean common mosaic, cowpea banding mosaic, chlorotic spot, aphid-borne mosaic, mosaic (southern bean mosaic virus), cowpea necrosis and cowpea yellow fleck. Fungal diseases are seed and seedling rot, root rot, wilt, *Phytophthora* blight, web blight, anthracnose, powdery mildew, *Cercospora* leaf spot and rust, whereas the major bacterial disease is bacterial blight. Economic losses also occur from nematode diseases such as root knot and root lesion. The only important mycoplasm disease is phyllody.

VIRAL DISEASES

Most cowpea viral diseases are transmitted by beetles, aphids and a few by thrips and whitefly. Cowpea mosaic virus and cowpea banding mosaic virus are readily transmitted by aphids in a non-persistent manner. A single viruliferous aphid appears sufficient to initiate an infection (Beningo and Favali-Hedayat, 1977). Acquisition, inoculation and transmission thresholds are 20, 25 and 50 seconds, respectively (Sharma and Varma, 1975a). Many cowpea viruses, however, are seed-borne; thus they are widely distributed. Cowpea aphid-borne mosaic virus is seed-borne but generally at a low level (Beningo and Favali-Hedayat, 1977; Khatri and Singh, 1974). There was a distinct correlation between the age when plants became infected and the percentage of seed transmission, with transmission decreasing with age of plants (Haque and Chenulu, 1972). Khatri and Chohan (1972) showed that transmission occurred in plants inoculated 10 days before flowering.

Cowpea banding mosaic virus (CpBMV) is highly seed-transmissible, whereas cowpea chlorotic spot (CpCSV) is not, according to studies by Gupta and Summanwar (1980) and Sharma and Varma (1975a). In the studies, virus-carrying seeds were randomly distributed in the pods and the virus was present in various concentrations in all parts of the seed, except the seed coat. CpBMV was located mostly in the plumule buds and cotyledon while CpCSV was located primarily in the testa (Gupta and Summanwar, 1980; Prakash and Joshi, 1980). Sharma and Varma's studies (1975a) showed seed transmission of CpBMV varied from 9 to 34 per cent in different cowpea varieties, with a direct correlation between time of infection and extent of seed transmission. Plants infected 13, 33 and 53 days after sowing had 32, 22 and 21 per cent infected seeds, respectively.

Some cowpea viruses are also transmitted by mechanical means or by whitefly or grafting.

Although cowpea mosaic virus is reported to be inactivated by sodium dodecyl sulphate (Rao and Raychaudhuri, 1979) and CpBMV by gallic acid (Prakash and Joshi, 1979), effectiveness of these chemicals in control programmes has not been verified. Control of the insect vectors may be an efficient and practical means of controlling the viral diseases.

Application of phorate and dimethoate has been shown to prevent the spread of CpBMV, CMV and CpCSV, and disulfotan significantly reduced the incidence of mosaic (Gay et al., 1973). Phorate increased cowpea seed yield by 13 per cent.

FUNGAL DISEASES

Cowpea is vulnerable to attack by fungal pathogens at all growth stages.

Singh and Chohan (1974) found that seed-borne fungi lowered cowpea seed germination by 5–20 per cent, and seedling mortality at 21 days after planting has been reported to be 17 per cent (Williams, 1975a). Grain yield reductions of about 20 and 40 per cent have been attributed to *Cercospora canescens* and *C. cruenta*, respectively (IITA, 1973), in Africa. *Cercospora* leaf spot is prevalent in tropical Asia, as well, although we are not aware of figures on the losses caused. Web blight is common in cowpeas planted after rice in a rainfed environment (Mew, unpublished data), as is wilt caused by *Fusarium oxysporum*. In a cowpea variety trial conducted at IRRI, fields previously grown with wetland rice harboured *Fusarium* sp., with the prevalence of wilt being as high as 97 per cent in variety PI 339587 (Table 1). Some varieties appeared resistant, with less than 1 per cent infection.

In Asia, powdery mildew caused by *Erysiphe polygoni* increases rapidly during the dry and cool season.

Table 1.   Incidence of *Fusarium* stem and root rot on some varieties of cowpea (data taken from a zero-tillage trial after wetland rice at 38 days after emergence), lot N5, IRRI, 1979

Pedigree	Percentage infection*
All season	1.50
TVx 6-4H	21.75
TVx 1836-187E	80.75
E.G. 2	2.56
TVu 2931	16.69
VITA 4	15.38
TVu 4-b-b-10-b	15.94
VITA 3	0.56
TVx 289-46	0.44
TVx 1836-19E	0.75
TVx 66-2H	10.31
PI 339587	97.00

* Based on estimated disease prevalence in a row.

DISEASE MANAGEMENT IN RICE-BASED CROPPING SYSTEMS

Many rice-based cropping systems in Asia comprise a rice crop in the rainy season for 4–5 months and an upland crop either before or after rice. For example in the Philippines, in lowland systems, cowpea is often the crop, grown as a source of compost. The salient characteristic of such lowland systems is field flooding which, no doubt, affects the disease complex. It has been said, and is often taken for granted, that continuous flooding reduces the inocula of destructive pathogens, especially those causing seedling blight and root rot. We attempted to test this old adage. As pathologists we were interested in analysing the impact of the disease interactions between the rice crop and the upland crops in the same field (Mew and Elazegui, 1982).

Cowpea, in tropical Asia, is often a component crop in the rice-based cropping systems, both rainfed lowland and upland. Thus, the pathogens that attack both rice and cowpea, such as *Rhizoctonia solani*, build up in the fields and can cause considerable losses.

We studied the fungi from three seasons in three cropping patterns: continuous cowpea; dryland rice–cowpea–cowpea, and dryland rice–maize–cowpea. We found that the prevalence of damping-off of cowpea did not vary much among the three patterns. *Fusarium* and *Rhizoctonia* prevailed in continuous cropping of cowpea, *Sclerotium* and *Rhizoctonia* in dryland rice–green maize–cowpea, and only *Rhizoctonia* in dryland rice–cowpea–cowpea (Figure 1). *Fusarium* infection increased in continuous cropping of cowpea and was dominant in all plants exhibiting rot.

For control of *R. solani, Trichoderma harzianum* offers some promise as it infects rice straw buried in the soil and is highly competitive (Mew and Rosales, in press). Its ability to decompose rice straw and limit the survival of *R. solani* needs to be further investigated.

Compost favours the growth of saprophytes that are antagonistic to soil pathogens, so we tested the effect of using rice-straw compost, as a means to control fungal diseases on cowpea. During the second growing season compost was incorporated at planting of plots that were being continuously cropped with cowpea, but infection of the third crop was actually increased in plots with compost.

In another study, cowpea stubble was compared with rice stubble. The influence of tillage was also investigated since disease surveys show that disease prevalence is increased in fields with zero tillage. Root infection or damping-off was low during the trial such that the treatment effects could not be determined. Shoot infection or *Fusarium* infection progressed rapidly up to 24 days after emergence in high-tillage plots, whereas throughout crop growth the prevalence did not increase in zero-tillage plots. In high-tillage

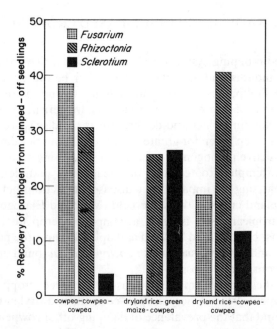

Figure 1.   Pathogen involved in damped-off cowpea as third crop in three cropping patterns

plots the stubbles, regardless of crop source, appeared to favour infection. For both stand count and grain yield, the two tillage levels were significantly different. The plots with zero tillage had significantly higher grain yields than did tilled plots.

These studies are a sample of work being done by IRRI, and they represent the focus of current efforts on controlling diseases that affect cowpea.

Cowpea Research, Production and Utilization
Edited by S. R. Singh and K. O. Rachie
© 1985 John Wiley & Sons Ltd

13

Cowpea Diseases and their Prevalence in Latin America

M. T. Lin* and G. P. Rios†
*University of Brasilia, Brasilia; and †EMBRAPA/CNPAF (Empresa Brasileira de Pesquisa Agropecuaria/Centro Nacional de Pesquisa—Arroz e Feijão), Goiânia, Goiâs, Brazil

Except for some viruses, diseases affecting cowpea in Latin America differ little from those found in other cowpea-growing areas of the world, although their significance, distribution and economic damage are more location-specific. Published information on the subject is reasonably complete for Brazil, whereas it is quite limited for most other countries in Latin America.

FUNGAL DISEASES

The major fungal diseases are scab, smut, leaf spot, powdery mildew, wilt and rots. Scab caused by *Elsinoe phaseoli* (synonym: *Sphaceloma* sp.) is common in Surinam, Central America (Singh and Allen, 1979; Rios and Neves, in press), and Brazil where recent surveys showed that as many as 16 per cent of cowpea fields had scab.

The pathogen is seed-borne so the most effective control method is to use resistant varieties. By using cultivars VITA 4, Kalkie and TVu 1888-C as sources of resistance, the cowpea team of EMBRAPA/CNPAF in Brazil have developed many resistant lines (Rios and Watt, 1980; Rios, 1983a; Rios and Neves, in press).

Smut (*Entyloma vignae*) occurs in the north and northeast of Brazil (Ponte, 1974, 1976; Ponte *et al.*, 1974; Santos, 1982; Rios and Neves, in press) and can cause 40 per cent loss of cowpea production (Ponte, 1966). Vakili (1977) reported 'carbon del frijol', a disease caused by *Entyloma* sp. in cowpea in Panama and in bean in other Central American countries. The cultivars

Cinzento, Ife Brown, V32-32, Serido, Carrapicho and Potomac are resistant to the pathogen (Ponte, 1974; Prabhu *et al.*, 1979).

Cercospora leaf spot (caused by *C. cruenta* and *C. canescens*) has been reported in Costa Rica (Araújo and Moreno, 1980), Puerto Rico (Vakili, 1977), Brazil (Ponte, 1976; Rios and Neves, in press), and other Central American countries (N. G. Vakili and W. J. Kaiser, unpublished). A survey carried out from 1978 to 1982 in Brazil showed 55–73 per cent of the cowpea fields are infected (Rios and Neves, in press), but the disease is considered minor because infection usually occurs when plants reach maturity. Neverthe-less, Fery *et al.* (1977) showed that *C. cruenta* can cause losses of up to 36 per cent of grain production. The incidence of this disease is higher in the rainy season than in the dry season (Araújo and Moreno, 1980; Rios, 1983b) but is not affected by the system of cultivation (Araújo and Moreno, 1980). Spraying with benomyl reduces the prevalence of the disease and increases significantly the yield from cowpea fields (Rios, 1983b).

Powdery mildew (caused by *Erysiphe polygoni*, synonym *Oidium* spp.) occurs from May to October in central Brazil (Rios and Neves, in press) and has been reported in Costa Rica and Puerto Rico as well (N. G. Vakili and W. J. Kaiser, unpublished). We believe it is found in all cowpea-growing areas in the world. Its development is favoured by cloudy conditions and wet weather, and intercropping of cowpea with corn increases the prevalence in the field (Rios, 1983b). Resistant cultivars are available, but the existence of races limits the practicality of attempting to control the disease by breeding for resistance.

Web blight (*Thanatephorus cucumeris*) occurs in hot and humid regions such as the north of Brazil where incidence of the disease is so high that growing bean is impossible. Cowpea, which is more tolerant to this fungus than bean, is planted to substitute for bean in this area (Cardoso and Mesquita, 1981). The fungus has a wide host range. In Brazil the control strategy is to plant resistant cultivars in the period when rainfall is limited.

Like powdery mildew and web blight, *Ascochyta* leaf spot (*A. phaseolor-um*) occurs more frequently in the hot, rainy season than in the dry season (Araújo and Moreno, 1980) but, in contrast with powdery mildew, intercrop-ping of cowpea with corn seems to result in lower incidence of the disease than is found in fields of cowpea intercropped with other crops (Araújo and Moreno, 1980; Rios *et al.*, 1982).

Fusarium wilt (*F. oxysporum* f. sp. *tracheiphilum*) is widely distributed on the American continent and causes considerable losses in Brazil (Rios *et al.*, 1982). The pathogen invades the plant systemically, causes necrosis and darkening of the vascular tissues, and results in the yellowing and premature fall of lower leaves. Young plants sometimes wilt and die. This disease and others caused by soil-borne fungi are becoming important in the irrigated areas in northeast Brazil where the soil is intensively cropped (Rios and Neves, in press).

Charcoal stalk rot (*Macrophomina phaseolina*), from 1980 to 1984, was responsible for up to 80 per cent of the premature deaths of plants in cowpea fields in northeast Brazil (Ponte, 1976; Rios and Neves, in press). Research data indicate that the disease is associated with stress, especially from lack of moisture (Magalhães *et al.*, 1982).

Sclerotium stem rot (*S. rolfsii*) is frequently found in north Brazil (Fulton and Allen, 1982; Rios and Neves, in press) where it is hot and humid. Control is difficult because the pathogen has a wide host range and can survive in the soil for a long time.

Other less important fungal diseases in Latin America include anthracnose (caused by *Colletotrichum lindemuthianum*) reported in Brazil (Ponte, 1976; Santos, 1982; Rios and Neves, in press) and Nicaragua (Singh and Allen, 1980); brown rust (*Uromyces appendiculatus*) in Brazil (Ponte, 1976); *Corynespora* leaf spot (*C. cassiicola*) in Brazil (Ponte, 1976; Rios and Neves, in press), Panama (N. G. Vakili and W. J. Kaiser, unpublished), and Venezuela (Subero, 1980); *Fusarium* collar and root rot (*F. solani*) in Brazil (Ponte, 1976; Santos, 1982; Rios and Neves, in press); lamb's tail pod rot (*Choanephora* sp.) in Brazil (Ponte, 1976; Rios and Neves, in press); *Pythium* stem rot (*P. aphanidermatum*) in Brazil (Rios and Neves, in press); damping-off of seedlings (*P. aphanidermatum* and *Rhizoctonia solani*) in Brazil (Santos, 1982; Rios and Neves, in press) and Nicaragua (Singh and Allen, 1979); and *Septoria* leaf spot (*S. vignae*) in Nicaragua (N. G. Vakili and W. J. Kaiser, unpublished).

BACTERIAL DISEASES

The key bacterial diseases are bacterial blight, bacterial wilt and halo blight.

Bacterial blight (*Xanthomonas vignicola*) is common in Puerto Rico during the heavy rainy seasons (Vakili *et al.*, 1975) and has been reported to occur in various parts of Brazil, although it has caused economic losses only in the humid areas and in areas using sprinklers for irrigation (Rios and Neves, in press).

Bacterial wilt (*Pseudomonas solanacearum*) was found by Ponte (1976) in the irrigated area of Piauí, Brazil, but apparently has no economic importance in Brazil.

Halo blight (*Pseudomonas syringae* pv. *tabaci*) was found, for the first time, recently in the states of Rio Grande do Sul and Goiás in Brazil (Rios and Neves, in press). Not much is known about this disease.

VIRAL DISEASES

Cowpea severe mosaic virus (CSMV) is the most widely distributed cowpea virus in Latin America. It is a member of the comovirus group with isometric particles of *ca.* 25 nm, transmitted by beetles and through seeds (de Jager,

1979). Symptoms induced by this virus include severe mosaic, blistering, colour breaking of the flowers and stunting. The beetle vectors in Brazil are *Cerotoma arcuata* (Costa *et al.*, 1978), *Diabrotica speciosa* (Costa *et al.*, 1981), and *Chalcodermus bimaculatus* (Lima and Goncalves, 1980). In Costa Rica the vectors are *Cerotoma ruficornis, C. atrofasciata, Gynandrobrotica variabilis, Diabrotica adelpha* and *Epilachna varivestis* (Valverde *et al.*, 1978). Several legumes including *Phaseolus lathyroides* (Lima and Nelson, 1977), *Canavalia ensiformis* (Lima and Souza, 1980), winged bean (Kitajima *et al.*, 1979), soybean (Anjos and Lin, 1984), *Calopogonium mucunoides, Centrosema pubescens, Vigna radiata, V. sesquipedalis* and *Crotalaria juncea* (Lin *et al.*, 1982a) have been reported as natural hosts of CSMV. Serological variation among the CSMV isolates has been reported (Fulton and Scott, 1979), and four serotypes have been identified (Lin *et al.*, 1981a, 1984). The cultivars or lines that are immune to this virus include Macaibo (Lima and Nelson, 1977), CNC 0434 (Rios and Neves, in press), TVu 612, TVu 1460-2, TVu 1948 and TVu 2480 (Fulton and Allen, 1982).

Blackeye cowpea mosaic virus (BlCMV) was found only recently in Brazil (Lin *et al.*, 1981b), with properties of the isolate (Lin *et al.*, 1981c) being similar to those of the Florida isolate (Lima *et al.*, 1979). The Brazilian isolate has a restricted host range, infecting only six of the 24 species in eight families tested (Lin *et al.*, 1981c). It is seed-borne (1.2 per cent in the cultivar Alagoas) and transmitted by the aphid *Myzus persicae*. Resistance to the Brazilian isolate was found in cultivars Canapu, Praiano, Pernambuco V-12, Mamoninha II, Bola de Ouro, Pendanga, Serido V-3 and Paraiba V-5 (Lin *et al.*, 1981c).

Cucumber mosaic virus, a member of the cucumovirus group, was recently found infecting cowpea in various parts of Brazil (Lin *et al.*, 1981b, 1982b), possibly because of its high transmissibility through seeds and by aphid vectors (Lin *et al.*, 1981c). In cowpea the virus causes vein-clearing and mosaic that usually disappear within 2–3 weeks after infection. The virus is frequently isolated from cowpea plants infected with other viruses (Lin *et al.*, 1981), and a synergistic effect by this virus and BlCMV in cowpea has been reported (Pio-Ribeiro *et al.*, 1978).

Cowpea green vein-banding virus (CGVBV) was first isolated in the state of Goiás, Brazil (Lin *et al.*, 1979), and was later found in the state of Piauí (Santos *et al.*, 1980). It is a seed-transmitted, aphid-borne potyvirus but is not serologically related to BlCMV (Lin *et al.*, 1979). In cowpea, CGVBV always induces the typical green vein-banding. Besides cowpea, it systemically infects bean, soybean, *Phaseolus arborigenus, P. acutifolius, P. lunatus, Cassia occidentalis* and *Lupinus albus* (Lin *et al.*, 1979). Resistance is found in cowpea cultivars such as Serido, TVu 91, Praiano, VITA 3, Pernambuco V-12, Carrapicho, 40 Días, Jatoba and Pitiuba.

Cowpea aphid-borne mosaic virus (CAbMV), a member of the potyvirus group, was originally isolated from cowpea in Italy in 1966 (Bock and Conti,

1974). Recently, Lima *et al.* (1981) reported the occurrence of this virus in the state of Ceará, Brazil, noting that the Brazilian isolate is transmitted by *Aphis craccivora*, infects bean, *Cassia tora*, *Centrosema brasilianum*, *Nicotiana benthamiana*, *Chenopodium amaranticolor* and 59 of the 60 cowpea cultivars tested. The virus is serologically related to BlCMV and bean common mosaic virus. Although the Italian isolate is transmitted through seeds (Bock and Conti, 1974), the Brazilian one apparently is not (Lima *et al.*, 1981). The cultivar CE-315 is immune to the Brazilian isolate.

Cowpea rugose mosaic virus (CRMV) and cowpea severe mottle virus (CSMtV) are both members of the potyvirus group. They were originally isolated from cowpea plants in the state of Piauí (Santos *et al.*, 1980; Santos, 1982) and are serologically related to each other but not identical (Santos *et al.*, in press). Together with CGVBV they constitute a serogroup that is not related to the group consisting of BlCMV and CAbMV. Besides cowpea, both CRMV and CSMtV infect bean, pea, *Calopogonium* sp., *Chenopodium* spp. and *Gomphrena globosa* but not plants of the families Solanaceae, Euphorbiaceae, Cruciferae, Compositae and Gramineae (Santos *et al.*, in press). CRMV, but not CSMtV, infects *Centrosema* sp. and *Cyamopsis tetragonolobus*, while *Cassia obtusifolia*, soybean, *Crotalaria paulina*, *C. retusa* and *C. mucronata* are hosts for CSMtV but not for CRMV (Santos *et al.*, in press). Of 30 cowpea cultivars tested, 16 showed similar reactions to the two viruses, five were immune to CRMV but susceptible to CSMtV and nine reacted differently to the two viruses. In the cowpea cultivar Serido, CSMtV, but not CRMV, could be transmitted by seeds (Santos *et al.*, in press). The aphid vectors for CRMV are *M. persicae* and *Aphis nerii*, while those for CSMtV are *M. persicae* and *A. citricola*. Sources of resistance are available; the cultivar Potomac is immune to both viruses and cultivars Pitiuba, Alagoas, Carrapicho, Oscariote and Mamoninha II are immune to CRMV (Santos *et al.*, in press).

Cowpea golden mosaic virus (CGMV), which is common in Africa and Pakistan (Singh and Allen, 1979), now may also occur in Latin America. Plants with symptoms similar to those reported for CGMV elsewhere were observed in northeast Brazil recently (Santos *et al.*, 1980; Santos, 1982; Lima *et al.*, 1983). Santos and Freire Filho (1984b) were able to transmit the disease by a whitefly species and by grafting, but not by mechanical means or by the seed. Cowpea cultivars with resistance to the disease in the field have been observed in Brazil (Lima *et al.*, 1983; Santos and Freire Filho, 1984b) and include the CSMV-resistant CNC 0434 (Santos and Freire Filho, 1984b) and the CAbMV-resistant CE-315.

CONCLUSIONS

Limited information is available on the diseases of cowpea in Latin America. Most of the studies reported here were carried out in Brazil after the establishment of the national cowpea programme in 1977 by EMBRAPA/ CNPAF in collaboration with the grain legume programme of IITA and the virology team at the University of Brasilia. It is our hope that this type of collaboration can be extended to other cowpea-growing countries in Latin America, so that we shall have a better picture about the disease problems affecting cowpea in this part of the world.

Cowpea Research, Production and Utilization
Edited by S. R. Singh and K. O. Rachie
© 1985 John Wiley & Sons Ltd

14

Fungal, Bacterial and Viral Diseases of Cowpeas in the USA

P. N. PATEL
University of California, Riverside, USA

At the beginning of this century in the southern United States, cowpea was grown for hay, forage, seed, and, more importantly, for its value as a fertilizer and soil renovator (Orton, 1902). Now it is predominantly grown for shelled green seeds for fresh market, canning or freezing. Disease pressure is generally high in these crops due to rains, warm weather, abundant vectors of viruses and continuous cropping. In contrast, the cowpea-production areas in California, where all the dry edible seeds (blackeye types only) are produced, receive practically no rains during the growing season, and the crops are largely free from the foliar diseases (Mackie, 1946).

Of the diseases reported in the USA (Table 1), only wilt and root knot are widely distributed and cause losses throughout production areas. Others, such as bacterial leaf spot, *Phytophthora* stem rot, *Chaetoseptoria* leaf spot and *Leptosphaerulina* leaf spot, have not been reported on subsequently. Sources of resistance to many of the prevalent diseases have been identified and bred into commercial cultivars. Compared with common beans, cowpeas receive relatively little research attention in the US.

FUNGAL DISEASES

Wilt

Wilt incited by *Fusarium oxysporum* f. sp. *tracheiphilum* was first reported from South Carolina (Orton, 1902) and is now prevalent in all the cowpea-growing areas. Many fields became wilt-sick due to a gradual build-up of the pathogen in soil and were abandoned until resistant cultivars became available.

Table 1.   Diseases of cowpeas in the USA

Disease	Pathogen	Reference
Seedling mortality	*Rhizoctonia solani, Pythium aphanidermatum, R. bataticola*	Mackie, 1946; Tompkins and Gardner, 1935
Red stem canker	*Phytophthora cactorum*	Weimer, 1949
Sclerotium stem rot, southern blight	*Sclerotium rolfsii*	Mackie, 1946; Toler *et al.*, 1963
Stem rot	*Diaporthe phaseoleorum*	Luttrell, 1947; Toler *et al.*, 1963
Ashy stem blight (charcoal rot)	*Macrophomina phaseolina*	Luttrell and Weimer, 1952
Sclerotinia wilt, stem canker	*Sclerotinia sclerotiorum*	Mackie, 1946; Toler *et al.*, 1963
Fusarium root rot	*Fusarium solani*	Weimer and Harter, 1926; P. D. Dukes (personal communication)
Fusarium wilt	*F. oxysporum* f. sp. *tracheiphilum*	Orton, 1902; Armstrong and Armstrong, 1950
Verticillium wilt	*V. alboatrum*	Baker *et al.*, 1940; 40; Strobel, 1961
Cercospora leaf spot	*C. cruenta*	Lathman, 1934; Fery *et al.*, 1977
Target spot	*Corynespora cassiicola*	Olive *et al.*, 1945; Spencer and Walters, 1969
Rust	*Uromyces appendiculatus*	Fromme, 1924; Toler *et al.*, 1963

Fusarium wilt, in contrast to typical vascular wilt, does not cause conspicuous wilting in cowpea. In the field the disease does not appear before plants are 6 weeks old when a few plants exhibit pale green, flaccid leaves that soon turn yellow and drop. These plants usually die prematurely. As the season advances, so does the destruction. Many plants, showing no external symptoms, show swelling in the lower stem and extensive vascular discoloration. These plants usually survive and produce some dry seeds. During threshing, the seeds get surface-contaminated with infected plant debris and act as a source for the disease in new areas. Systemic infection of the seeds, if it occurs at all, is rare (Kendrick, 1931). The pathogen, as chlamydospores, survives in

Disease	Pathogen	Reference
Powdery mildew	*Erysiphe polygoni*	Fennell, 1948
White leaf spot	*Aristatoma oeconomicum*	Tehon, 1933; Lefebvre and Stevenson, 1945; Brown and Dukes, 1982
Cladosporium spot	*Cladosporium vignae*	Gardner, 1925; Toler *et al.*, 1963
Chaetoseptoria leaf spot	*Chaetoseptoria wellmanii*	Tehon, 1937
Leptosphaerulina leaf spot	*Leptosphaerulina vignae*	Tehon and Stout, 1928
Choanephora pod rot	*Choanephora cucurbitarum*	Lefebvre and Weimer, 1939; Toler and Dukes, 1965
Bacterial leaf spot	*Pseudomonas syringae* pv. *syringae*	Gardner and Kendrick, 1925
Bacterial stem canker	*Xanthomonas campestris* pv. *vignicola*	Burkholder, 1944
Mosaic	Blackeye cowpea mosaic virus	Taiwo *et al.*, 1982
	Cucumber mosaic virus	Kuhn *et al.*, 1966
	Tobacco mosaic virus	Kuhn *et al.*, 1966
	Southern bean mosaic virus	Kuhn *et al.*, 1966
	Cowpea severe mosaic virus	Shepherd, 1964
Stunt	Mixed infection of blackeye and cucumber mosaic virus	Pio-Ribeiro *et al.*, 1978
Chlorotic mottle	Cowpea chlorotic mottle virus	Kuhn *et al.*, 1966

soil for a long time. Anaerobic conditions may inhibit the growth of the pathogen but do not adversely affect the survival in soil (Toler *et al.*, 1966).

Control through breeding was started soon after 1902, when the cultivar Iron was identified as resistant to wilt and root-knot nematode (Orton, 1902; Webber and Orton, 1902). The first successful breeding programme was started in California in 1929, with crossing of Iron and susceptible California Blackeye and screening of segregating populations in naturally infested, wilt-sick fields. Several resistant breeding lines were developed, and California Blackeye 5 (CB5) was released about 1936. A wilt- and root-knot-resistant cultivar, Calva, was developed from the resistant parent, Virginian

Blackeye, but was never released (Kendrick, 1936). CB5 is still the most popular blackeye variety in California, in spite of breakdown of its wilt resistance. Mackie (1946) noted that none of the breeding lines were as resistant as Iron. This was attributed, by him, to the influence of backcrossing. Another wilt-resistant cultivar selected from a farmer's field in the Chino area in the 1950s, CB3, is currently grown in badly wilt-sick fields; but, because of lodging at maturity and low yields, it has never completely replaced the agronomically superior CB5. A new wilt-resistant blackeye (CB7), developed at the University of California, Davis, is now being released. Its wilt resistance is derived from a germ-plasm line, PI 166146.

In the southern states an active breeding programme to instil resistance to root knot and wilt has been pursued at Mississippi State University by W. W. Hare since the early 1950s and has resulted in the release of Mississippi Crowder and Mississippi Silver. Mississippi Silver, with resistance to all three races of the wilt pathogen and *Meloidogyne* spp., as well as tolerance to several viruses, soon became the most widely cultivated variety for fresh pick, canning and freezing in the south (Hare, 1957b, 1967). Other wilt-resistant cultivars released during this period were Calhoun Wonder in Texas (George, 1948) and Lagreen in Louisiana (Miller *et al.*, 1962).

Three races of the wilt pathogen have been recognized on the basis of ability of an isolate to cause death in a differential cultivar. Race 1 kills the cultivar Groit but not Chinese Red and Arlington. Race 2 kills Chinese Red but not others, whereas race 3 kills Chinese Red and Arlington but not Groit. All are prevalent in the cowpea areas in the southern states. An isolate from California was identified as race 2 (Armstrong and Armstrong, 1950; Hare, 1953). Variability in wilt fungus in California has not been studied. Selection of wilt-resistant CB5 and its subsequent breakdown suggests predominance of race 1 in the wilt-sick fields used for screening by Mackie and gradual build-up of races 2 or 3. However, in a recent survey of races in California, none of the 25 isolates obtained in 1983 or 1984, which were inoculated in 25 cowpeas by the root–clip–dip method, behaved like race 1; they could kill all the race differentials as well as the lines reported as highly resistant in Nigeria (Oyekan, 1977). Known wilt-resistant cultivars, Mississippi Silver, CB3, Magnolia and others, behaved as highly resistant. In wilt-sick fields many plants of Arlington and Groit did not show visible wilt symptoms, but roots and stems were extensively invaded, as indicated by vascular discoloration. It has been found that reactions in race-differential cultivars were influenced by temperature, age of the plants (Swanson and Van Gundy, in press), method of inoculation, inoculum load and virulence of the isolates. There is a need to establish race differentiation on a more stable host–pathogen genetic system.

Three dominant genes have been implicated in wilt resistance in Iron (Hare, 1957a), but how each gene interacts, singly or in combination, with each race has not been reported. No race has yet been found that can overcome resistance found in Iron.

A root-knot nematode species, *Meloidogyne javanica*, has been shown in artificial inoculations to increase wilt severity in a tolerant cultivar, Grant, believed to be the same as CB3 (Thomason *et al.*, 1959). In soils where nematodes were controlled, the severity of wilt was markedly reduced in both Grant and a susceptible cultivar, Chino-3, and seed yields increased nearly threefold (Thomason, 1958). In surveys in California, my colleagues and I have not encountered a conspicuous breakdown in wilt resistance of CB3. Breakdown may depend on genetic background of a wilt-resistant cowpea cultivar or the degree of its susceptibility to *M. javanica*. In artificial inoculations of breeding lines we have noted that some resistant lines show extensive vascular discoloration in stems and roots. Whether such lines are more vulnerable to breakdown by root-knot nematode is now being examined.

Cercospora leaf spot

The disease caused by *Cercospora cruenta* (synonym *Mycosphaerella cruenta*) was first recorded from Mississippi in 1891 and occurs in most of the southern states (Lathman, 1934; Toler *et al.*, 1963). The spots first appear as a chlorosis on the upper surface of the leaves, enlarge, and gradually become necrotic with profuse masses of conidiophores and spores on the under-surface. Leaves with many lesions defoliate prematurely. The pathogen is seed-transmitted and can survive in plant debris. Recently this leaf spot has become a major foliar disease in fall crops in the southeastern states and can reduce seed yields up to 36 per cent in a susceptible cultivar. Frequent epiphytotics are attributed to susceptibility of the present cultivars, inadequate crop rotation and sequential plantings (Fery *et al.*, 1977). Cowpea accessions with high levels of resistance were identified and used to develop the cultivar Colossus 80, now released as a replacement for a widely adapted and popular home garden and fresh market cultivar, Colossus (Fery *et al.*, 1982). Inheritance studies have shown that the resistance is governed by two distinct genes. One, designated as *Cls-1* and present in a breeding line CR 17-1-34, was completely dominant, and its expression was not influenced by environment. The other gene, *cls-2*, present in Ala 963-8, was incompletely dominant and highly influenced by the environment when in heterozygous conditions. These genes are neither allelic nor linked (Fery *et al.*, 1976; Fery and Dukes, 1977a).

Charcoal rot

Rhizoctonia bataticola (synonym: *Macrophomina phaseolina*), a pathogen inciting charcoal rot in seedlings and leaf spot and ashy stem blight in adult

plants, is prevalent in most of the cowpea-growing areas in the USA (Tompkins and Gardner, 1935; Mackie, 1946; Luttrell and Weimer, 1952; Toler *et al.*, 1963). Cowpea seedlings infected by inoculum in the seed or soil develop charcoal rot, whereas leaf spot, stem canker, and ashy stem blight result from secondary infection from conidia or sclerotia (Luttrell and Weimer, 1952). In California, Mackie (1946) found that charcoal rot was favoured by dry soils and high temperatures and, therefore, caused more damage in blackeyes in unirrigated or lightly irrigated soils. He found Iron to be resistant and incorporated this single gene-controlled resistance in several breeding lines.

Other diseases

Scab, pod rot and fungal diseases of seedlings also cause economic losses.

Scab, first observed in Indiana, was described by Gardner in 1925 and the causal fungus was designated as *Cladosporium vignae*. Subsequently it was encountered in Georgia, Florida, Delaware, California, North Carolina and South Carolina. It occurs in crops planted in cool, humid weather (P. D. Dukes, personal communication). The disease is characterized by blackened or purplish, sunken, scabby spots on the pods, stems, peduncles, leaves and bracts. Only young tissues are susceptible to infection. The fungus is transmitted through seeds. Cultivars Louisiana Purchase and Florida 133-0110 are immune to the disease (Strider and Toler, 1963).

Pod rot was first reported in cowpeas in 1920 (Wolf and Lehman, 1920). The fungus (*Choanephora cucurbitarum*) is an opportunistic polyphagous pathogen. In cowpeas, Toler and Dukes (1965) estimated 1–15 per cent loss in the spring-planted crops in Georgia under conditions of continuous high humidity and rainfall. The first symptom on mature pods is a water-soaked appearance, followed by a soft rot, accompanied by abundant fruiting of the fungus. Infection could spread to peduncles. The nutrients leached from flowers in highly humid environments provide an ideal base for proliferation of the fungus. Cultivars with a tendency to retain dried corolla on the pods are more likely to be damaged severely. High plant density and insect damage to pods also favour this desease (Cuthbert and Fery, 1975b); Carolina Cream is resistant (P. D. Dukes, personal communication).

Several fungal pathogens damage cowpeas in the seedling stages. During early planting when soil temperatures are low and there is abundant moisture, *Rhizoctonia solani* infections in hypocotyls can reduce seedling vigour or cause mortality and affect the plant stand. Excessive irrigation can also favour *Pythium aphanidermatum*. However, fungicide treatment of the seeds provides adequate protection to the seedlings until they develop natural resistance (Mackie, 1946; Gay, 1970).

Among the remaining fungal diseases, powdery mildew (*Erysiphe polygoni*) occurs widely but appears late in the growing season and does not cause much loss in yield. Resistance is available and was shown to be controlled by a single dominant gene (Dundas, 1939) or multiple recessive genes (Fennell, 1948). Rust (*Uromyces appendiculatus*) frequently occurs in fall plantings in South Carolina and Georgia, but many of the cultivars grown are resistant (P. D. Dukes, personal communication; Gay, 1971). Sources of resistance to white leaf spot (*Aristatoma oeconomicum*) are being identified (Brown and Dukes, 1982).

BACTERIAL DISEASES

Leaf spot and blight are the major bacterial diseases of cowpea in the USA.

Bacterial leaf spot, caused by *Pseudomonas syringae* pv. *syringae*, was first described in 1923 by Gardner and Kendrick and has been reported to be present in Indiana, Delaware, Kentucky, Florida other adjoining states (Gardner and Kendrick, 1925), and Australia (Wilson, 1936). Cool, humid environments favour this pathogen. It does not cause economic losses currently, but the same pathogen, inciting bacterial brown spot disease in *Phaseolus* beans, remained unimportant for 34 years after its first report (Burkholder, 1930) before becoming severe again (Patel *et al.*, 1964). In cowpea the disease is characterized by reddish-brown lesions on the leaves, stems and pods. Young spots are greasy and water-soaked. Severe symptoms in young growing organs result in distortion and shattering of the leaves, deformity in the pods, and discoloration and shrivelling of the seeds. The pathogen is seed-borne and has an extensive host range (Gardner and Kendrick, 1925).

Bacterial blight, incited by *Xanthomonas campestris* pv. *vignicola*, was first reported from Oklahoma in 1931 and became important in Texas and other states in the south (Burkholder, 1944; Hoffmaster, 1944; Lefebvre and Sherwin, 1945), but not in California (Mackie, 1946). Recently its prevalence has increased in Georgia, and the pathogen has been detected in seeds produced in Florida, Georgia, Texas and California (Gitaitis and Nilakhe, 1982). It is also prevalent in Puerto Rico (Vakili *et al.*, 1975). Bacterial pustule disease, earlier considered to be caused by a strain of the blight organism but recently suggested as a distinct pv. *vignaeunguiculatae* (Patel and Jindal, 1982), does not occur in the USA.

Watkins (1943) recorded 60 per cent plant mortality from blight in a California Blackeye variety in Texas. The disease first appears as small water-soaked spots on the under-surface of leaves, and surrounding tissues gradually become necrotic with a yellow border. The bacterium invades the cortex as well as the vascular tissue and moves systemically into stems, peduncles and pods. Infected stems and peduncles develop cankers and split.

Early infection from seed-borne inoculum can cause seedling mortality, whereas secondary infection at the tip of the peduncles can destroy young pods or infect seeds through the funiculus (Shekhawat *et al.*, 1977).,

The blight pathogen also infects certain varieties of common beans (Sherwin and Lefebvre, 1951; Vakili *et al.*, 1975) and *Vigna umbellata* (Jindal and Patel, 1980). In infiltration inoculations of bean pods it produces distinct reactions, useful in differentiating it from the pathogens of bacterial pustule of cowpea (Patel and Jindal, 1982) and common and fuscous blights of bean (Jindal and Patel, 1980).

The primary source of inoculum is the infected seeds so the use of clean seed is a potential control measure. Also, hot-water treatment of infected seeds reduces seed-borne inoculum (Boettinger and Bowers, 1975). Cowpeas resistant to blight were identified long ago, as was the mechanism of inheritance (Lefebvre and Sherwin, 1950; Sherwin and Lefebvre, 1951), but resistance has not been bred into present cultivars in the USA. Two types of resistant responses, which are stable against the prevailing strains of the blight organism, have been identified by artificial inoculations (Gitaitis, 1983).

VIRAL DISEASES

Viral diseases in cowpeas are more widespread and economically important in the southern states than in California. Identification, symptomatology and sources of resistance for blackeye cowpea mosaic virus (BlCMV), cucumber mosaic virus (CuMV), cowpea chlorotic mottle virus (CCMV), cowpea strain of the southern bean mosaic virus (SBMV-CS) and stunt disease caused by a mixed infection of CuMV and BlCMV were described in the intensive investigations of C. W. Kuhn and associates at Georgia (Kuhn *et al.*, 1966; Pio-Ribeiro *et al.*, 1978; Pio-Ribeiro and Kuhn, 1980). Cowpea severe mosaic virus (CSMV) occurs in Arkansas (Shepherd, 1964) and Louisiana (Gabriel *et al.*, 1981). In cowpea-growing areas of California, isolated occurrences of CuMV, SBMV and curly-top diseases have been noted (Mackie, 1946; personal observations).

Though sources of resistance to most of the cowpea viruses have been identified and inheritance has been studied, progress towards improving present cultivars for virus resistance has been slow (Kuhn *et al.*, 1966; Sowell *et al.*, 1965; Fery, 1980; Meiners, 1981; Taiwo *et al.*, 1981; Walker and Chambliss, 1981; Fulton and Allen, 1982; Allen, 1983). Mississippi Silver, though not directly bred for virus resistance, is reported to have tolerance to several viruses. Its hybridization with Pinkeye Purple Hull led to the selection of Worthmore, which combines resistance to BlCMV, CCMV and SBMV (Gay, 1974). Worthmore is now being used to improve Knuckle Purple Hull, released in 1959 in Alabama, for BlCMV resistance (O. L. Chambliss, personal communication). Since resistance to CuMV is not available in the

USA, resistance to BlCMV should reduce the damaging synergistic effects of stunt disease. Three BlCMV-resistant varieties were recently released. Pink-eye Purple Hull-BVR, White Acre-BVR and Corona were selected from original commercial cultivars, during evaluations of disease reactions to BlCMV and CuMV in single and mixed inoculations (Kuhn *et al.*, 1984).

My participation in the conference that produced this publication was funded by grant MSU/AID/DSAN-XII-G-0261 from the United States Agency for International Development to the Bean/Cowpea Collaborative Research Support Program.

PART 6

Entomology

Cowpea Research, Production and Utilization
Edited by S. R. Singh and K. O. Rachie
© 1985 John Wiley & Sons Ltd

15

Insect Pests of Cowpeas in Africa: Their Life Cycle, Economic Importance and Potential for Control

S. R. SINGH and L. E. N. JACKAI
International Institute of Tropical Agriculture, Ibadan, Nigeria

In Africa, insect pests are often responsible for 100 per cent losses of cowpea yields, and, if not controlled, virtually always limit yields to less than 300 kg/ha. Although mixed cropping reduces pest populations of some species, unless combined with other protection measures, it is not possible to obtain high yields. A realistic method of control appears to be cultivation of insect-resistant varieties in combination with applications of insecticide in minimal amounts and use of cultural-control methods.

Insects covering the main phytophagous taxa attack the crop from seedling to harvest and can cause economic damage at all stages of plant growth (Table 1). Not all pests are found at all locations; however, some are consistently found in large numbers and are considered major pests.

HOMOPTERA

Aphids

There are two *Aphis* spp. (Homoptera: Aphididae) reported as pests of cowpeas in Africa: *A. craccivora* Koch, which is the main aphid infesting cowpeas throughout Africa and Asia, and *A. fabae* (Scopoli), which has been reported as a minor pest in East Africa (Singh and van Emden, 1979).

Mostly females are found, and they reproduce parthenogenetically. If food is abundant and climatic conditions are optimal, apterous forms are produced. Alates are produced whenever crowding occurs and food is in short supply, as well as during adverse climatic conditions. The biology of *A.*

Table 1.   Insect pests of cowpea in Africa

Common name	Scientific name	Pest status	Damage
Cowpea aphid	*Aphis craccivora*	Major	Feeds on foliage, pods, at seedling stage; virus vector
Pea aphid	*Aphis fabae*	Minor	Feeds on foliage at seedling stage; virus vector
Leafhoppers	*Empoasca signata, E. dolichi*	Minor	Feed on foliage at seedling stage
Beanfly	*Melanagromyza phaseoli*	Minor	Feeds on stem at seedling stage
Foliage beetles	*Ootheca mutabilis, O. bennigseni*	Sporadic	Feed on foliage at seedling stage; virus vector
Striped foliage beetles	*Medythia quaterna* (syn. *Luperodes lineata*)	Minor	Feeds on foliage at seedling stage; virus vector
Blister beetles	*Mylabris farquharsoni, M. amplectens, Coryna apicicornis*	Sporadic	Flower feeders
Other beetles	*Lagria villosa, Chrysolagria* spp.	Minor	Foliage feeders
Striped bean weevil	*Alcidodes leucogrammus*	Sporadic	Feeds inside stem and on foliage
Pod weevil	*Piezotrachelus varius* (syn. *Apion varius*)	Minor	Feeds on seeds within green pods
Legume pod borer	*Maruca testulalis*	Major	Feeds on flower buds, flowers, green pods

Pod borer	*Cydia ptychora* (syn. *Laspeyresia ptychora*)	Minor	Feeds inside green pods
Egyptian leaf worm	*Spodoptera littoralis* (syn. *Prodenia litura*)	Sporadic	Feeds on green pods, flowers, foliage
African bollworm	*Heliothis armigera*	Sporadic	Feeds on green pods, flowers, foliage
Lima bean pod-borer	*Etiella zinckenella*	Minor	Feeds on green pods
Legume bud thrips	*Megalurothrips sjostedti* (syn. *Taeniothrips sjostedti*)	Major	Damages flower buds, flowers
Foliage thrips	*Sericothrips occipitalis*	Minor	Feeds on foliage at seedling stage
Green stink bugs	*Nezara viridula, Aspavia armigera*	Minor	Feed on green pods
Coreid bugs	*Anoplecnemis curvipes, Riptortus dentipes, Clavigralla tomentosicollis* (syn. *Acanthomia tomentosicollis*), *C. shadabi* (syn. *A. horrida*), *C. elongata*	Major	Feed on green pods
Cowpea storage weevil	*Callosobruchus maculatus, C. chinensis*	Major	Damage seeds in storage

craccivora has been studied extensively at IITA. It varies a great deal with the host plant, soil fertility, soil moisture and temperature. The adults live from 5 to 15 days and have a fecundity of over 100. Daily progeny production may vary from 2 to 20. A generation can be completed within 10–20 days, and there are four nymphal instars (Singh, 1980). The biology of *A. fabae*, though not extensively studied, appears to be similar to that of *A. craccivora*.

Aphids primarily infest seedlings, although large populations also infest the pods. They cause direct damage to the plant by removal of its sap. Small populations may have little impact on the plant, but large populations can cause distortion of leaves, stunting of plants and poor nodulation of the root systems. Yield is reduced, and in extreme cases the plant dies (Singh and van Emden, 1979). Indirect, and often more serious, damage is through transmission of aphid-borne viruses (Bock and Conti, 1974).

Several insecticides such as phosphamidon and dimethoate are effective against aphids, but pirimicarb is probably the best (Singh and Allen, 1980). In nature, aphid populations are often kept under control by natural enemies and adverse climatic conditions, and cowpea varieties resistant to aphids are being developed.

For example, at IITA, screening for aphid resistance has been conducted and several resistant lines have been identified, including TVu 36, TVu 408, TVu 410, TVu 801 and TVu 3000 (Singh, 1977b, 1980). Initial screening was in the greenhouse, and selected lines were also tested in the field. In the greenhouse, test materials were planted in wooden trays 54 × 40 × 11 cm filled with soil to 8 cm. The materials were planted in single rows—six rows, 40 cm long, 9 cm apart. The distance between plants was 4 cm. A known locally improved variety Prima was included as a susceptible check row. Later, after identification of resistant lines, a resistant check row was also included. When the plants were about 10 days old each plant was infested with five 4th-instar aphids by a camel's-hair brush from an aphid culture maintained in the greenhouse. The infested trays were then introduced into cages (wooden frames covered with saran mesh) kept in the greenhouse. The dimensions of the cages were 140 × 81 × 66 cm. About 15 days after infestation the plants were rated for aphid damage on a scale of 1–5. Plants rated as 1 were almost immune to aphid damage, and those rated 5 were highly susceptible and died within 15 days after infestation. Whenever a line had a score of 1 or 2, it was tested at least twice. The resistant plants were transplanted in pots and further tested for resistance at flowering and podding.

The segregating plant populations were also screened in the greenhouse. Field screening was conducted only during the dry season, December–February, when aphid populations peak at IITA. Application of DDT on the cowpea seedlings increased aphid populations and resulted in uniform aphid infestations. The materials were planted as single rows, 1 m apart and 3 m long, with a susceptible check every 10th row.

Leafhoppers

Several species belonging to the genus *Empoasca* (Homoptera: Cicadellidae) attack cowpeas and are ubiquitous in the tropics. *Empoasca dolichi* Paoli is common but rarely causes economic damage to the crop. The leafhoppers are considered minor, and cowpea varieties grown on farms appear to be resistant to this pest. In East Africa, *E. signata* Haust is reported from Ethiopia, Sudan and Egypt, and other species have been reported in Kenya, Tanzania and Uganda (Singh, 1980).

The different leafhopper species infesting cowpea have similar biology (Singh and van Emden, 1979). Mostly females are found, and they reproduce parthenogenetically. The females lay their eggs in the leaf veins on the underside of the young leaves; when populations are dense, eggs are also laid in the stems of young seedlings. The number of eggs laid varies with the species and the host plant. On average, 50–150 eggs are laid by a single female. Depending on temperature, they hatch within 5–9 days. Five nymphal instars are found, with a nymphal duration of about 7–10 days. Adults live from 30–50 days (Parh, 1983).

Leafhoppers are seedling pests and during drought can seriously damage the crop (Raman *et al.*, 1978). Both adults and nymphs infest leaves and suck the plant sap from the underside. The leaves yellow and begin cupping. Frequently, however, severe damage occurs without reduction in yield. Leafhoppers are active insects, easily disturbed, and are particularly mobile during the warm part of the day. When the temperature is high, for instance during bright sunshine, leafhoppers hide in the bordering grasses. Population counts on plant surfaces should be taken early in the day when the hoppers are present on the plant and are less mobile (Singh, 1980).

Leafhoppers on cowpeas are easy to control. Several insecticides are effective, and a single insecticide application is often sufficient. Some of the more effective insecticides are chlorpyrifos, methomyl, entrimfos and dimethoate (Singh and Allen, 1979). Also, resistance to damage by leafhoppers has been identified, and the development of resistant varieties is under way.

Screening for resistance to leafhoppers at IITA involved both field and greenhouse tests. Field screening was conducted during late September and October. Fifteen days before the plantings, a susceptible cowpea cultivar, Prima, was planted at the border of the field plots to ensure that adequate leafhopper populations were present. Test cultivars were then planted 1 m apart in single rows 5 m long, and the susceptible check planted every 10th row. The check rows were useful in rating the lines for resistance and as indicators for evaluating the leafhopper populations in the field. When the test materials were about 15 days old the susceptible border rows were uprooted and the tall uncut grass around the field was also cut. This resulted in a mass movement of the leafhopper populations to the test cultivars. Plants were rated for damage on a 1–5 scale, similar to that for aphids, when they were about 25–30 days old. Materials rated as resistant were further tested in replicated field trials and later in the greenhouse for confirmation. Four cultivars, TVu 59, TVu 123, TVu 662 and TVu 1190E were rated as resistant (Singh, 1977a, 1980). TVu 1190E (VITA 3) has multiple disease resistance, tolerance for drought, and resistance to root-knot nematodes (Singh *et al.*, 1975). In the greenhouse, techniques and cages were similar to those used for aphids.

HEMIPTERA

Pod bugs

Several species of pod bugs infest cowpeas at the podding stage and do considerable damage. Normally their populations are high because the adults constantly migrate from wild host plants to cultivated fields. Control is often difficult. They breed throughout the year if food is available and the climate favourable. Adults and nymphs suck the sap from developing pods and can cause serious yield losses through premature drying of pods and lack of normal seed formation.

Among the major pod bugs, four species in the family Coreidae cause economic losses in Africa: *Clavigralla tomentosicollis* (Stal.) synonym: *Acanthomia tomentosicollis* Stal. found in East and West Africa; *C. shadabi* Dolling synonym: *A. horrida* (Germar) found in West Africa; and *C. elongata* Signoret in East Africa. *Anoplocnemis curvipes* (Fabricius) is found in East, West and Central Africa.

Clavigralla tomentosicollis is medium-sized, hairy, and grey. Nymphs form large colonies on cowpea pods and peduncles and are not easily disturbed. Adults are not strong fliers and have a longevity of 100–150 days. Eggs are laid in batches of 10–70, and, on average, about 200 eggs are laid by each female. Each instar lasts about 2 days, but the last instar is about 6 days. The total nymphal period is about 14 days (Singh and Taylor, 1978).

Clavigralla shadabi and *C. elongata* are smaller than *C. tomentosicollis* and are grey. The former has a spiny dorsal thorax, and the latter is distinguishable by its long cylindrical body. The two pests feed on pods and they have similar biology. Adult longevity is from 40 to 80 days. Eggs are laid singly, about 250 eggs per single female, and they hatch in about 6 days. There are five nymphal instars, and the total nymphal period is about 20 days (Singh and Taylor, 1978).

At present, insecticides such as endosulfan and dimethoate offer the only effective control method, although manipulating the planting date holds some promise.

Anoplocnemis curvipes (Fabricius) is a major pest of cowpeas. Found throughout Africa, it has a large host range, including leguminous trees and several other wild host plants.

The adults are strong fliers, dark black and fairly large; they live from 24 to 84 days, but unmated males and females survive up to 150 days. Eggs are laid in batches, normally in chains, and are dark grey. They are laid on leguminous trees and wild host plants rather than on cowpeas. Each batch contains 10–40 eggs, and a single female normally lays 6–12 batches. The eggs hatch in about 7–11 days. There are five nymphal instars, and the early instars resemble ants. The total nymphal period is about 30–60 days, depending on the host plant and climatic conditions.

Control is difficult when populations are dense. Endosulfan, fenitrothion and dimethoate are effective, and several egg parasites have been recorded that appear capable of checking the host populations.

Several species of *Riptortus* (Hemiptera: Alydidae) are found on grain legumes (Singh and van Emden, 1979), and, of these, *R. dentipes* (Fabricius) is most commonly found on cowpeas, feeding on pods or resting on foliage.

The adult is light brown, with whitish-yellow lines on the side of the body. Fertilized females have a large black abdomen. The adults are strong fliers; their life span is about 10–20 days. Eggs are laid in small batches every 5–15 days, with an average 50 eggs being laid by a single female, and they hatch in about 6 days. There are five nymphal instars, the first few lasting about 3 days and the last about 6 days.

Because of the constant invasion by adults from adjacent areas, control of this pest is difficult, although endosulfan, fenitrothion and dimethoate are reasonably effective.

Screening for cowpea resistance to pod-sucking bugs is being carried out at IITA: test lines are planted in the field with a known susceptible check every 10 entries. The plants are sprayed with deltamethrin (12.5 g a.i./ha) first at bud formation and then 8–10 days later for control of thrips and pod borers.

A sample of pods (100–150/m row) is collected and examined for seed damage. A resistance index is calculated: the percentage seed damage in the test cultivar expressed as a fraction of the susceptible check. Only cultivars with less than 60 per cent of the damage found in the susceptible check are further tested.

Selected lines are screened in cages in replicated entries. Test lines are planted in pots and transferred to cages at the onset of flowering. Pod bugs (10–12 pairs) are released in the cages for 15 days.

The plants are removed, and damage to seeds and peduncles is measured. So far the cultivars tested have all been susceptible.

Nezara viridula (Linnaeus) (Hemiptera: Pentatomidae) is a minor pest of cowpea with a wide host range.

Its biology varies a great deal with host plant and climatic conditions. A single female lays 100–400 eggs in four to six batches of 30–80 on the underside of young, tender leaves. There are five nymphal instars, and the entire life cycle takes 40–100 days. The early instar nymphs are bright, coloured with spots, and are found in batches. They can be located easily. Adults are green, live about 30–60 days and are fairly strong fliers. Damage is caused primarily by adults.

The pest is difficult to control in areas where populations are large, but endosulfan, fenitrothion and dimethoate are effective. The frequency of application depends on reinfestation levels. Manipulation of planting time also offers a means to minimize damage from the stink bug.

Of the other pentatomids, *Aspavia* deserves mention. *Aspavia armigera* (Fabricius) adults are dark brown with large scutellum and three white or orange angular spots. Several species are known, but *A. armigera* is by far the most common, its eggs frequently being found on cowpea.

THYSANOPTERA

Among Thysanoptera, several species of thrips damage cowpea in Africa, but only a few are important.

Legume bud thrips, *Megalurothrips sjostedti* (Trybom) synonym: *Taeniothrips sjostedti* (Trybom) (Thysanoptera: Thripidae), are a major pest of cowpeas and often cause up to 60 per cent damage to the crop (Singh and Taylor, 1978).

The biology of this pest is not completely known, although the entire life cycle takes about 18 days. Eggs are laid in the flower buds, and nymphs and adults develop. Adults are shiny black and are easily noticed on the flowers where they feed on pollen.

The nymphs and adults feed on flower buds and can cause complete loss of flower production. The racemes of severely infested plants do not have any flower buds and existing flowers appear diseased (Singh, 1978).

Methomyl, monocrotophos and cypermethrin are effective against this pest.

Screening for resistance to thrips has been practical only in the field. Simple techniques were developed at IITA to ensure uniform and high infestation. First, a mixture of dwarf pigeon peas and *Crotalaria juncea* is planted all around the cowpea field in advance so that they flower long before the cowpeas and attract the thrips. *Crotalaria* serves as a trap for the pod-borer moth and minimizes the infestation and competition between the borers and the thrips. Then, germ-plasm lines are planted in single rows, 4 m long. Every 10th row, VITA 7, a multiple disease-resistant line from IITA, is planted as a susceptible check. After about 35 days, endosulfan is sprayed at the rate of 200 g a.i./ha to reduce the borer populations further and prevent them from interfering with the thrips' infestation. Plants are rated visually on a scale of 1–5 for thrips' damage at about 45 and 55 days after planting. The plants that do not produce any flowers and have severe thrips damage on buds are rated as susceptible (4–5 score) and those that produce normally with little or no thrips damage are rated as resistant (1–2 score). Cultivars with the least damage are further tested in replicated field trials for confirmation. From the entire germ plasm tested so far only one cultivar, TVu 1509, has been found to be moderately resistant; the rest were susceptible (Singh, 1977b). TVu 1509 has been extensively utilized in the breeding programme at IITA. TVx 3236, a line derived from TVu 1509, combines thrips resistance, high-yield potential and other desirable agronomic characters.

Foliage thrips, *Sericothrips occipitalis* Hood (Thysanoptera: Thripidae), have been noticed as a minor pest of cowpea seedlings during drought stress. The infestation declines with the onset of the rains, and the plants appear to recover fully. Foliage thrips are often found on seedlings in greenhouses or on irrigated crops in the dry season.

DIPTERA

Beanfly, *Ophiomyia phaseoli* (Tryon) synonym: *Melanagromyza phaseoli* Tryon (Diptera: Agromyzidae), has been reported to infest cowpeas in East Africa. It is minute, shiny black, with light-brown legs and antennae. Females are about 2 mm, and males are slightly shorter. The fly oviposits white, elongate–oval eggs about 0.3 mm long, mostly on the upper surface of the leaf beginning at the two-leaf stage. The eggs hatch in 2–3 days. The maggot behaves like a leafminer, penetrating underneath the epidermis of the leaf until it reaches a vein. It then tunnels to the midrib and to the leaf stalk, finally reaching the stem near the ground surface or in the petiole in older plants. The maggot stage lasts about 10 days. In young plants the larvae pupate in the root zone. The pupation period lasts for about 10 days. The entire life cycle takes about 21–24 days (Singh and van Emden, 1979).

Leaves of infested plants have punctures that become visible as irregular lines; the plants are stunted and yellow and finally die. In most cases the stem is swollen below or near the ground level; it cracks and turns dark brown as the tissues rot. Heavy infestation occurs in dry conditions and results in major losses in yield.

Soil insecticides such as carbofuran and thimet have been found effective in control, as has the adjustment of planting date (Swaine, 1969). Development of beanfly-resistant varieties also offers good potential for control.

LEPIDOPTERA

Pod borers

Numerous Lepidoptera cause damage to cowpea in Africa. For example, the legume pod borer, *Maruca testulalis* (Geyer) (Lepidoptera: Pyralidae), is found throughout the continent and can cause serious losses in cowpea yield (Singh and van Emden, 1979).

The adult moth is dull white, with light-brown markings on the forewings and at the edges of the hind wings. The female moth lays up to 200 eggs on flower buds, flowers and tender leaves (Jackai, 1982). Eggs hatch in 2–3 days, and there are five larval instars. Larval development takes about 8–14 days (Jackai, 1981a,b). The late larval instars can be easily identified by the characteristic black dots on their body (Usua and Singh, 1978). A 2-day prepupal period follows the larval period, during which feeding ceases. The pupal stage takes 6–9 days, and the pupae are initially green or pale yellow but later darken to greyish-brown. Pupation occurs on the soil in a double-walled pupal cell, and adults emerge after about 5–10 days and have a life span of 5–15 days (Taylor, 1967).

The early larvae, in the absence of flower buds and flowers, feed on young tender shoots and peduncles. Later, when the flower buds and flowers are formed, they prefer to feed on floral parts and the green pods. They hide during the day in the flowers or pods and are active during the night, wandering around the host plant and invading uninfested flowers and pods. The larvae usually web together leaves, flowers and pods (Taylor, 1967).

Several insecticides including methomyl, endosulfan and cypermethrin are effective against this pest (Singh and Allen, 1980). Screening for pod-borer resistance has also been conducted on IITA fields. The germ-plasm collection was planted in single rows, 5 m long, and the plants were rated for percentage damage to stems, flowers and pods. To reduce damage from thrips on flower buds, staff sprayed the crop with monocrotophos at bud initiation, about 35 days after planting, at the rate of 200 g a.i./ha. This application drastically reduced the thrips population without affecting the borer population. This method identified TVu 946 as being resistant (Singh, 1977a). The mechanism of resistance was studied by one of us (Jackai, 1982) who also assessed the damage to seeds in the pod.

Another Lepidoptera, African bollworm, *Heliothis armigera* (Hubner) (Lepidoptera: Noctuidae) is a sporadic pest and in recent years has been increasingly observed damaging the cowpea crop, particularly in regions where maize is cultivated (Singh and van Emden, 1979).

The female lays eggs singly, mostly on tender leaves, flowers and shoots. The eggs hatch in about 3–8 days. There are six larval instars, and the larval period lasts from 15 to 30 days. Fully grown larvae are about 40 mm long and have characteristic longitudinal markings on each side of the body: a pale band, an almost black band and another light band. The larvae are often dark brown or green. Pupation occurs in the soil, about 40 mm deep, and the pupal period lasts from 10 to 25 days (Swaine, 1969). The adult moth is grey to brown.

This is a polyphagous pest. Besides cowpeas, it attacks other cultivated crops. The larvae feed on the leaves, flower buds, flowers and green pods and cause serious damage to the crop. The larvae of early instars make a hole in the pod and move inside to feed on the grains. The older larvae feed on the grain from outside the pod and can be easily detected.

For effective control, insecticides should be applied when the larvae are at an early stage of development. Endosulfan, chlorpyrifos, monocrotophos, methomyl and cypermethrin are effective against this pest. Several larval parasites have also been recorded (Robertson, 1963).

Cowpea seed moth, *Cydia ptychora* (Meyrick) synonym: *Laspeyresia ptychora* (Meyrick) (Lepidoptera: Tortricidae), is often found in the humid tropics of Africa, infesting the green pods of cowpea.

The adult moths are characteristically cylindrical, small and grey. On average, each female lays 45 eggs during its life span (about 5 days). Eggs are flat and oval, approximately 0.5 mm long. White and translucent when laid, they become reddish when the embryo matures and can be easily seen. Eggs are normally laid on the green pods at the point of attachment. Incubation is 3 days. There are five larval instars and the larval period lasts about 12 days. The first-instar larvae enter the pod and feed on developing seeds and do extensive damage. Sometimes, under moist conditions, damage to the seed continues even after harvest. Early instars are white, whereas the later ones are pink to dark red. At the time of pupation the larvae move from the pod to the soil and pupate for about 5 days.

Endosulfan and cypermethrin applied during the pod-filling stage are effective in control of this pest.

Lima bean pod-borer, *Etiella zinckenella* (Treitschke) (Lepidoptera: Pyralidae), is a minor pest reported from East Africa. The first-instar larvae enter cowpea pods and feed on the developing seeds. Mature larvae emerge and pupate in the soil. The entire life cycle may take 20–30 days. Endosulfan and cypermethrin are effective against this pest.

Foliage feeders

Egyptian leaf worm, *Spodoptera littoralis* (*Boisdural*) synonym: *Prodenia litura* (Fabricius) (Lepidoptera: Noctuidae), has been reported as a serious pest of cowpeas in Egypt (Singh and van Emden, 1979).

The adult moth is brownish and lays eggs on the underside of young leaves in clusters of 50–250, normally covered by hairy scales from the body of the female. Each female lays 4–8 egg clusters, so the total number of eggs can be up to 2000. The early-instar larvae are greenish with black dots on the abdomen or sometimes only on the first and last abdominal segments. The older larvae are brown, with five pale-yellow or greenish longitudinal lines along the body with black spots laterally on each segment. The older larvae hide in the soil by day but are active, voracious feeders in the night, cutwormlike in habit. Depending on climatic conditions, the larval period can vary from 15 to 30 days (Singh and van Emden, 1979). The pest pupates in soil for about 10 days when there is no winter diapause.

This is a polyphagous insect that attacks a wide range of crops. Primarily it is a defoliator but also bores into green pods or feeds inside the pod from the outside. When large numbers are present, complete crop loss is possible within a short time.

It is easy to control early instars, but not so the older instars. Insecticides endosulfan, methomyl, chlorpyrifos and cypermethrin, as well as other synthetic pyrethroids, are effective against the pest. Parasites appear to play a role in keeping the pest population at a low level, and in Egypt a polyhedrosis virus is effective.

COLEOPTERA

Beetles

A large number of coleopterous beetles feed on cowpea foliage and flowers, and some are effective vectors for viruses (Singh and Taylor, 1978). In general they are sporadic pests, and unless their populations are high, the damage by direct feeding is insignificant.

Cowpea leaf beetle, *Ootheca mutabilis* (Shalberg) (Coleoptera: Chrysomelidae), is probably the most damaging of the foliage-feeding beetles that attack cowpea in East and West Africa (Booker, 1965; Halteren, 1971). A related species, *O. bennigseni* Weise, has been reported from East Africa (Le Pelley, 1959).

Ootheca mutabilis adults are normally shiny, light brown or orange; however, light-black or brown adults are also found. Adults are about 6 mm long, oval and can live up to 3 months. Eggs are laid in the soil (about 60 eggs/egg mass), and the total number of eggs laid by a single female varies from 200 to 500. The eggs, which are elliptical, light yellow and translucent, are held together in a mass by a sticky substance secreted by the female; they hatch in about 13 days. The larvae develop in the soil, and there are three larval instars. The first and second instars last about 6 days each, and the third lasts about 18 days followed by a prepupal stage, which lasts about 15 days. The pupal stage is about 16 days. The life cycle of this pest is greatly affected by the season and ranges from 60 to 250 days. In southern Nigeria, where the rainfall is bimodal, during the second season, adults undergo obligatory diapause (Ochieng, 1977). After emergence they remain inactive in the soil for almost 60 days while reproductive parts are not fully developed and they are incapable of flight.

The damage is done by the adults, feeding between veins on the leaves. Dense populations can totally defoliate cowpea seedlings, resulting in the death of the plant. The larvae feed on the cowpea roots but seldom cause serious damage (Singh and Taylor, 1978). The adult beetles cause indirect damage by transmitting cowpea yellow mosaic virus (Bock, 1971).

Striped foliage beetle, *Medythia quaterna* (Fairmaire) synonym: *Luperodes lineata* (Kars) (Coleoptera: Chrysomelidae), is found mainly in the forest zone of West Africa but also in the derived savanna. Compared with *O. mutabilis* this pest has rather limited distribution.

The biology is not fully known, but egg, larval and pupal stages are found in the soil. The adults are about 3–4 mm long and striped longitudinally with white and light-brown markings. The adults attack young cowpea seedlings by feeding on newly emerged leaves, mostly at the margins, and can completely defoliate and kill the plants. The adults are also vectors of cowpea yellow mosaic virus.

These beetles are easy to control, and several insecticides such as endosulfan, methomyl and chlorpyrifos are effective. Normally, a single application is sufficient.

Blister beetles, *Mylabris* spp. (Coleoptera: Meloidae), are pests of cowpea flowers. Those that are commonly found are *M. farquharsoni* Blair and *M. amplectens* Gerstaecker. A few beetles of the genus *Coryna*, especially *C. apicicornis* (Guerin), are also important flower feeders.

They are often found in cowpeas planted with or near maize. The life history of blister beetles is rather complex: the larvae undergo hypermetamorphosis, and each larval instar is different. Eggs are laid in the soil where larvae and pupae are usually found. The adults are strong fliers and feed on flowers and pollen. These beetles are often difficult to control because the adults are found only on flowers, whereas the larvae are scattered in the soil.

Several other beetles belonging to the Lagriidae family often feed on cowpea foliage, but their populations are seldom high enough to cause economic damage. In Africa, *Lagria villosa* Fabricius and *Chrysolagria* spp. have been reported; their feeding causes characteristic holes in the leaves. *Lagria villosa* is a comparatively large, blackish beetle, and *Chrysolagria* spp. are smaller bluish beetles.

Insecticides such as endosulfan, methomyl and chlorpyrifos are effective against these pests, and a single application is often sufficient.

Weevils

Other Coleoptera that cause damage are the striped bean weevil and the pod weevil.

Striped bean weevil, *Alcidodes leucogrammus* (Erichson) (Coleoptera: Curculionidae), is a sporadic pest found in parts of East and West Africa. The adults feed on the leaves and lay eggs on the stem. The larvae tunnel and feed inside the stem, causing stunted growth; later the stem splits and the plant dies. Pupae are found inside the stem. Adult weevils are 7–9 mm long, dark brown with white markings on the elytra. Several insecticides are effective against this pest, including endosulfan, methomyl and monocrotophos.

Pod weevil, *Piezotrachelus varius* Wagner synonym: *Apion varius* Wagner (Coleoptera: Curculionidae), infests cowpea pods. It is a minor pest, seldom causing serious damage to the crop. Eggs are laid in the green pods, and larvae feed inside the seed and pupate; then tiny, black weevils emerge when the pods are dry. The life cycle takes about 20 days. Several insecticides are effective against this pest.

Bruchids

Of the storage weevils, several species of *Callosobruchus* (Coleoptera: Bruchidae) cause damage but the two most important are *C. maculatus* (Fabricius) and *C. chinensis* (Linnaeus).

Infestation occurs in the field when the pods are nearly mature. Eggs are laid on the pods, but weevils prefer to slip inside the pods through holes made by the other pests and lay eggs directly on the seed. After the crop is harvested the bruchids multiply and do considerable damage to stored cowpeas. The adult life span is from 5 to 7 days, and each female lays 50–80 eggs. The eggs are glued on the top of the seed in storage; they are glossy and oval when fresh and hatch in about 3–5 days. After moulting three times and reaching the fourth instar, the larvae construct a cell inside the seed and pupate in it. The life cycle is completed in about 30 days.

The damage is done by larvae feeding inside the seed. Often farm storage for 6 months is accompanied by about 30 per cent loss in weight with up to 70 per cent of the seeds being infested and virtually unfit for consumption.

For large-scale storage, efficient methods to control bruchids have been developed, and these involve the use of proper storage facilities and various fumigants. On subsistence farms, treatments with pirimiphos-methyl are suitable. Wherever the produce can be fumigated under airtight conditions, phostoxin has been found effective. Phostoxin treatment is effective against the eggs and larvae inside the seed and does not have any residue. Therefore, produce has to be protected from reinfestation in storage. For subsistence farmers who produce less than 10 kg seeds for their own consumption, mixing the seed with groundnut oil (about 5 ml/kg of seed) was found practical and effective (Singh *et al.*, 1979).

The entire germ-plasm collection at IITA (12,000 accessions) was screened for resistance to this pest: 40 seeds of each accession at about 13 per cent moisture were held in a plastic box 5 × 5 × 2 cm. Three pairs of 1-day-old bruchids were introduced into the box and were left for 24 h for egg laying. The adults were removed next day. Bruchid cultures were maintained in Kelnar jars that had mesh lids. A filter paper dipped in Kelthane® MF was kept between the mesh layers which kept the bruchid culture free of mite infestation. The infested boxes and the bruchid culture were kept in a room at about 28°C and relative humidity of 78–80 per cent. Five days after infestation, when the eggs had hatched and were easily visible, egg count was made. If any seeds had no eggs they were removed. After 25 days the number of adults that had emerged had been recorded. Only one accession, TVu 2027, was found resistant (Singh, 1977b); it held adult emergence to less than 20 per cent, whereas in all the other accessions the emergence was more than 60 per cent. The accessions or breeding lines showing promise were included in replicated trials for confirmation of resistance.

Cowpea Research, Production and Utilization
Edited by S. R. Singh and K. O. Rachie
© 1985 John Wiley & Sons Ltd

16

Recent Trends in the Control of Cowpea Pests in Africa

L. E. N. Jackai,* S. R. Singh,* A. K. Raheja† and F. Wiedijk*
*International Institute of Tropical Agriculture, Ibadan, Nigeria; and
† Institute for Agricultural Research, Ahmadu Bello University, Zaria, Nigeria

Cowpea is attacked by many insect pests throughout its geographical range, although the number and their status vary from one region to another (Singh and Allen, 1980). The pests include aphids, beanfly, leafhoppers, thrips, pod borers, pod-sucking bugs, cowpea curculio and the storage beetle. The key pests in the various geographical regions are reviewed by Singh and Allen (1980).

Losses in grain or foliage attributable to cowpea field pests are from 20 per cent to almost 100 per cent (Raheja, 1976a; Singh and Allen, 1980; IITA, 1983a). Few other important food crops suffer such losses from insects. These high losses have contributed immensely to cowpea's subsidiary position in the farming systems of several countries in the tropics. To increase the hectarage of this food crop, researchers and farmers must minimize the damage caused by insect pests. The peasant farmer who produces most of the cowpea available in the world market obtains about 200–350 kg/ha. No conscious attempt is made to control the pests, except possibly in a few Far Eastern countries where insecticides are used routinely by farmers (Litsinger et al., 1983). In most tropical countries cowpea cannot be grown successfully without at least one or two insecticide sprays. It is grown in most countries as a mixed crop with cereals (maize, sorghum, millet), root crops (yam and cassava), okra, cotton, etc. (Okigbo and Greenland, 1976). These associated crops perform various roles—some serve as hosts for the cowpea pests (Ochieng, 1977); others serve as barriers restricting movement of the pests in the field (Jackai, 1983); others play roles yet undetermined. Some farmers plant with the first sign of rain to enable the crop to escape damaging populations of certain pests, and harvest before the pest population peaks.

In the past, 6–10 insecticide sprays were required for complete protection (Booker, 1965), but today an optimal number is 2–4 sprays with resulting increases in yield of 5–10 times (Raheja, 1976b; Singh and Allen, 1980; Matteson, 1982). The progress is a reflection of research on other methods of control that aim not at replacing insecticides but at complementing them (Jackai and Singh, 1983).

This paper discusses the changing concepts and the latest developments in chemical control of cowpea field pests, using examples mainly from Nigeria but also from other countries. The present and potential use of cultural methods in insect pest suppression is also discussed, and an optimization model is proposed: it incorporates the two methods of control and host-plant resistance for maintaining pest populations below damage thresholds.

CHEMICAL CONTROL

More research has been done in chemical control than in other methods of cowpea insect pest control. New insecticides flood the markets every year, and insecticide-application technology, even though much slower, is also changing. Recent studies in Nigeria at IITA and the Institute for Agricultural Research (IAR) have shown that there is a new range of insecticide formulations that provide excellent control of the entire insect spectrum on cowpea (Table 1). The recently developed hand-held Electrodyn® spraying system and the combined insecticide formulations are some of the new ideas in the chemical-control arena.

Even though chemical control of cowpea insect pests is not widely practised by peasant farmers, it is not new to them. Cowpea growers have for long recognized the usefulness of insecticides, but factors such as availability and cost have kept this technology beyond their reach. What is new is the changing attitudes of growers who, having realized the dramatic results of insecticide application on cowpea, and having become more sensitive to the dangers and costs involved in chemical protection, are starting to express concern about the continued and increasing use of these chemicals. Although opinions on these concerns differ, there is one area of general agreement: in most of its geographical range cowpea production cannot become a successful venture (i.e. production cannot be both sustained and economic) without the use of insecticides. What then becomes important is the need to develop or formulate an algorithm (or strategy) for safer and more economic and ecologically sound use of insecticides.

In the past, insecticides were used by everyone in Africa except traditional farmers with smallholdings. Today, all categories of farmers are beginning to use insecticides. To optimize production without damaging the environment or endangering consumers, those who use insecticides or are in a position to advise on their use, must consider:

○ the feeding behaviour of the target pest;
○ the activity cycle of the pest, especially the damaging stage;
○ the part of the plant attacked;
○ the mode of action of the chemical (whether contact, systemic, stomach poison, fumigant, or translaminar);
○ the residual activity of the compound;
○ the phytotoxicity of the chemical at effective dosages; and
○ the efficacy of the compound.

Getting this information is the responsibility of the researcher and the extension agent who should educate growers. Choosing a wrong chemical can lead to unexpected and costly results—for example, applying monocrotophos to control *Maruca* borer would not control the pest (Jackai, 1983).

Research efforts in chemical control have tended to be more environmentally conscious in recent years than in the past. Researchers are collecting increasing amounts of biological data when evaluating insecticide efficacy. For example, at IITA they study the behaviour of the pest population—onset, fluctuation and composition—as well as the efficiency and safety of the method of application. This information allows one to determine when the population of a target insect will peak, and to synchronize this with insecticide application.

Nine insecticides were screened for use on cowpea in Nigeria and Ivory Coast in 1983. They included single, conventional insecticides and new, combined formulations involving two insecticides, generally a synthetic pyrethroid and an organophosphate. The combined formulations are the outcome of continuing collaboration between chemical companies and research institutions. The results show that a range of insecticides is effective against all the pests currently attacking cowpeas in Africa; the formulations are safer than those available in the past (Table 1). Although yields did not differ significantly for the top seven compounds, the control of pests was quite variable. In this respect Sherpa Plus® stood out as the best formulation but was followed closely by mixtures of deltamethrin and dimethoate or endosulfan. The combined formulations use smaller amounts of each active ingredient than are normally recommended.

The need to make chemical control accessible to all farmers (often uneducated) in the tropics is the impetus behind the new technical developments. The combined formulations have eliminated the need for farmers to purchase more than one chemical and the need for them to decide whether, and when, to change from one chemical to another. Also, even though the dosages used in combined formulations are still somewhat high, they are substantially lower than earlier recommended dosages, with reduced risks of dermal and chronic poisoning to the user and consumer. Their cost is still prohibitive but likely to become affordable in the future.

Insecticides have normally been applied with hydraulic, high-volume (HV)

Table 1. Field evaluation of foliar insecticides for their efficacy in the control of three major cowpea pests in Nigeria, IITA, second season, 1982*

	Application dosage (g a.i./ha)	Mean no. of					Yield (kg/ha)
		Flowers/ plant	Thrips/10 racemes	Thrips/20 flowers	Pod bugs/ 10-m row	Maruca/20 flowers	
Sherpa + dimethoate	30 + 250	17.6	0.0	19.1	2.7	0.0	1622
Endosulfan + deltamethrin	500 + 10	24.3	2.7	54.8	18.0	0.1	1613
Cybolt (flucythrinate)	80	15.5	6.5	121.1	55.5	0.1	1603
Oftanol	500	12.9	0.1	23.4	39.2	4.6	2589
Zolone (phosalone)	700	14.3	0.9	112.3	111.7	3.0	1587
Endosulfan + deltamethrin	650 + 10	15.5	1.2	39.5	17.0	0.2	1586
Cypermethrin + chlorpyrifos	25 + 500	20.3	1.4	44.9	29.5	0.0	1579
Deltamethrin + dimethoate	12.5 + 400	18.8	1.6	37.1	21.7	1.4	1486
Cypermethrin + chlorpyrifos	50 + 500	20.1	0.0	32.7	22.2	0.1	1440
Cybolt (flucythrinate)	40	15.7	8.4	131.1	28.9	0.3	1433
Hostaquick	550	12.4	6.0	170.0	38.5	2.2	1397
Tamaron	600	14.7	4.5	43.4	10.7	3.1	1397
Control (unsprayed)		4.1	81.1	423.6	16.0	2.4	268
LSD at 5%		6.7	4.2	80.7	47.7	2.2	366
Standard deviation		4.6	2.5	56.3	33.3	1.5	256

* Insecticide applied at 35, 62, 49 and 56 days after planting (DAP); racemes sampled at 38, 40 and 45 DAP; flowers sampled at 45, 47, 52 and 54 DAP; flower count made at 45, 47, 52 and 54 DAP; pod bug counts made at 50, 57, 61, 64, 68 and 71 DAP.

sprayers or spinning-disc sprayers that are ultra-low volume (ULV), with predictably good results. High-volume sprayers require water for mixing and are bulky to carry around. This makes them inappropriate for use in the dry savannas where water is scarce. Their use is also highly labour-intensive, requiring several hours to spray 1 ha. HV insecticide application is also prone to errors of dilution and high risk of spillage, and therefore contamination, even among researchers. These shortcomings notwithstanding, the hydraulic (HV) sprayer gives excellent results where it is properly used.

The ULV sprayer, unlike the HV sprayer, requires no mixing. It is prone to excessive drift and requires the use of batteries that are difficult to find in remote locations. Its effective use is highly dependent on wind direction and velocity. The ULV sprayer has the advantage of being highly time-efficient and is probably the most popular spraying method used at present in the African savanna and sahelian regions where water economy is important and farms tend to be a little larger than in other parts of the tropics. When properly used, this sprayer provides good results in every cropping system.

The latest in the line of spray technology innovations is the Electrodyn® sprayer. This sprayer is a hand-held battery-operated machine that releases negatively charged droplets of the insecticide. The droplets are attracted to the nearest grounded object—the plant. The sprayer has no moving parts and no mixing is required as the chemical comes in a ready-to-use bozzle, which completely eliminates the risk of accidental spillage. Although much more labour-intensive than the ULV sprayer (a maximum of two rows can be sprayed at a time) and possibly not appropriate for use in intercropping situations, the Electrodyn is the safest and the cheapest method currently available. Little or no drift is associated with this method of spraying. Unfortunately, the Electrodyn has a rather narrow range of chemicals at present but has become relatively popular in recent years in parts of Nigeria because farmers found it easy to transport and use. Furthermore, two of the available ED formulations, Cymbush Super® and Pirimor® are among the most effective compounds for controlling the pests of cowpea. Pirimor is the best aphicide available on the African market, and Cymbush Super is among the most effective broad-spectrum insecticides. Once farmers master the use of the Electrodyn sprayer, they obtain excellent control of pests.

With the increasing use of pesticides, concern is being expressed about the health hazards associated with insecticide residues. Little work has been done on residue analysis in cowpea, and this is an area needing urgent attention. Fortunately for the consumer, most of the chemicals used in cowpea-pest control in many countries are synthetic pyrethroids, but other compounds, organophosphates as well as organochlorides, such as dimethoate, monocrotophos and endosulfan, are also used extensively (Ayoade, 1969; Dina, 1976; Raheja, 1978; Ezueh, 1980; Litsinger et al., 1983). The non-pyrethroid insecticides are generally toxic to arthropods and mammals alike, thus underscoring the need for residue studies. Dutta and Lal (1980) found that the residue on leaves after 7–8 days from time of application of monocrotophos was below tolerance levels; findings were similar for disulfotan and dimethoate (Satpathy et al., 1975; Rajukanno et al., 1977). When used as directed, the insecticides cause only rare cases of consumer contamination or poisoning.

Whereas in the past the priorities were pesticide efficacy on target pests, today the growers have concerns about the effects on environment and safety. The shift is also apparent in research efforts, and priorities are being tailored to include features of safety of formulations and sprayer technologies as well as the economics of chemical vs cultural and other methods of control.

CULTURAL CONTROL

Pest problems on cowpeas can be reduced through the use of methods that alter the microenvironment of the pest—for example, species diversification, manipulation of planting date and pest diversion (or trap cropping).

The population dynamics of insects are regulated by several factors. The most important is changes in the microenvironment (Wilken, 1972), which can be manipulated by increasing or changing the sequence or pattern of crops in the ecosystem. Companion cropping, which increases crop diversity, changes the insect's habitat (habitat modification) and interferes with the insect's identification of, and responses to, its host plant (Tahvanainen and Root, 1972; Southwood and Way, 1980). Modifications that lead to the reduction of populations of a pest have been referred to as 'cultural control' or 'association resistance' (Root, 1973).

Many farmers in the tropics practise companion cropping (see e.g. Okigbo and Greenland, 1976) involving a few to several crops. To date, the choice of crops has been governed primarily by the crops' contribution to diets and subsistence rather than their effects on insect control. In other words cultural control of insects has not been consciously practised by most tropical farmers and, we believe, has not received the attention it deserves from entomologists (Kayumbo, 1976).

Intercrops

Cowpeas are mostly intercropped with maize, sorghum, millet and cassava, but occasionally with cotton and groundnuts. Studies on the effects of companion cropping on insect pests have, for the most part, been conducted in systems involving cowpeas–maize, cowpeas–sorghum and cowpeas–cassava.

When cowpea is grown as a monocrop it is subjected to heavy depredation, and yields are low. However, several workers have shown that when cowpea is intercropped the populations of several pests are reduced and yields increased. Work conducted in Nigeria (Perfect *et al.*, 1978) and in Tanzania (Karel *et al.*, 1982) showed that the populations of leafhoppers, *Empoasca dolichi*, *Sericothrips occipitalis* and *Callosobruchus maculatus* were reduced in cowpea–maize intercrops. Similar trends were reported for flower thrips by the same authors and Matteson (1982), Ezueh and Taylor (1983), and Wiedijk (unpublished). Several reports have shown that the pod borer, *Maruca testulalis*, are not affected by the cropping system (Taylor, 1977; Perfect *et al.*, 1978; IITA, 1982), even though, with sorghum as the companion crop, there is a net reduction in damage and infestation (Amoako-Atta *et al.*, 1983). Taylor (1977) also showed that intra-row mixing resulted in less *Maruca* damage to flowers than either monocrop or inter-row association, and Karel *et al.* (1982) reported a higher incidence of *Maruca* in monocrops of cowpea than when cowpea was intercropped with maize. These findings were supported by those of Ezueh and Taylor (1983) and Aidar and Kluthcouski (1983), among others.

For pod-sucking bugs the reports have been mixed: Perfect *et al.* (1978) and Matteson (1982) indicated a decrease in numbers in cowpea–maize plots at locations in southwestern Nigeria, whereas at other locations increased numbers have been associated with cowpea–maize and cowpea–sorghum intercrops (Kayumbo, 1976; Ochieng, 1977; Perfect *et al.*, 1978; Matteson, 1982).

There are other documented cases of pest-population increases with intercropping, a classic example being that of the foliage beetle, *Ootheca mutabilis*. This insect feeds on cowpea foliage, and its incidence is dependent on the onset and distribution of rainfall during the cropping season (Kayumbo, 1976). This means that the population of *O. mutabilis* can be quite variable. In intercrops involving either cowpea and maize or cowpea and sorghum in Tanzania the population of this insect has been found to increase (Kayumbo, 1976; Karel *et al.*, 1982). Increased shading is thought to be partly responsible for this observation (Kayumbo, 1976). In northern Nigeria, Matteson (1982) also recorded higher populations of meloid beetles (*Mylabris* sp. and *Coryna* sp.) on cowpea–maize intercrops than on monocrop cowpeas, and suggested that the insects normally fed on maize pollen and then infested and damaged cowpea flowers as the maize was drying.

These apparent contradictions on the effects of intercropping on insects underscore the need for standard crop varieties, sampling techniques, methods of analyses and experimental designs in investigations to evaluate insect responses to cropping systems. Studies in cultural control are made for the prevalent local systems of crop associations. Therefore, parochial biases cannot be avoided, but sampling techniques and methods of analyses need to be standardized if valid decisions are to be made on a comparative basis.

Intercropping cowpea with cassava, even though commonly practised, has been given inadequate attention in investigations of the dynamics of insect pests. In studies in Ibadan (Jackai, 1983), however, intercropping of two rows of cowpea with one of cassava reduced the populations of thrips and pod-sucking bugs. It did not affect *Maruca* borers. Similarly, results from Brazil showed an increase in *Maruca* populations but reductions in flower thrips, foliage beetles (*Diabrotica speciosa*) and leafhoppers (*Empoasca kraemeri*) (Aidar and Kluthcouski, 1983).

Cassava may be acting as a physical barrier to movement of thrips or an alternative food source for the pod bugs. However, the possibility that cassava is emanating chemicals, or that the cowpea host becomes less apparent to the insects, deserves investigation. The shading from associated crops adversely affects cowpea performance and may also be responsible for the observed population changes. In this system the humidity is known to increase and insolation is reduced inside the crop canopy (IITA, 1982).

Several explanations have been suggested for the changes in pest populations under different conditions of intercropping. First, decreases have been generally attributed to a disruption of the insect's perception of its host plant. Plant species diversity alters the host-selection repertoire involving vision, olfaction and contact (Southwood and Way, 1970; Tahvanainen and Root, 1972; Altieri *et al.*, 1978).

In addition, in cowpea–maize or cowpea–sorghum intercrops, the factors that are suspected of playing a vital (combined) role in reducing pest populations include:

○ restriction of movement (barrier effect) of thrips, leafhoppers, aphids;
○ increase in canopy closure leading to increased humidity, reduction in temperature and the provision of greater shading (for shelter); and
○ increase in natural enemy populations leading to a net reduction in pest populations.

The population increases that have been observed for a number of pests in intercrops seem to nullify the assumptions; they have been attributed to:

○ Reduction of overall effort required for movement of insects that prefer different hosts, for example, for oviposition, for feeding, etc. For instance, the meloid beetles (reported by Matteson, 1982) and some pod bugs (e.g., *Anoplocnemis curvipes*) oviposit on maize but feed (mostly) on cowpea (Ochieng, 1977). Intercropping therefore effectively reduces the time and energy required by the insect to move from one host to another.
○ Increased shading and humidity and reduced temperature brought about by some crop mixtures favour higher populations of some foliage beetles (e.g. Karel *et al.*, 1982).

Host evasion

When temporal asynchrony is created between an insect and its host plant, the dynamics of some insects change. In the United States, for example, temporal asynchrony through manipulation of planting dates has provided a good measure of control of the Hessian fly on wheat (Knippling, 1979), and planting date has been shown to be important in damage to southernpeas by *Chalcodermus aeneus* (Sherman and Todd, 1938).

Researchers in the tropics have confirmed the value of manipulating the planting date, thus giving scientific credence to the traditional practice of planting early in the rainy season. Results from Mokwa in Nigeria showed that when cowpeas are planted in July they are likely to mature before the peak of infestation of *Clavigralla tomentosicollis* (IITA, 1982). If an early-maturing cowpea variety were planted early, it should be able to avoid damage, although this possibility has not been tested. A number of pests such as thrips and *Maruca* borer cannot be effectively controlled (or avoided) by planting early or using early-maturing varieties, but the prevalence of pod-sucking bugs seems to be more dependent on environmental triggers than on the phenologic stage of the plant. Host evasion is probably the most promising approach in control of pod-sucking bugs at present because of the difficulty in finding resistance to this group of pests (Jackai, 1983).

Trap cropping

Trap cropping is the use of plant species to divert pests from the principal crop. This strategy has been successfully used in several crops (McGovran, 1969; Hill and Mayo, 1974; Newson and Herzog, 1977; Jackai, 1983).

The underlying assumptions in trap cropping have been outlined (Jackai, 1983) earlier in a preliminary report on the use of *Crotalaria juncea* as a trap crop for *Maruca* borer. The risk in the use of trap crops is that other pest species will increase. However, the potential for trap cropping in cowpea pest management needs to be investigated.

PEST-MANAGEMENT PACKAGES

Each method of cultural control can contribute to cowpea-pest suppression, but no one method provides satisfactory results. Cowpea resistance to insect pests offers the greatest hope for the peasant farmer but may take several years to develop. Also it may never be available at reasonable levels to all major pests, and its durability cannot be guaranteed. The state of the art in host-plant resistance in cowpeas is fully discussed by Singh and Jackai (this volume). Chemical control is the most readily available technology for the suppression of cowpea pests, but it has disturbing consequences (e.g. Luck *et al.*, 1977; Van den Bosch, 1978).

In recent years entomologists in several countries have stressed the need for developing integrated-management strategies for insect control to optimize agricultural production without upsetting the balance of nature. We know:

○ there are effective insecticides for every insect pest on cowpea;
○ there are cropping patterns that reduce populations of some pests and others that increase populations;
○ some damage by insects can be minimized through manipulation of planting dates; and
○ there are cowpea varieties with resistance to one or two insect pests.

With this in mind, one can proceed to formulate a package that will optimize production and at the same time minimize insecticide uses and, thus, overall production costs. A cowpea variety with resistance to thrips planted early and in association with maize or cassava will need two, or in exceptional cases three, insecticide sprays to suppress insect populations. The same variety planted in the same location as a monocrop at the normal planting time will require up to four sprays to achieve the same result, and a variety of similar maturity but with no resistance will require even more sprays under monocropped conditions. An algorithm can be used to represent schematically the pattern that would gain maximal benefit from the varietal characteristics, cropping practices and other components of the overall system (Figure 1).

Obviously, when several control methods are included in a control programme, the amount of insecticide needed is reduced. However, to date no integrated pest-management programmes have been developed for cowpea

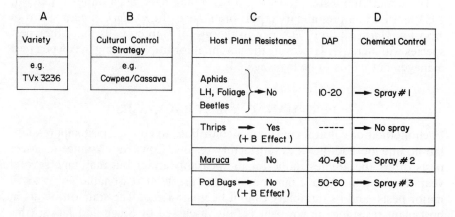

Figure 1. Schematic representation showing the use of resistance, cultural and chemical control in IPM of cowpea pests

Table 2.   Yield (kg/ha) of cowpeas sprayed regularly for insect control (Jackai and Singh, 1983)*

Location in Nigeria	Variety	Insecticide treatment	Yield (kg/ha)
Ayangba	TVx 3236	A	1144
		B	1156
Ayangba	TVx 7-5H	A	300
		B	383
Ilorin	TVx 3236	A	1500
		B	1589
Ilorin	Ife Brown	A	956
		B	1667

* Insecticide treatment: A = 2 sprays; Cymbush at 35 days after planting; dimethoate at 60 days after planting; B = 4 sprays: Cymbush at 30 and 40 days after planting; dimethoate at 50 and 60 days after planting.

pests in most countries. Because the individual components have to be developed before they can be consolidated into a management package, researchers have hesitated to devote much effort to integrated control packages, although an effective package does not necessarily need to include all the methods of control. For example, we reported earlier (1983) a study in which a cowpea variety with resistance to flower thrips was compared with two other (susceptible) varieties under two levels of protection: two sprays and four sprays. No significant yield differences were found between the two levels of protection for the resistant variety, whereas marked differences were observed for the other varieties (Table 2). Traditional cultural control does not lend itself to effective insecticide use, particularly late during crop growth when crop canopies are closing and insect pressure is increasing. Unless the planting pattern is changed from alternating rows to 'strip cropping' (two to four rows of cowpea alternated with two to four rows of the other crop), insecticides will not be effective in the system. At present only host evasion, natural enemies and resistant varieties can be used in concert with intercropping in any meaningful way. Our experience is that where insect problems are serious, as is the case in most African countries, the most meaningful strategy will have to include the use of insecticides.

Cowpea Research, Production and Utilization
Edited by S. R. Singh and K. O. Rachie
© 1985 John Wiley & Sons Ltd

17

Insects Damaging Cowpeas in Asia

S. R. SINGH
International Institute of Tropical Agriculture, Ibadan, Nigeria

Cowpeas in Asia suffer less damage from insects than do other grain legumes. Leafhoppers, aphids and beanfly appear to be the serious pests to date. It should be possible to control these pests by the development of resistant varieties.

BEANFLY

Ophiomyia phaseoli (Tryon) synonym: *Melanagromyza phaseoli* (Tryon) (Diptera: Agromyzidae) is a sporadic pest and can cause serious damage to the crop. It infests the seedlings, causing the stem to swell and crack at ground level. If the plants survive the attack they are stunted and produce only a few pods.

The shiny black insect (about 2.0 mm long) lays eggs on the upper surface of leaves, and the larvae tunnel to the stem near the ground or near the petiole of older leaves and pupate (Agrawal and Pandey, 1961).

Control measures, besides the development of resistant lines (Singh and van Emden, 1979), include adjusting the date of planting (Ho, 1967), planting in zero-tillage fields with long (60 cm) rice stubble (Ruhendi and Litsinger, 1982), and applying carbofuran or phorate granules to the soil (Pandey, 1962; Wickramasinghe and Fernando, 1962).

APHIDS

Aphis craccivora Koch (Homoptera: Aphidae) is a major pest of cowpea seedlings (Kobayashi *et al.*, 1972; Saxena, 1978; Singh and van Emden, 1979), with 5–10 aphids being born each day under favourable conditions and a single aphid producing more than 100 offspring (Bernado, 1969). The nymphs and adults suck the plant sap and are also found on green pods. When the populations are dense some damage to pods occurs because of honeydew deposits and mould formation.

The methods for control of aphids are similar to those for beanfly: use of resistant varieties (Singh and van Emden, 1979), zero tillage in fields with rice stubble (Ruhendi and Litsinger, 1982) and application of insecticides, including pirimicarb, phosphamidon and dimethoate as foliar sprays (Sarup *et al.*, 1961; Saxena, 1978) or phorate and carbofuran as granules.

LEAFHOPPERS

Empoasca spp. (Homoptera: Cicadellidae) are the other major pest of cowpeas in Asia. Like beanfly and aphid, they infest the crop seedlings and suck the sap. The most common species are *E. kerri* Purthi (Saxena, 1978) and *Amrasca biguttula* (Shiroki) synonym: *E. biguttula* (Shiroki), although there are many species and their biology is similar. The females lay eggs in the leaf veins on the underside of the cowpea leaves, each female laying about 100 eggs that hatch in a week.

The cowpea variety VITA 3 appears to be resistant (Singh, 1980), and leafhopper populations have been reported to be lower in zero-tillage plots with rice stubble than in high-tillage plots with no stubble (Ruhendi and Litsinger, 1982). Several insecticides, including phosphamidon and dimethoate, have been found effective as foliar sprays, and granular application of phorate and carbofuran at planting also gives excellent control.

MINOR PESTS

The other pests in Asia are of minor importance at present. They include leaf miners, leaf beetles, lima bean pod borers, blue gram butterflies, thrips, African bollworms, Egyptian leafworms and pod bugs.

Leaf miners, *Acrocercops phaeospora* Meyrick and *A. caerulea* Meyrick (Lepidoptera: Gracillariidae), are sporadic pests with a life cycle of about 15–20 days. The larvae tunnel under the epidermal cells of the leaves and pupate under the transparent leaf sheath. A few cowpea varieties seem to suffer less damage from leaf miners than do others (Saxena, 1978), and infestation is reported to be significantly reduced on zero-tillage plots with rice stubble (Ruhendi and Litsinger, 1982). Methomyl and endosulfan have been found effective as foliar sprays.

Leaf beetles, *Madurasia obscurella* Jocoby (Coleoptera: Chrysomelidae), have been reported from India (Saxena, 1978). The adults feed on leaves at seedling stage and produce round holes on the leaves. The life cycle of this pest has not been fully studied, but it is known that the adults are nocturnal (Saxena, 1978).

Saxena *et al.* (1971) reported systemic insecticides (phorate, disulfotan, and aldicarb basally applied as granules at planting) protected the seedlings up to 3–4 weeks.

Pod borer, *Maruca testulalis* (Geyer) (Lepidoptera: Pyralidae), is common throughout Asia (Rejesus, 1978; Saxena, 1978; Subasinghe and Fellowes, 1978) and can cause serious damage to crops in humid regions and during the rainy season. The larvae feed on stems, peduncles, flower buds, flowers and green pods, characteristically producing frass and webbing when they feed. The adult is a nocturnal moth, light brown with whitish markings on its forewings. It lays eggs on flower buds and flowers. The larva is light brown with irregular brown–black spots and a black head. The larval period is about 10–15 days, with pupation taking place in the soil under debris (Singh and Allen, 1979).

Damage by the pod borer is minimized in cowpea lines that have solid rather than hollow stem and peduncles and that have pods held above the plant canopy on long peduncles at a wide angle from each other. Methomyl, endosulfan, cypermethrin and decamethrin are effective against the borer (Singh and Allen, 1980).

The lima bean pod borer, *Etiella zinckenella* (Treitschke) (Lepidoptera: Pyralidae), is found throughout the tropics and subtropics of Asia, sporadically infesting a wide range of grain legumes (Srivastava, 1964). The tiny larvae enter the young, green pods and feed on seeds. The adult, a tiny moth, lays eggs on the growing tips of the plant, on flowers, or on developing pods. The first-instar larvae penetrate the pods and feed on the developing seeds. The mature larvae emerge from the pod and pupate in the soil for 7–10 days, depending on the moisture in the soil (Singh and Dhooria, 1971). It is easy to control this pest by the application of endosulfan, methomyl or cypermethrin early in the podding stage.

Blue gram butterfly, *Euchrysops cnejus* (Fabricius) synonym: *Catochrysops cnejus* (Fabricius) (Lepidoptera: Lycaenidae), is a sporadic pest during the rainy season (Saxena, 1978). The greenish larvae feed on flowers and on the developing grains inside green pods. They are easily controlled by the application of several insecticides, including endosulfan, methomyl and synthetic pyrethroids.

Thrips spp. are found in rice fallows mostly during drought stress. The adults lay eggs in the epidermal cells of leaf buds, and the life cycle, which has not been fully studied, appears to take about 20–30 days. In rice-fallow plots with long stubble, thrips have been observed to be fewer than in high-tillage plots without stubble (Ruhendi and Litsinger, 1982). The pests are easily controlled by the application of any of the synthetic pyrethroids.

African bollworm, *Heliothis armigera* (Hubner) (Lepidoptera: Noctuidae), is a minor pest of cowpea but does extensive damage to other grain legumes (Srivastava, 1964). The early-instar larvae bore inside the pod and feed on the seeds, the older larvae feeding on the grain from outside the pod. Several insecticides, including endosulfan, have been found effective against this pest (Agpad-Verzola and Cortado, 1969).

The Egyptian leafworm, *Spodoptera littoralis* (Fabricius) synonym: *Prodenia litura* (Fabricius) (Lepidoptera: Noctuidae), sporadically infests cowpea in Asia. It is primarily a defoliator but also feeds on young pods (Saxena, 1971). The older larvae are not easy to eliminate, but early instars can be controlled by insecticides such as endosulfan and cypermethrin.

Clavigralla gibbosa (Spinola) and *Riptortus clavatus* (Thunberg) (Hemiptera: Coreidae) are the two most common pod bugs found on cowpea in Asia. They suck the sap from the pods, causing the seeds to become shrivelled. *Clavigralla gibbosa* are grey to brown and are often found in large numbers on the cowpea plant; *R. clavatus* are cylindrical and light brown. The latter are easily disturbed and are fast fliers. Pod bugs can be controlled by the application of monocrotophos and endosulfan, whereas synthetic pyrethroids are generally not too effective.

STORAGE PESTS

Callosobruchus maculatus (Fabricius) and *C. chinensis* (Linnaeus) are the two major storage pests, with 100 per cent infestation common within 6–9 months of storage at the subsistence-farm level. Infestation often starts in the field and multiplies in storage. The adults, which are brown, lay eggs (up to 80/female) on cowpea seeds. The larvae hatch in about 3–5 days and bore into the seeds where the larval and pupal stages are found. The adult emerges after about 20 days.

It is not difficult to control this pest chemically: small quantities of seed (up to 10 kg) can be protected by a covering of edible oil (for example, groundnut oil mixed at the rate of 5 mℓ/kg seed—Singh *et al.*, 1979); large quantities for commercial use can be dusted with pirimiphos-methyl or, in airtight containers, fumigated with phostoxin.

Cowpea Research, Production and Utilization
Edited by S. R. Singh and K. O. Rachie
© 1985 John Wiley & Sons Ltd

18

Distribution, Biology and Control of Cowpea Pests in Latin America

RICHARD A. DAOUST,*† DONALD W. ROBERTS† and BELMIRO P. DAS NEVES*
*Centro Nacional de Pesquisa—Arroz e Feijão (CNPAF), Goiânia, Goiás, Brazil, and †Boyce Thompson Institute for Plant Research, Cornell University, Ithaca, New York, USA

In Latin America, as elsewhere, insects attack cowpeas at nearly all growth stages and in storage, and are the leading cause of low yields, which generally average between 400 and 600 kg/ha (Watt, 1978). Insect control could reduce yield losses by 75 per cent or more. In storage, losses caused by bruchids sometimes reach 100 per cent.

DISTRIBUTION

The cowpea-pest complex has not been extensively studied in most regions of Latin America. In Brazil, however, owing to the importance of the crop, published literature is available on the principal pest species and on economic damage caused by some of the pests in several regions. In other Latin American countries the pest complex frequently appears to reflect that of other leguminous crops, particularly dry beans (*Phaseolus vulgaris*), which are often grown in proximity to cowpeas. For example, insect pests known to attack cowpeas in Peru are normally the same species as those found on dry beans or soybeans. In Haiti, pod-sucking bugs have been observed moving from pigeon peas to cowpeas.

The most important cowpea pests can be categorized into seven damage groups in five orders, including pod and stem borers, leaf-feeding Lepidoptera, leaf-feeding chrysomelid beetles, storage bruchids, leafhoppers, pod-sucking bugs and leaf miners (Table 1). Chrysomelid leaf-feeding beetles, leafhoppers in the genus *Empoasca* and noctuid pod and leaf feeders in several genera are the most widely distributed pests in eight Latin American

Table 1. Major insects occurring in 17 Latin American countries and in the Commonwealth Caribbean Islands on cowpeas*

Insect group	Number of countries in which pest group is a major problem	Countries in which pest group occurs		
		Caribbean	Central America	South America
Pod borers, stem borers, cutworms	10	CC, PR, JA	HO, EL, NI	CO, VE, BR, SU, PE
Leaf-feeding Lepidoptera	5	CC	HO, NI	CO, VE, BR, SU, PE
Leaf-feeding Chrysomelidae	12	CC, PR, JA, HD	ME, EL, CR, PA	CO, VE, BR, EC, GU, SU, PE
Grain-storage bruchids	11	CC, PR, JA, HD	ME, EL, HO, PA	CO, VE, SU, BR
Leafhoppers	9	CC, PR, JA	HO, CR, PA	CO, VE, BR, EC
Pod-sucking bugs	7	CC, JA, HD	ME, HO, EL, CR	SU, BR, PE
Leaf miners	6	CC, PR	EL, NI	VE, GU, BR

* Countries include Caribbean—Commonwealth Caribbean (CC) (except Dominican Republic, Jamaica and Guyana); Puerto Rico (PR); Haiti/Dominican Republic (HD); and Jamaica (JA). Central America—Mexico (ME); Honduras (HO); El Salvador (EL); Nicaragua (NI); Costa Rica (CR); and Panama (PA). South America—Brazil (BR); Colombia (CO); Venezuela (VE); Peru (PE); Surinam (SU); Guyana (GU); and Ecuador (EC). The information presented was obtained from published and unpublished reports and survey trips and from personal communication with Latin American scientists (see acknowledgements for some of the contributors). The information from the Caribbean was from Buckmire (1978) and others.

Table 2. Principal insect pests of cowpea in eight Latin American countries*

Insect species or group	Country							
	Brazil	Panama	El Salvador	Venezuela	Colombia	Peru	Jamaica	Nicaragua
Elasmopalpus sp.	P	—	—	—	—	P	—	P
Spodoptera sp.	S	—	—	P	P	P	—	P
Heliothis sp.	—	—	—	—	P	—	P	P
Maruca sp.	S	—	—	—	—	—	S	—
Chrysomelidae	P	P	P	P	P	P	P	—
Chalcodermus sp.	P	—	—	—	—	—	P	—
Callosobruchus sp.	P	—	—	—	P	—	P	—
Empoasca sp.	P	P	—	P	—	—	P	—
Aphids	S	S	—	P	—	—	S	—
Coreidae	P	—	P	—	—	—	—	—
Pentatomidae	P	—	—	—	—	P	P	—
Liriomyza sp.	S	—	P	P	—	—	—	P

* P = principal pest species; S = secondary pest or pest that occurs in sporadic outbreaks only; — = pest species has not been reported or is rarely reported.

Table 3. Principal cowpea pests and their relative importance in 11 northern and northeastern states of Brazil*

Pest species	States of Brazil in which pest species causes economic damage to cowpea†										
	North			Northeast							
	PA	AM	AC	PE	MA	CE	PI	RN	PB	AL	BA
Lepidoptera											
Pyralidae											
Elasmopalpus lignosellus	—	—	—	P	P	P	P	P	S	S	P
Hedylepta indicata	P	S	—	P	—	S	S	—	—	—	—
Etiella zinckenella	—	—	—	S	—	P	PS	P	—	—	S
Maruca testulalis	S	S	—	S	—	S	S	S	—	—	S
Hesperidae											
Urbanus proteus	P	P	—	S	—	S	—	S	—	—	—
Noctuidae											
Spodoptera frugiperda	S	—	—	S	—	—	P	—	—	—	—
Spodoptera latifascia	P	P	—	—	P	—	—	—	—	—	—
Agrotis ipsilon	—	S	—	—	—	S	PS	S	—	—	S
Coleoptera											
Curculionidae											
Chalcodermus sp.	S	S	—	P	S	P	P	P	S	S	P
Aracanthus sp.	—	—	—	S	—	PS	—	S	S	S	S

	PA	AM	AC	PE	MA	CE	PI	RN	PB	AL	BA
Bruchidae											
Callosobruchus maculatus	P	PS	P	P	P	P	PS	S	S	P	P
Chrysomelidae											
Diabrotica speciosa	P	P	P	S	S	S	P	S	S	S	S
Cerotoma sp.	P	S	P	P	P	P	P	P	S	S	S
Andrector sp.	P	P	—	—	—	—	—	—	—	—	—
Hemiptera											
Pentatomidae											
Nezara viridula	S	S	S	S	S	S	P	P	S	S	S
Piezodorus guildinii	S	P	P	P	S	S	P	P	S	S	S
Edessa sp., *Acrosternum* sp.	—	P	P	S	S	S	S	S	S	—	S
Coreidae											
Crinocerus sanctus	P	P	P	S	P	P	P	P	S	S	S
Homoptera											
Cicadellidae											
Empoasca kraemeri	P	PS	S	P	P	P	P	P	P	P	P
Aphidae											
Aphis craccivora	—	PS	—	S	P	P	PS	S	S	—	S
Diptera											
Agromyzidae											
Liriomyza sativae	—	—	P	—	P	PS	PS	—	—	—	P

* This information was compiled from published accounts, including Assis, 1976; Bastos, 1973a; Carneiro, 1983; Castro et al., 1975; Cavalcante et al., 1975; Morães and Ramalho, 1980; Nogueira, 1981; Sales et al., 1979; Santos et al., 1977c, 1982; and Silva and Magalhães, 1980, and from unpublished reports, survey trips and personal communications with entomologists in Brazil.

† P = principal pest species; S = secondary pest or pest that occurs in sporadic outbreaks only; — = pest species has not been reported or is rarely reported in the state; PA = Pará; AM = Amazonas; AC = Acre; PE = Pernambuco; MA = Maranhão; CE = Ceará; PI = Piauí; RN = Rio Grande do Norte; PB = Paraíba; AL = Alagoas; BA = Bahía.

countries (Table 2). Other species such as *Elasmopalpus lignosellus, Chalcodermus* sp., *Callosobruchus maculatus* and several species of pod-sucking bugs have a more restricted distribution but are principal pest species in several countries. In Brazil at least 20 pest species, varying greatly between states and regions of the country, are documented as primary pests (Table 3). The most important pests in Brazil are leafhoppers (*Empoasca kraemeri*), the cowpea curculio (*Chalcodermus* sp.), leaf-feeding chrysomelid beetles (*Cerotoma* sp. and *Diabrotica speciosa*), the cowpea weevil (*C. maculatus*) and the lesser cornstalk borer (*E. lignosellus*).

BIOLOGY AND ECONOMIC IMPORTANCE

Homopterous and hemipterous pests

The green leafhopper, *E. kraemeri*, is the most important piercing insect on cowpeas throughout Latin America, and in Brazil it is among the principal pest species on both dry beans (Ramalho and Ramos, 1979) and cowpeas (Santos *et al.*, 1977c). On *Vigna unguiculata* the insect causes the most damage in the semiarid and often drought-stricken 'Sertão' region of northeast Brazil, while less severe attacks occur in the humid tropical regions of the north and along the northeast coastal region referred to as the 'litoral'. Although *E. kraemeri* is not known as a vector of viral diseases of cowpeas, heavy populations can cause production losses of 60 per cent or more as a result of direct sucking of plant juices and the injection of toxic substances into cowpea plants during feeding (Santos *et al.*, 1977c). The reduced productivity is generally related to a reduced number of pods/plant.

When densities of leafhopper are high, a yellowing with subsequent drying occurs at the foliar margins. Leafhopper attack is most critical from the time trifoliate leaves form until the flowering phase of the crop. If protection is applied during this period, significant increases in crop yield can be obtained. For example, Moraes *et al.* (1980) showed that the average yield (1250 kg/ha) from plants protected for 8–76 days after germination was approximately 2.5 times that (498 kg/ha) produced when no protection was applied. Yield reductions caused by leafhopper damage vary according to pest-population densities as well as cultivar: for example, Moraes and Oliveira (1981) reported that at population densities between 0.24 and 0.34 nymphs/leaflet, yield losses, depending upon cultivar, were 26–42 per cent higher in untreated plots than in treated plots. At a higher population density (2.09–2.86 nymphs/leaflet), production losses in unsprayed plots were even greater (66–79 per cent). In other studies, reductions in productivity of up to 78 per cent (501 kg/ha *vs.* 2230 kg/ha) have occurred when control measures were not applied (EMBRAPA, 1984). Moraes *et al.* (1982) calculated the economic injury caused by leafhoppers for two cowpea cultivars, with losses of productivity in the susceptible cultivar varying from 32 per cent with 0.5 nymphs/leaf to 100 per cent when 3 or more nymphs were present on each leaf.

Aphids are also important pests of cowpeas in several Latin American countries (Table 2). The most important species, *Aphis craccivora* (the black cowpea aphid), attacks the crop most severely in the wet or irrigated regions of northeast Brazil, principally infesting seedlings (Santos *et al.*, 1977a). Aphid populations generally remain dense for 30–35 days after which infestations decline naturally (Santos *et al.*, 1977c). In the state of Ceará, which has the highest cowpea production in Brazil, *A. craccivora* is known to cause serious damage to cowpeas and, according to Santos *et al.* (1977c), is the most widely distributed pest after *E. kraemeri* and *Chalcodermus* sp. Other aphids present on cowpeas in Brazil include *A. gossypii* in Pará and *Picturaphis brasiliensis* in Piauí, but these species do not generally cause economic losses (Santos *et al.*, 1977c; Silva and Magalhães, 1980; Santos *et al.*, 1982).

Aphids typically develop by sucking plant sap from the terminal shoots and petioles of new foliage. If climatic conditions are favourable they can also attack cowpea pods. Infested plants often suffer from retarded development, with the initiation of flowering being delayed for 1 week or more. In addition, *A. craccivora* serves as a vector in the transmission of potyviruses (Araújo *et al.*, in press).

Cowpea yield losses incurred by aphids have not been well documented for most locations. However, production losses in one field study in northeast Brazil ranged from 11.6 to 25.4 per cent when mean populations of aphids in a defined 3-cm area on petioles were between 7.4 and 16.3 aphids, respectively (Santos *et al.*, 1977a).

Pod-sucking Pentatomidae and Coreidae cause various degrees of damage throughout Latin America. For example, the stink bug *Nezara viridula* is widespread in the Caribbean, Honduras and Brazil but is often a secondary pest. Other pentatomids, e.g. *Piezodorus guildinii* and *Acrosternum* sp., cause serious damage to cowpeas in the highland jungle areas of Peru and in humid tropical regions of Brazil; *Edessa* sp. is important in north Brazil. The coreid bug *Leptoglossus* sp. causes major damage to cowpeas in Mexico. In Brazil, *P. guildinii* and *Crinocerus sanctus* are the two most important species of pod bugs on cowpeas (Araújo *et al.*, in press). *Piezodorus guildinii* adults cause a reduction in size and darkening of attacked seeds, while *C. sanctus* sucks juices from stems, new foliage and pods. In general, published research is not available on economic damage levels caused by pod-sucking bugs of cowpeas.

Other minor hemipterous pests in Brazil include the tingid *Gargaphia* sp. and several other coreid species.

Coleopterous pests

Coleopterous pests are widespread, and the most widely distributed groups, the storage bruchids and chrysomelid leaf-feeding beetles, cause economic losses in the Caribbean (Jamaica, Puerto Rico, Dominican Republic and several smaller islands), Central America (Mexico, Honduras, El Salvador and Panama), and South America (Venezuela, Surinam, Colombia and Brazil). While the storage bruchids cause direct losses to seeds in storage, chrysomelid beetles cause damage to plants by feeding on roots, nodules and stems as larvae and by defoliation and mechanical transmission of the cowpea severe mosaic virus (CSMV) as adults (Costa et al., 1981).

The species complex of Chrysomelidae in Latin America varies considerably from one region to another. Throughout most regions of Central America and the Caribbean, Diabrotica balteata, D. adelpha and Cerotoma ruficornis are widespread on grain legumes (Risch, 1976). Cerotoma ruficornis is primarily a legume feeder while the two Diabrotica species eat from a much wider range of plant families (Risch, 1976). In a study conducted in Costa Rica, Risch (1976) showed that C. ruficornis prefers cowpeas to beans while the Diabrotica species ate virutally no cowpeas, suggesting that cowpeas introduced into new areas of Central America would suffer less damage from diabroticite beetles than is currently suffered by P. vulgaris. Although research has not been conducted on the biology of these species on cowpeas, Gonzales et al. (1982b) studied the morphology, biology and habits of D. balteata and Cerotoma facialis on beans. They found that these species have three larval instars lasting from 10.6 to 14 days, a pupal period in the soil 6–7 days and adult longevities from 37 days (Diabrotica) to 52 days (Cerotoma).

In Brazil, D. speciosa, Cerotoma sp. and Andrector arcuatus are the predominant pest species on cowpeas in both the dry, firm regions and the irrigated riverbeds in north Brazil (Carneiro, 1983). In the Amazon region Chrysomelidae are present all year, although peak incidence is from August to September (Carneiro, 1983). In riverbed plantings leaf beetles consume foliage, leaving only stalks with naked leaf veins and, in heavy infestations, cause damage to the stems and pods (Castro et al., 1975).

Economic levels of damage by chrysomelid beetles on cowpeas have not been assessed. In dry beans in Costa Rica, however, mixed and pure adult populations of leaf beetles at the rate of two and four adults/plant, respectively, resulted in yield losses of up to 60 per cent during the first 15 days of plant growth (Gonzalez et al., 1982a). In cowpeas, losses are probably similar, although high pest populations in the Amazon can cause even greater damage to the crop (up to 100 per cent).

The cowpea weevil, also referred to as the cowpea seed beetle, is the only important bruchid pest of cowpeas in Latin America. Callosobruchus maculatus is the principal species, although C. chinensis may be more important in some locations in the Caribbean region. *Acanthoscelides obtectus*, the bean weevil, has been reported to attack cowpea seeds in Mexico, but it is probably a secondary pest (Jarry and Bonet, 1982).

Bruchids damage cowpeas almost exclusively as postharvest pests, although infestations can occur during the drying stage of the pods in the field. *Callosobruchus maculatus* females deposit eggs on the exterior surface of cowpea seeds, sticking them firmly to the grains. In 6–8 days larvae hatch and penetrate directly into seeds through small holes (*ca.* 1 mm in diameter). Most of the seed is destroyed during larval development, with several individuals often developing in each seed. Oviposition does not occur on broken or damaged seeds or on flour derived from cowpeas or other grain legumes (Bastos, 1969).

The biology of bruchids on cowpeas in Brazil has been extensively studied by Santos (1971). At 25°C and 75 per cent RH the durations of oviposition, incubation of eggs, and larval and pupal development are 8, 8, 16 and 7 days, respectively (Bastos, 1981). Adult longevity under the same conditions varies from 9 to 11 days. In northeast Brazil the best temperatures for development of weevils are between 20°C and 30°C, which are common in this region (Bastos, 1981).

Santos and Vieira (1971) showed that the loss in seed weight caused by the cowpea weevil is directly proportional to the number of holes produced in seeds. Based on this information, seed losses can be predicted with regression analysis for different levels of *C. maculatus* attack (Santos *et al.*, 1978; Oliveira *et al.*, 1984). From an economic standpoint cowpea seeds with 5 per cent bruchid damage lose up to 53 per cent of their value on the market, as was shown in an analysis in Fortaleza, Ceará (Bastos, 1973a). In addition to direct seed losses and economic losses due to reduced seed quality, bruchid attack also results in lowered seed germination (from 90 per cent for healthy seeds to 0 per cent for grains with four holes/seed) (Santos and Vieira, 1971).

Among Curculionidae attacking cowpeas, *Chalcodermus* sp. is by far the most important. In Brazil it is considered the most destructive field pest of cowpeas in the northeast, particularly in Ceará (Santos *et al.*, 1977c). Its distribution in other Latin American countries is unknown; however, it has been reported in several Caribbean countries, e.g. Jamaica and Puerto Rico.

Other curculionid species are also occasionally reported to cause damage to cowpeas by attacking pods and foliage. For example, *Pantomorus glaucus* causes defoliation in the state of Ceará (Cavalcante *et al.*, 1974). However, Santos *et al.* (1977c) consider this species an accidental pest only. Other curculionid species that cause damage, but whose economic statuses are still unknown, include *Aracanthus* sp. and *Promecops* sp. (Cavalcante *et al.*, 1979; Moraes and Ramalho, 1980).

The cowpea curculio was first described as economically important in northeast Brazil by Aciola (1971). The species status is still unclear, having been reported in much of the earlier literature as *C. bimaculatus* and more recently as *Chalcodermus* sp. or *C. aeneus*. Adult curculios typically attack developing green pods. As pods begin to mature and dry, the pests form progressively fewer external feeding and oviposition punctures (Bastos, 1974c). Adults perforate pods with their rostrum, and females deposit eggs into the holes created. Adults also feed on plant stems in proximity to pods (Bastos, 1974c). Larvae, upon hatching (2–3 days), penetrate and develop within seeds until they reach the fifth instar. Fully developed, they drop from pods onto the ground, penetrate the soil and pupate at a depth of between 1 cm and 6 cm (Bastos, 1974e). At favourable temperatures (24–37°C), pupal development in the field lasts from 12.9 to 13.3 days, but pupae are unable to develop at temperatures below 15°C and above 40°C (Bastos, 1974d).

Although both quantitative and qualitative losses occur because of curculio damage to cowpea seeds, economic damage assessments have not been attempted. Santos and Bastos (1977) have calculated the production losses that would occur by relating seed weight losses caused by curculio damage to reductions in yield, given the accurate prediction of both the number of pods/hectare and the number of seeds/pod. They estimated that when 20 larvae/30 pods develop in cowpea seeds, a production loss of 20 kg/ha results. In addition to direct production losses, the seed quality and germination capacity are significantly reduced by curculio damage (Vieira *et al.*, 1975).

Lepidopterous pests

Throughout much of Latin America the lepidopterous pest complex on cowpeas is diverse and highly damaging, with leaf- and pod-feeding noctuids and pyralids as the most commonly encountered families. *Elasmopalpus lignosellus*, the lesser cornstalk borer, is the most important lepidopterous species on cowpeas in Brazil. *Elasmopalpus lignosellus* larvae attack the stems of cowpea seedlings at the soil surface or slightly below. Larvae quickly bore into the cortical region of the plant and move up the stem through a tunnel formed. As larval development proceeds, plant stems begin to rot and foliage becomes yellow. Soon thereafter plants die and fall over. Normally, plants more than 20 days old are not susceptible to attack by lesser cornstalk borer larvae (Santos *et al.*, 1977c). Larvae form only one channel inside stems but do not stay inside the stems. A small shelter made of webs, soil and other materials is formed at the opening of the gallery, and larvae frequent these shelters (Santos *et al.*, 1977c).

Economic losses caused by lepidopterous insects have not been evaluated for cowpeas in Latin America, but, since these species are generally polyphagous, more published research exists in relation to other crops. In most cases *E. lignosellus* causes stand losses of only between 5 and 10 per cent on cowpeas (Santos *et al.*, 1977c). However, in dry areas with sandy soils, such as are found in Bahia, yield losses due to a reduction in plant stand can be severe.

Other lepidopterous species, e.g. *Agrotis ipsilon, Hedylepta indicata, Urbanus proteus, Spodoptera frugiperda, S. latifascia* and *Etiella zinckenella* in Brazil and *Heliothis zea* and *Spodoptera* spp. in Colombia, Peru, Nicaragua and Jamaica, also cause sporadic production losses, but research is not available on the damage levels or the yield losses that these species cause. The legume pod borer, *Maruca testulalis*, although frequently encountered in the Caribbean and Brazil, is of minor importance on cowpeas in contrast to its economic status in many African nations.

Other pests

Most other pests are of secondary importance. However, the leaf miner, *Liriomyza* sp., is reported as an important pest in El Salvador, Nicaragua and Venezuela and is a secondary pest in several states in Brazil. Outbreaks of *Thrips tabaci* are occasionally reported to cause severe damage to cowpeas in several northeast states in Brazil, but this species has not been reported from other Latin American countries. Mites cause some damage to cowpeas in Peru and Venezuela, and ants (*Atta* and *Ectatoma*) as well as grasshoppers (*Gryllus assimilis, Gryllotalpa* sp. and *Schistocerca pallens*) are reported as occasional pests in several regions of Brazil.

CONTROL MEASURES

In spite of the fact that insect pests take a heavy toll on cowpea production throughout most of Latin America, control measures have generally not been developed. In a few locations chemical control strategies have been adapted for cowpea pest regulation, but have not been implemented on a large scale. One reason is that cowpeas are generally produced on small farms as a subsistence crop. Even farmers who can afford insecticides are unwilling to make financial investments for insect control because of the low productivity and high risk of crop production in tropical semiarid regions with poor soils. The prevailing agronomic practice among farmers is to intercrop cowpeas with other subsistence crops or low-value cash crops with the expectation that they will realize at least a small yield from one of the crops planted.

Nevertheless, research has shown that significantly higher cowpea yields are obtained when insecticides are applied (Juárez *et al.*, 1982). To date, the use of chemical insecticides represents the only control strategy that has been implemented in the field, even though recommendations have generally not been developed for insecticide usage on cowpeas. Several studies have been conducted, and they demonstrate the successful control of *Chalcodermus* sp. and *C. maculatus* on cowpeas in northeast Brazil (Bastos, 1974a,b; Bastos

Table 4.   Chemical control of cowpea pests in Brazil

Pest species	Damage level for treatment	Critical stage to apply treatment	Product	
			Technical name	Dosage (a.i.)
Leafhopper *Empoasca kraemeri*	1 nymph/leaf	20–55 days post-emergence	Monocrotophos, 60% Carbaryl	100 mℓ/100 ℓ 0.5–1 kg/ha
Lesser cornstalk borer *Elasmopalpus lignosellus*	1 dead plant/m	Emergence until 25 days	Monocrotophos, 40% Carbofuran, 5% 35% Dimethoate, 50%	200 mℓ/100 ℓ 20 kg/ha 1 ℓ/75 kg seed 100–200 mℓ//100 ℓ
Cowpea curculio *Chalcodermus* sp.	Punctures on new pods	Initiation of fruit	Monocrotophos, 40% Endrin, 20%	300 mℓ/ha 5 ℓ/ha
Cowpea weevil *Callosobruchus maculatus*	Eggs on seeds	At time of storage	Methylbromide Phostoxin Phosphine	20 mℓ/600 kg seed 0.6 g/300 kg seed 0.6 g/240 kg seed
Leaf beetles *Diabrotica speciosa* *Cerotoma* sp.	15 adults/m	Emergence until flowering	Carbaryl, 85% Endosulfan, 35%	850 g/ha 350 g/ha
Pod bugs *Crinocerus sanctus* *Piezodorus guildinii*	—	Initiation of fruit	Dimethoate, 60% Diazinon, 60% Methyl parathion, 60%	200 mℓ/100 ℓ 100 mℓ/100 ℓ 80 mℓ/100 ℓ

and Assunção, 1975). A general list of chemical control recommendations for cowpea pests in Brazil is shown in Table 4.

Little research on nonchemical strategies has been conducted in Latin America; however, in Brazil, various levels of varietal resistance have been shown against several cowpea pests (Santos et al., 1977b; Neves and Watt, 1981; Quinderá and Barreto, 1983). At CNPAF in Brazil studies have been under way for several years to evaluate the varietal resistance of cowpea cultivars to several important pests including Chalcodermus sp., E. kraemeri and E. lignosellus. In these studies, preliminary evaluations of promising cultivars were made in screenhouses. The more resistant lines were entered into advanced trials throughout north and northeast Brazil. Cultivars showing sources of resistance were utilized in the breeding programme. Preliminary results have shown that some varietal resistance exists against several major pests. Several cowpea lines were identified as having mechanical resistance and two lines showed antibiosis against Chalcodermus sp. (Neves, 1983). In other studies, sources of resistance have been identified for both E. kraemeri and E. lignosellus (Neves, 1983). Preliminary studies have also shown that several lines introduced into Brazil sustain less damage from M. testulalis than do local cultivars (Neves et al., 1982), and cowpea cultivars showing reduced damage from leaf miner L. sativae were identified in northeast Brazil (Moraes et al., 1981).

The use of natural enemies for the control of cowpea pests is another alternative to chemical insecticides. Some predators and parasites have been identified from cowpea pests in Latin America (Table 5), but virtually no research has been conducted with these agents. The greatest range of natural enemies has been observed on the lesser cornstalk borer (Silva, 1983).

Entomopathogens may possess an even greater potential than do other natural enemies for cowpea pest regulation. Entomogenous fungi infecting cowpea pests in Brazil are particularly widespread (Table 6), and epizootics of fungal diseases have been observed occurring on cowpea pests in several regions of Brazil (Daoust et al., 1983a,b; Wraight et al., 1983). Among the fungal species recorded as causing epizootics were Erynia radicans and Hirsutella guyana on E. kraemeri, Beauveria bassiana attacking D. speciosa and Paecilomyces fumosoroseus infecting Lagria villosa.

Through a cooperative research effort between scientists from the Boyce Thompson Institute and CNPAF, research has been initiated to assess the potential for using entomopathogens to control cowpea pests. Preliminary results indicate that the greatest potential exists for the use of B. bassiana and Metarhizium anisopliae for the control of chrysomelid leaf beetles, the cowpea curculio, and the lesser cornstalk borer (Daoust et al., 1984; Lima et al., 1984). Also, the phycomycetous fungus E. radicans has potential to control E. kraemeri.

The cost of a system of biological control using entomopathogens has to be

Table 5.   Some known parasites and predators of cowpea pests in Latin America

Pest species	Parasite or predator	Country	Reference
Empoasca kraemeri	*Anagrus flaveolus*	Brazil	Pizzamiglio, 1979
	Aphelinoidea pluetella	Brazil	Pizzamiglio, 1979
	Anagrus sp.	Colombia	Pizzamiglio, 1979
Chalcodermus bimaculatus	*Urosigalphus* sp.	Brazil	Bastos, 1973b
Maruca testulalis	*Sarcophaga* sp.	Puerto Rico	Leonard, 1931
	Microbracon thurberiphagae	Cuba	Leonard, 1931
	Apanteles n. sp.	Cuba	Leonard, 1931
	Argyrophyla albincisa	Cuba	Leonard, 1931
Etiella zinckenella	*Eurytoma* sp.	Puerto Rico	Leonard and Mills, 1931
	Heterospilus etiellae	Puerto Rico	Leonard and Mills, 1931
Liriomyza sativae	*Chrysocharis* sp.	Brazil	Haji, 1984
	Chrysonotomia sp.	Brazil	Haji, 1984
	Diglyphus sp.	Brazil	Haji, 1984
	Euilophidae sp.	Brazil	Haji, 1984
Aphis craccivora	*Eriopis connexa*	Brazil	Haji, 1984
	Cycloneda sanguinea	Brazil	Haji, 1984
	Coleomegilla maculata	Brazil	Haji, 1984
	Pseudodorus clavatus	Brazil	Haji, 1984
Urbanus proteus	*Apanteles* sp.	Brazil	Haji, 1984
Nezara viridula	*Trissolcus basalis*	Brazil	Ferreira, 1980
Elasmopalpus lignosellus	Approximately 25 species listed in Latin America	Cuba, Puerto Rico, Trinidad, Jamaica, Brazil	Silva, 1983

Table 6.   Entomopathogenic fungi isolated from cowpea pests in Brazil, 1981–84

Pest species	Entomopathogen species	Isolates
Chrysomelidae		
Diabrotica speciosa	*Beauveria bassiana*	36
Cerotoma sp.	*Beauveria bassiana,*	8
	Metarhizium anisopliae	2
Podisus sp.	*Beauveria bassiana*	1
Curculionidae		
Chalcodermus sp.	*Beauveria bassiana,*	9
	Metarhizium anisopliae,	2
	Paecilomyces sp.	1
Aracanthus sp.	*Beauveria bassiana*	1
Lagriidae		
Lagria villosa	*Paecilomyces fumosoroseus*	3
Bruchidae		
Unidentified	*Metarhizium anisopliae*	3
Cicadellidae		
Empoasca kraemeri	*Entomophaga* sp.,	1
	Entomophaga australiensis,	1
	Hirsutella guyana,	7
	Erynia radicans	13
Unidentified	*Erynia radicans,*	1
	Conidiobolus sp.,	1
	Erynia delphacis,	1
	Beauveria bassiana	1
Pentatomidae		
Piezodorus guildinii	*Metarhizium anisopliae*	1
Nezara viridula	*Beauveria bassiana*	1
Unidentified	*Beauveria bassiana,*	4
	Metarhizium anisopliae	2
Noctuidae		
Spodoptera sp.	*Nomuraea rileyi,*	3
	Beauveria bassiana	1
Heliothis zea	*Entomophthorales*	1

kept low if it is to be realized. We (Daoust *et al.*, 1984) have shown that such a system could be implemented with the fungus *B. bassiana* against *Cerotoma* sp. in north Brazil using a chrysomelid beetle-attracting cucurbitaceous root.

This work was supported in part by a grant to BTI/CNPAF from the Bean/Cowpea Collaborative Research Support Program. We wish to thank several scientists who provided information on cowpea entomology in their respective countries: Drs J. R. de L. Maycotte and M. C. J. Ramirez Choza in Mexico; Dr A. Cruz O. in Honduras; Dr A. Barrios G. in Venezuela; Dr G. Bastidas in Colombia; Drs H. Tapia B. and R. Obando in Nicaragua; Dr C. Cruz in Puerto Rico; Dr C. F. Burgos in Costa Rica; Dr G. Hernández-Bravo in Peru and Drs F. N. P. Haji, P. H. S. da Silva, S. M. de L. Barbosa, R. Cavalcante and U. M. Soares in Brazil. We also wish to express our appreciation to CNPAF and IITA scientists who provided internal research and trip reports with information on cowpea pests in Latin America.

Cowpea Research, Production and Utilization
Edited by S. R. Singh and K. O. Rachie
© 185 John Wiley & Sons Ltd

19

Entomological Research on Cowpea Pests in the USA

RICHARD B. CHALFANT
Coastal Plain Experiment Station, Tifton, Georgia, USA

Cowpea, or southernpea, is well adapted to the warm climate of the south but sustains a succession of insect pests (Chalfant, 1976). Early in the planting season (May, June), a complex of foliage thrips (*Frankliniella* spp. and *Sericothrips variabilis*) feed on the unfolding leaves, causing chlorosis, wrinkling and distortion. During this time and later, the cowpea aphid, *Aphis craccivora*, feeds on seedlings and pods, causing stunting, loss of vigour and spread of viruses. Also, various lepidopterous defoliators feed on the young plants.

The cowpea curculio, *Chalcodermus aeneus*, is the most serious insect pest in the southeast throughout the growing season. Adults occasionally feed on young seedlings, but the oviposition activities of the female are responsible for economic damage. Both sexes chew through the pod wall into the developing seed where the female deposits an egg or 'stings' the seed. The resultant grub then consumes one or more peas. The presence of a single grub or stung pea in a standard 300-g sample of shelled peas for processing results in condemnation of the whole lot. Growers must treat the crop with insecticides to produce a marketable yield.

The corn earworm, *Heliothis zea*, often infests cowpeas during midsummer, usually after the maize crop has begun to dry. The larvae chew round holes in several pods and usually consume one pea per pod. They seldom remain within the pod. The damage is more apparent in peas grown for fresh market where peas are sold in the pod. For processing, peas are shelled in the field, and the damaged peas and larvae are discarded.

Late in the season (September and October), larvae of the European corn borer, *Ostrinia nubilialis*, tunnel into the main stem, branches, peduncles and pods, causing poor growth and contamination. The larvae remain within the pods and render the peas unfit for consumption.

The larvae of the lesser cornstalk borer, *Elasmopalpus lignosellus*, tunnel up the stem of seedling peas causing wilting and death. Infestations are prevalent in sandy soils during the hottest months after periods of drought.

Stink bugs and related species, *Nezara viridula, Euschistus* spp. and leaf-footed bug, *Leptoglossus phyllopus*, cause distorted pods and blemished peas, pod abscission, seed abortion and reduced yield. They are prevalent during late summer and fall.

COWPEA CURCULIO

Because of the curculio's severity in Georgia, my colleagues and I at the University of Georgia have directed our efforts towards its control. In the early 1970s, studies (Chalfant, 1972) showed that four applications of toxaphene during pod production protected the peas against attack and ensured a marketable crop. In 1978–80 the curculio showed resistance to toxaphene so effective substitutes were sought. In laboratory and field studies, Stacey and I (unpublished) found that encapsulated methyl parathion (Penncap-M®) and the pyrethroids permethrin and fenvalerate could replace toxaphene. They are more expensive and because pyrethroids are contact insecticides they present difficulties in proper coverage. Most commercial growers rely on aerial application, but dense crop foliage and careless application often result in control failures.

New methods of application have been developed to ensure complete coverage. One method a colleague and I have been studying (Chalfant and Young, 1982, 1984) is chemigation by which the insecticide is injected into irrigation water delivered by a centre-pivot system. Research includes evaluation of selected insecticides, different formulations, water volume, droplet size and water pressure.

A breakthrough in the identification of host-plant resistance occurred when Todd and Canerday (1968) identified resistance in Ala 963-8. Subsequent studies in Georgia revealed pod-wall thickness as a resistance factor and reported a water-soluble arrestant and feeding stimulant for the cowpea curculio (Canerday and Chalfant, 1969; Chalfant and Canerday, 1972; Chalfant *et al.*, 1972). Field screening of commercial cultivars, breeding lines and plant accessions has indicated resistance to various insects, but heritability has not been determined (Nilakhe *et al.*, 1981b; Nilakhe and Chalfant, 1982a). We are continuing to screen plant introductions and experimental lines in the field as they occur.

At the US Department of Agriculture Vegetable Breeding Laboratory in Charleston, South Carolina, high levels of plant resistance to the cowpea curculio were located in several Florida breeding lines as well as Ala 963-8 (Cuthbert and Chambliss, 1972). Using these lines as resistant sources and California Blackeye 5 (CB5) as a susceptible source, researchers identified three independent resistance mechanisms (Cuthbert et al., 1974). These were non-preference (measured by oviposition and feeding punctures), pod factor (measured by the percentage of penetrations through the pod wall into the pea) and antibiosis (measured by larval survival). No single factor was dominant in all lines produced. Later, three resistant lines were developed (Cuthbert and Fery, 1975a). CR 17-1-13 has pod-factor resistance inhibiting penetration, and CR 18-13-1 and CR 22-2-21 have non-preference resistance. The resistant cultivar Carolina Cream was recently released (Fery and Chambliss, 1984).

Ala 963-8 was termed resistant to the cowpea curculio because only 10–25 per cent of the pod punctures penetrated the pod wall compared with 61–64 per cent in the highly susceptible CB5 (Cuthbert and Fery, 1975a). Inheritance is simple, not quantitative. Further investigations of the pod factor (Fery and Cuthbert, 1979) indicate that seeds of CB5 are more accessible to the adult curculio than are those of Ala 963-8. This relationship is called the pod:seed (PS) ratio [(pod weight − seed weight)/seed weight] and is negatively correlated with resistance. Varieties with high PS ratios have walls that are effective barriers to penetration. However, these varieties also produce a low percentage shell-out, and resistance is only an adjunct to chemical control (Cuthbert and Fery, 1979).

Progress toward development of non-preference resistance is slow under field conditions because of low heritability and large variations in insect distribution (Fery and Cuthbert, 1978). For this reason Fery and Cuthbert suggested using the seedling stage for screening in the greenhouse and laboratory.

In Alabama, Chambliss, Rymal and their associates at Auburn University have also focused on physiology of resistance to the curculio. They noted the presence of thickened cell walls in the fibre layer of the pods of Ala 963-8 as well as increased tannin, which was associated with non-preference resistance (Ennis and Chambliss, 1976; Grundlach and Chambliss, 1977). They measured pod-wall toughness with a pressure-testing instrument (Chambliss and Rymal, 1980) and have estimated broad and narrow sense heritability for pod-wall strength (56.6 and 44.8 per cent, respectively) (Chambliss et al., 1982). They have derived 13 lines from crosses between the susceptible CB5 and Ala 963-8.

In studies of the curculio itself they found that eicosenoic fatty acid in cowpeas stimulated the insect to feed (Rymal and Chambliss, 1976). Volatile extracts were isolated from pods in studies to determine factors repelling curculios in resistant varieties and attracting them in susceptible lines (Rymal et al., 1981; Chambliss and Rymal, 1982).

Currently, Chambliss and Rymal are planning to incorporate resistance to mosaic viruses as well as resistance to the cowpea curculio into commercial cultivars, and to characterize the chemical nature of host-plant resistance.

POD BUGS

Other studies at the University of Georgia include the effect of foliar and systemic insecticides on yield as well as the interactions among systemic insecticides, nematicides and varietal resistance to nematodes (Chalfant and Johnson, 1972; Chalfant et al., 1982). Nilakhe and co-workers have developed an economic threshold for stink bugs and sampling procedures and distribution patterns for lepidopterous and soil pests (Nilakhe et al., 1981b, 1982; Nilakhe and Chalfant, 1982b).

At present, research at Charleston is also directed towards finding varieties resistant to pod bugs such as the southern green stink bug and leaf-footed bug, which reduce pod set and seeds per pod as well as damaging seeds (Nilakhe et al., 1981a; Schalk and Fery, 1982). Fery and Schalk (1984) have identified several cowpea lines with resistance to southern green stink bug, including PI 293476, PI 293557, PI 293570, PI 353074 and PI 354580.

Lygus bugs

In California, cowpeas are grown for dried grain. The major pests are both species of Lygus—L. hesparus and L. elisus, which cause pitting of the seeds and loss of blossoms and pods.

A programme for screening cowpeas for resistance to Lygus is in progress, and 22 lines have been identified as having antibiosis (Moshy et al., 1983). Three lines that reduce the rate of nymphal growth have been crossed and backcrossed to CB5 and show much promise for high yield and quality. The cowpea curculio-resistant line CR 17-1-13 has also demonstrated resistance to seed damage by Lygus and has been entered into the breeding programme.

COWPEA WEEVIL

In Florida, J. R. McLaughlin and associates have focused on the chemical ecology of host plants in relation to insect behaviour and reproduction and research on oviposition stimulants for the cowpea weevil, Callosobruchus maculatus. They have measured oviposition response to cowpea seed extracts placed on glass beads. They plan to continue exploring chemical and physical cues involved in weevil host selection. The oviposition stimulant study will focus on identification of the stimulus and on the effect of variety and host maturity, and will attempt to characterize the associated oviposition behaviour and sensory structures in the insect.

At the Boyce Thompson Institute in Ithaca, New York, studies are now in progress on the biology of the cowpea weevil and on seed resistance.

One line of research on C. maculatus is aimed at discovering the mechanisms involved in dispersal. Low-density populations consist of sedentary, 'normal' adults, and larval crowding within seeds triggers the development of 'active' (dispersing) morphs, although the ratio of active to normal beetles declines with intense crowding. Five geographically separate beetle strains have been found to differ genetically in their propensity to produce active morphs. In the laboratory, most active beetles fail to reproduce. Plans are to determine the environmental conditions (such as flight exercise, host-plant stimulation, or diapause development) that are needed for ovarian development and successful oviposition by the beetles (Messina, in press; Messina and Renwick, submitted for publication).

Research has also focused on the initial infestation of cowpeas by weevils in the field. To determine when plants become vulnerable to *Callosobruchus*, researchers put plants at different stages of growth in cages with active beetles. The beetles preferred to deposit eggs on the walls of large, green pods and on exposed seeds, avoiding the intervening mature-pod stage. Eggs laid directly on seeds were most likely to survive to become adults. On green or mature pods, newly hatched larvae often failed to penetrate the seed after they drilled through the pod wall. Nevertheless, since 20–50 per cent of eggs laid on green pods develop into adults, cowpea plants are susceptible to the weevils before pods mature. Plant traits likely to reduce infestation include rough pod walls and seed coats as well as the tendency for the seed to separate from the pod wall as it dries.

Research on oviposition behaviour of 'normal' *Callosobruchus* in storage is aimed at explaining how beetles are able to disperse their eggs evenly among seeds. When clean seeds are available the beetles avoid seeds that already have eggs on them.

Varietal resistance to the cowpea weevil is also under study. Several improved varieties derived from a single resistant accession from IITA are being tested to determine whether varieties are resistant to all five geographical strains of *Callosobruchus* maintained at the Boyce Thompson Institute. The stability of resistance is being monitored to determine the number of generations elapsing before beetles overcome the resistance.

The researchers have found that, unlike the parental resistant variety, the improved varieties do not increase mortality of larvae; rather they delay larval development in all strains. Six generations of beetles reared on resistant varieties showed only slightly shorter delays in larval development.

Wendell Burkholder and his group at the University of Wisconsin have isolated and identified pheromones from many stored-product pests. They are close to identifying the sex pheromones of *C. maculatus* and will provide them to the Boyce Thompson researchers for testing on the active form of the beetle and in field trials in northern Cameroon.

In Savannah, Georgia, Helen Su and associates have a programme using naturally occurring materials to protect cowpeas and beans from bruchid infestation in storage. They are especially interested in indigenous plant

270 COWPEA RESEARCH, PRODUCTION AND UTILIZATION

materials that serve as toxicants, repellents, attractants or antifeedants.

Black pepper, *Piper nigrum*, both the dried and brine-preserved unripe fruit, protects peas and bean seeds against *Callosobruchus* sp. (Su, 1977; Myikado *et al.*, 1979; Su and Horvat, 1981). On contact it is highly toxic to adults and to newly hatched larvae before they penetrate seeds. West African black pepper, *P. guineense*, is slightly less toxic (Su, 1984). Toxicity of *Piper* spp. is mainly caused by *N*-isobutylamides (Su *et al.*, 1972b).

Several citrus oils used in bakery and flavour industries are toxic to weevils on contact, and the volatile component of citrus oils, mainly limonene, has been shown to be a fumigant (Taylor and Vickery, 1974). Even the non-volatile component, especially of lemon oil, provides strong and long-lasting protection against adults and larvae (Su *et al.*, 1972a,b; Su, 1976).

Researchers in other locations have also been investigating the use of vegetable oils as a surface protectant for peas and beans against bruchids. For instance, at Boyce Thompson Institute, studies on the use of the various oils to protect cowpea seeds against weevil attack have been completed. The oils were found not only to kill the eggs but also to deter oviposition (Messina and Renwick, 1983). The oil surface also interferes with respiration of the insects. However, many of these oils contain unsaturated fatty acids and fats that turn rancid due to oxidation. The result is a disagreeable odour and reduced protection against the insects (H. C. F. Su, personal communication, 1984).

In California, very light infestations of bruchids are perennial, and major infestations are associated with residual seeds in the hoppers of planters, in harvesters, and, rarely, carelessly stored beans (T. F. Leigh, personal communication, 1984).

APHIDS

Recently the cowpea aphid has gained attention. Screening for resistance to aphids has been conducted with 12 varieties developed at IITA and 200 varieties assembled by the Plant Introduction Station in Georgia. Thus far, the varieties resistant to *A. craccivora* in Africa appear not to be resistant to an aphid population from Georgia. Five varieties from the Georgia station seem to be resistant, however, and small-scale field experiments are being used to confirm the results. Experiments to determine the exact nature of resistance are planned.

I wish to acknowledge the contributions by T. F. Leigh, University of California, Shafter; J. A. A. Renwick and F. J. Messina, Boyce Thompson Institute, Ithaca, New York; J. M. Schalk and R. L. Fery, USDA (United States Department of Agriculture), ARS (Agricultural Research Service),

Vegetable Laboratory, Charleston, South Carolina; O. L. Chambliss and K. S. Rymal, Department of Horticulture, Auburn University, Alabama; Helen C. F. Su, USDA, ARS, Stored Product Insects Research Laboratory, Savannah, Georgia; and J. R. McLaughlin, USDA, ARS, Insect Attractants Laboratory, Gainesville, Florida.

Vegetable Research, Geneva, 1985.

20

Nematological Studies Worldwide

F. E. Caveness* and A. O. Ogunfowora†
*International Institute of Tropical Agriculture and †Institute of Agricultural Research and Training, Ibadan, Nigeria

Soil-inhabiting nematodes, perhaps being more numerous than any other animal of similar size, must be considered an important segment of the soil fauna. They are a well-defined group of invertebrates ranked as a phylum or a class in the animal kingdom. Nematode is a word derived from nematoid, meaning 'like a thread', and is used with other common terms such as eelworm, threadworm or roundworm. Nematodes are widely spread worldwide and in all climatic zones, occurring often in great numbers wherever food, moisture and conditions are favourable. They can be grouped according to their life style as parasites of animals, insects, plants, fungi or as free-living in the soil or in fresh or marine water. Nematodes, or the diseases they cause in humans, animals and plants, have been known for centuries and are mentioned in some biblical accounts.

As nematode diseases tend to debilitate rather than devastate crops, their study has lagged behind that of other pathogens.

Cowpea was mentioned with some of the earliest reports of plant-parasitic nematodes (George, 1922; Feder and Feldmesser, 1956; Thomason, 1959; Klein, 1960; Sprau, 1960; Thomason and McKinney, 1960; Ayoub, 1961; Gundy and Rackham, 1961; Robinson, 1961; Toler et al., 1963; Prasad et al., 1964; Goodey et al., 1965; Prasad et al., 1965), and there are numerous reports in recent literature of nematodes associated with and attacking cowpea (Ichinohe and Asal, 1956; Janarthanan, 1972; Germani and Dhery, 1973; Gupta and Edward, 1973; Heyns, 1976; Razak and Evans, 1976; Ibrahim, 1978; Gupta, 1979; Rushdi et al., 1980; Kheir and Farahat, 1981; Sharma, 1981). The attack of nematodes on cowpea can be severe or inconsequential. Plant-parasitic nematodes associated with cowpea and their distribution are given in Table 1.

Table 1.   Plant-parasitic nematodes reported associated with cowpea

Nematode	Location (reference)
Aphasmatylenchus straturatus	Burkina Faso (Germani and Dhery, 1973; Dhery *et al.*, 1975)
Belonolaimus gracilis	Florida (Christie, 1952), South Carolina (Graham and Holdeman, (1953)
B. longicaudatus	Florida (Standifer, 1959)
Criconomoides mutabile	California (Siddiqui *et al.*, 1973)
C. obtusicaudatum	South Africa (Heyns, 1962)
Helicotylenchus cavenessi	Nigeria (Caveness, 1967a,b, 1968, 1971)
H. dihystera	California (Siddiqui *et al.*, 1973), Nigeria (Caveness, 1967a), Texas (Brown, 1972)
H. pseudorobustus	California (Siddiqui *et al.*, 1973), East Africa (Jensen, 1972), Nigeria (Caveness, 1967a,b)
Hemicriconemoides cocophilus	Nigeria (Caveness, 1967a; Prasad *et al.*, 1964)
Hemicycliophora arenaria	California (Gundy and Rackham, 1961)
Heterodera cajani	Egypt (Diab, 1968; Aboul-Eid and Ghorab, 1974), India (Koshy and Swarup, 1972; Sharma and Sethi, 1975)
H. glycines	Arkansas (Riggs and Hamblen, 1962), Egypt (Diab, 1968)
H. schactii	California (Raski, 1952)
H. vigni	India (Edward and Misra, 1968; Gupta and Edward, 1973; Gupta *et al.*, 1975)
Hoplolaimus columbus	Georgia (Hogger and Bird, 1976), South Carolina (Fassuliotis, 1974)
H. indicus	India (Das *et al.*, 1970)
H. pararobustus	Nigeria (Caveness, 1967a, 1968)
H. seinhorsti	Nigeria (Bridge, 1973a,b)
Macroposthonia ornata	Georgia (Johnson *et al.*, 1981)
Meloidogyne africana	East Africa (Whitehead, 1968; Jensen, 1972)
M. arenaria	Nigeria (Caveness, 1967a,b, 1968), Philippines (Madamba, 1981)
M. ethiopicum	East Africa (Whitehead, 1968), Tanzania (Jensen 1972)

continued

Nematode	Location (reference)
M. incognita	East Africa (Jensen, 1972), India (Bala-subramanian and Sivakumar, 1982), Nigeria (Caveness, 1967b, 1972, 1973; Amosu, 1974), Philippines (Madamba, 1981)
M. javanica	Australia (Jensen, 1972), California (K. Williams, 1972, 1973), East Germany (Salem, 1978), Egypt (Taha and Kassab, 1979), Nigeria (Caveness, 1967a,b, 1972, 1973)
M. kikuyensis	East Africa (Jensen, 1972)
Meloidogyne spp.	Arizona (George, 1922), Brazil, California (Siddiqui *et al.*, 1973), Hawaii (Linford *et al.*, 1938), Kenya (Ngundo, 1972; Ngundo and Taylor, 1974), Nigeria (Ogunfowora, 1976; Olowe, 1978)
Merlinius brevidens	California (Siddiqui *et al.*, 1973), Pakistan (Pakistani Botanical Society, 1982)
Paralongidorus maximus	East Germany (Decker, 1969)
Paratrichodorus minor (*P. christiei*)	California (Siddiqui *et al.*, 1973)
P. porosus	California (Siddiqui *et al.*, 1973)
Peltamigratus nigeriensis	Nigeria (Caveness, 1967b; Olowe, 1978)
Pratylenchus brachyurus	Australia (Jensen, 1972), California (Siddiqui *et al.*, 1973), Georgia (Hogger and Bird, 1972), Louisiana (Heyns, 1976), Nigeria (Bridge, 1973b; Caveness, 1967b, 1971)
P. neglectus	California (Siddiqui *et al.*, 1973)
P. thornei	California (Siddiqui *et al.*, 1973)
P. vulnus	California (Jensen, 1953; Siddiqui *et al.*, 1973)
P. zeae	California (Siddiqui *et al.*, 1973)
Pratylenchus spp.	California (Siddiqui *et al.*, 1973)
Quinisulcius acutus	California (Siddiqui *et al.*, 1973)
Radopholus similis	East Africa (Jensen, 1972)
Rotylenchulus parvus	California (Dasgupta and Raski, 1968; Heyns, 1976), South Africa (Furstenburg, 1972; Heyns, 1976)

continued

Nematode	Location (reference)
R. reniformis	Egypt (Badra and Yousif, 1979; Taha and Kassab, 1979), Hawaii (Linford and Oliveira, 1940; Linford and Yap, 1940), India (Dasgupta and Seshadri, 1971; Gupta and Yadav, 1980), Nigeria (Caveness, 1967b, 1968), Poland (Kaminski, 1969), Texas (Heald, 1978)
Scutellonema aberrans	Nigeria (Caveness, 1967a, 1968)
S. brachyurum	California (Siddiqui et al., 1973)
S. bradys	Nigeria (Olowe, 1978)
S. cavenessi	Nigeria (Sher, 1964)
S. clathricaudatum	East Africa (Jensen, 1972), Nigeria (Caveness, 1967a,b; Ogunfowora, 1976; Olowe, 1978)
Telotylenchus spp.	Nigeria (Caveness, 1967a)
Trichodorus spp.	California (Siddiqui et al., 1973)
Tylenchorhynchus clarus	California (Siddiqui et al., 1973)
Xiphinema americanum	California (Teliz et al., 1966; Siddiqui et al., 1973), Nigeria (Caveness, 1972)
X. ebriense	Nigeria (Caveness, 1967a)
X. ifacolum	Nigeria (Caveness, 1967b, 1971)
X. nigeriense	Nigeria (Caveness, 1967a, 1971), South Africa (Heyns, 1962)
X. setariae	Nigeria (Caveness, 1967a)
X. vanderlinde	South Africa (Heyns, 1962)

LOSSES IN COWPEA PRODUCTION

Data on yield losses are scarce, and not available for most developing countries, so an economic appraisal is impractical. The lack of data reflects the minor role given to cowpea, the ignorance of the nematode problem and the lack of trained staff to make the necessary evaluations. The reports that are available on nematode damage to the cowpea industry leave no room for doubt about the destructiveness of plant-parasitic nematodes and the importance of their role in agricultural production. Nematodes are frequently subtle, insidious crop pests so that yield reductions of a few to 20–30 per cent can pass undetected unless carefully managed control plots are introduced to observe differences.

SYMPTOMS OF INFECTION AND INJURY

Plant-parasitic nematodes are obligate parasites, as they are unable to reproduce without sustained feeding on live cells of a host plant. In the absence of a suitable host plant they will, over time, gradually deplete the stored-energy reserves within their bodies. A more favourable environment will encourage nematode activity and food reserve consumption while the stress of drought or cold will restrict nematode activity. Once food reserves are exhausted, nematodes die, but the time needed for this to occur can be a few months to several years, depending on the nematode species involved and the environment. Some species have built-in mechanisms that help preserve their populations during periods of stress.

Roots of cowpea plants appearing unhealthy suggest nematode damage, but laboratory analysis of soil and roots is essential for confirmation. Nematodes feeding on cowpea roots may cause root lesions, surface necrosis or reductions in total root volume by causing stubby root, root galling or the killing of fine feeder roots resulting in a coarse root appearance (Germani and Dhery, 1973; Amosu, 1974; Fassuliotis, 1974).

Above-ground symptoms are essentially those caused by other conditions that deprive the plant of an adequate and properly functioning root system. This may be expressed as reduced top growth, attendant lower yields, general lack of vigour, less resistance to drought conditions and early wilting during the heat of the day. In soils deficient in some necessary elements, the plants tend to develop symptoms of mineral deficiency.

The degree of stunting, chlorosis, wilting or plant death is dependent on nematode numbers, plant susceptibility, soil and climatic effects and the interplay of other pests and pathogens, particularly fungus root pathogens. Good weather and a high soil fertility ameliorate nematode damage, while drought and low soil fertility accentuate it. Under favourable conditions of moisture, temperature and food supply to roots, female nematodes can produce several dozen to several hundred eggs in each generation. With three or four generations often occurring in one season, nematode densities can reach high levels in soil.

Underground plant symptoms caused by nematode feeding include a general reduction in the root system; root galls (or knots); lesions on the root; excessive root branching; injured root tips causing short, stubby clusters of roots; an open-root system devoid of fine feeder roots; and curling of the root tip.

Gross symptom expression is generally related to the number of nematodes and the influence of the environment.

The injury extends from simple mechanical damage (caused by nematodes' directly feeding on protoplasm or moving in, out, or through plant tissue) to highly involved nematode–plant interactions caused by chemicals introduced by the nematode. The protoplasm-rich root tip is often the preferred feeding and invasion site. The overall result is often a reduction or cessation in tip growth, an alteration of tissue, or death (Jensen, 1972; Bridge, 1973a; Aboul-Eid and Ghorab, 1974; Sharma and Sethi, 1975; Wilson and Caveness, 1980).

THE FEEDING SITES AND LIFE CYCLE OF THE PARASITES

Most plant-parasitic nematodes are root feeders and live and reproduce entirely within the soil or root tissue. Some forms, endoparasites, wholly enter the root where they further develop and reproduce. Some soil-inhabiting nematodes feed on roots without bodily penetrating, and are known as ectoparasitic nematodes. Nematode forms with attenuated stylets can feed on cells of the cortex or stele while the body remains outside in the soil. The feeding activities of some nematodes lie between endo- and ectoparasites: they partially enter the root tissue with the anterior part of their bodies and are rarely found wholly within root tissue.

Typically, in plant-parasitic nematode development, there are four juvenile stages, with each being terminated by a moult. The first-stage juvenile develops within the egg shell and the first moult takes place within the shell. The second-stage juvenile, which is the infective stage of some species, leaves the egg shell and is free in the soil or plant tissue. With a suitable host plant as a food source, the mature female lays eggs and the life cycle is repeated. The male nematode is essential in some species as reproduction occurs only after copulation and fertilization. With other species the male form does not occur or is rare, and eggs develop without fertilization by the male. In nematode species where the male is necessary, the sex ratio is generally 1:1.

NEMATODE–*RHIZOBIUM* RELATIONSHIPS

There are several reports that nematodes cause reduced nodulation on leguminous plants, e.g. cyst nematodes on soybean (Ross, 1959; Ichinohe, 1961; Epps and Chambers, 1962; Epps, 1969; Lehman *et al.*, 1970), white clover and peas (Oostenbrink, 1955; Jones and Moriarty, 1956; Wardojo *et al.*, 1963), root-knot nematodes on peanuts, hairy vetch, alfalfa and cowpea (Miller, 1951; Malek and Jenkins, 1964; Nigh, 1966; Brown, 1971; Sharma and Sethi, 1975; Embabi *et al.*, 1982). Other studies reported no significant effect on cowpea nodulation by nematodes (Brown, 1972; Gupta and Yadav, 1980; Taha and Kassab, 1980). In-depth studies are needed to clarify the causes of different results.

In one study, Ali *et al.* (1981) reported that when both root-knot nematode, *M. incognita* and *Rhizobium leguminosarum* were used jointly as inocula on cowpea, infected plants showed more severe nitrogen deficiency and retarded growth than did plants with nematodes only or no nematodes. Nematode invasion reduced the numbers of nodules and inhibited nitrogen fixation by 63 per cent in nodular tissue. Infected nodules contained all developmental stages of root-knot nematodes. Infected nodules were not altered in structure, but bacteroids did not develop adjacent to infection sites. Nodules with nematodes deteriorated earlier than uninfected nodules.

In another study, the root-knot nematode, *M. javanica*, and the reniform nematode, *R. reniformis*, did not decrease nodulation (Taha and Kassab, 1980) except when reniform nematodes infected the plants before rhizobial inoculation. Nitrogen nodules formed on root-knot galls and served as infection sites for both types of nematodes. Similar results have been reported in other studies (Brown, 1971; Gupta and Yadav, 1980).

CONTROL METHODS

Control of nematodes gains importance as population growth places an increased demand on arable land. Nematodes are persistent in the soil, and shortened crop rotations favour the build-up of high levels of nematode populations. Put another way, greater nematode problems are coming with the future.

The principle underlying all control measures is to impose stress on nematode populations, impeding the nematode's ability to feed, reproduce and survive. Some control measures limit or reduce nematode populations over various periods in contrast to a quick kill by the use of heat or chemicals. Control methods can be placed in categories such as cultural control, physical control, chemical control, biological control and legal restriction (quarantine). These control measures need to be modified to fit the problem; crop involved; facilities available; and the inputs of labour, equipment and funds deemed appropriate.

Where the nematode species is relatively host-specific, crop rotation can be effective and is the only profitable method of control where inputs are limited or crop value is low. However, a number of nematode species often occur in the same field, complicating crop-rotation schemes and making it necessary to collect accurate information on cultivar resistance and nematode biology.

Amending the soil with organic materials and mulches has often resulted in improved plant growth. However, this may be due as much to improved plant-growing conditions as to biotic competition with the nematodes. Further development of alley cropping methods could help make soil amendments and mulching much more attractive as the hauling distance is drastically reduced.

Nematode populations tend to be reduced under fallow periods of weeds, grass or bush, as the preferred food source for the nematodes is reduced on a unit area basis. The eradication of volunteer crop plants is important for the success of this practice.

Sanitation involves the use of clean planting stock in addition to other control methods, especially where the transplanted crops in a rotation scheme come from a communal seedbed and are moved to a number of fields.

Heat treatment can be used for small quantities of soil. Wood fires over seedbeds will kill most nematodes to a depth of 30–40 cm. Recent research utilizing clear plastic sheets to trap solar heat makes nematode control over larger areas more feasible (Katan, 1981).

Where nematode-resistant crops are available they complement crop rotation. Resistant cultivars increase and stabilize yields, but at no extra cost to the grower and with the additional advantage of mitigating nematode attack on the following crop. Thus, resistant cultivars in combination with improved rotations could contribute greatly towards achieving optimum economic and long-term land use within traditional farming systems as well as more sophisticated and technical farming systems.

The principle in using a trap crop to control nematodes is the planting of a favourable host crop where the roots are invaded by the second-stage juvenile of the nematode. The trap crop is allowed to remain on the land just long enough for the second-stage juveniles to begin development and become immobile within the root tissue. The trap crop is then destroyed (ploughed under, uprooted or killed by herbicide) along with the 'trapped' juvenile nematodes. This practice is effective against those genera that become sessile when they develop beyond the second stage. The root-knot, reniform and cyst-forming nematodes are examples.

Certain crops can be effectively used as trap crops as they are highly susceptible to invasion by second-stage juveniles but are unsuitable host plants; thus, the nematodes are unable to mature and they die without laying eggs. *Crotolaria* and, in some areas, groundnuts have been used successfully to reduce populations of certain root-knot nematode species.

Chemical sterilizers for the soil are effective and can be used where economic and technologically safe (Rohde *et al.*, 1980; Balasubramanian and Sivakumar, 1982; Prasad *et al.*, 1982; Reddy and Singh, 1983). In most developing countries their use is restricted by the lack of sufficient infrastructure to develop and sustain an agricultural chemical control industry.

Plant quarantine measures are instituted by governments to restrict international movement of plant pests and diseases. These measures are effective when respected by the populace.

MAJOR NEMATODE DISEASES

Root-knot nematodes

Meloidogyne spp., root-knot nematodes, cause major economic losses of cowpea (K. Williams, 1972, 1973, 1974, 1975; Zannou, 1981; Ponte and Santos, 1982; Duncan and Ferris, 1983). They are distributed throughout the tropical and temperate zones and attack virtually all crops. Also, root-knot nematodes have been shown to predispose cowpea to infection by fungi and bacteria.

Meloidogyne incognita, M. javanica, M. arenaria and *M. hapla*, generally in that order of importance, are the cause of root-knot disease in cowpea production, with some cultivars being susceptible to one or more root-knot species and generally resistant to other species.

As a result of their ability to attack most food crops, root-knot nematodes are liable to cause losses wherever there is intensive cultivation of susceptible crops, especially on light soils, and where precautions against population build-up are not taken. Losses may be direct due to decreased yields or reduced stands or indirect from increased production costs incurred by costly control measures, the need to grow less-productive but resistant cultivars, or less frequent appearance of high-yielding but susceptible cultivars in the rotation scheme. Yield losses in cowpea of 5–10 per cent for California (Toler *et al.*, 1963); 10 per cent for Mexico, Central America and the Caribbean; and 43 per cent for West Africa have been reported (Sasser, 1979). From field studies in Nigeria, yield losses of 20–30 per cent (Caveness, 1967a), 40 per cent (Bridge, 1972) and 59 per cent (Ogunfowora, 1976) have been reported. In inoculated pot experiments Olowe (1978) reported grain-yield reductions of 25 and 94 per cent at inoculation levels of 1000 and 80,000 second-stage juveniles/kg soil respectively.

Second-stage juveniles of root-knot nematode occur in the soil and are vermiform in shape. They are the infective forms, and they penetrate the root near the tip. Plant roots can become infected within 2–6 days after germination. After becoming located at a feeding site in root tissue, the juveniles become sedentary parasites and cannot change their position. Nematode feeding induces gall formation by initiating hypertrophy and hyperplasia of infected and adjacent tissues. The main and branch roots become infected. The vermiform juveniles feed and develop to the fusiform stage by the 7th day. Rapid saccate growth commences about the 11th day, and fully developed rounded females with matrix can be found by the 19th day after penetration.

Gall sizes range from minute to 20 mm depending on the nematode species and number, as well as the cultivar's susceptibility. In infected tissues the vascular elements are disrupted, and water and nutrient movement to the top of the plant is restricted. In heavy infections the root system is reduced, and feeder roots do not develop, thus inhibiting the plant's ability to absorb needed water and nutrients. Compared with uninfected plants, infected plants have heavier roots, because of gall formation, and reduced top weight. The nematode infection acts as an energy sink, absorbing photosynthates needed by the plant for growth and grain production. An accumulation of metabolites coupled with membrane changes allows nutrients to leak into the soil. This enriched rhizosphere favours growth of bacteria and fungi, increasing root rot.

At maturity the female is 650 μm long and 400 μm wide, pear-shaped or globose. During her life span she produces 500–2000 eggs, which are laid outside the body and root in a gelatinous matrix. There are instances where the egg mass is within a gall, and the second-stage juveniles from these eggs may just start a new cycle of infection within the same gall. The eggs hatch within a few days, and the juveniles may infect the same gall or seek a new root tip. The generation time from egg to egg is about 30 days at 25°C so that a number of generations can occur during one growing season.

Symptoms of root-knot nematode injury on cowpea can be divided into aboveground and underground symptoms. These above ground include stunting, general lack of vigour, early wilting, death of seedlings and expression of mineral deficiency. Infected cowpeas often mature earlier than do healthy ones, while fertilization, especially with nitrogen and phosphorus, minimizes root galling and gives higher yields. Those below ground include root galls (these occur singly or coalesce to form typically knotted roots), excessive lateral roots, lack of lateral roots, stunted root system and root-rot complexes.

The root-knot nematodes have extensive host ranges and often occur as mixed populations in the same field. The differential-host test has shown certain plant species and cultivars to be poor or non-hosts for the common root-knot species.

Root-knot nematodes are ubiquitous in the tropics and subtropics, and cool-weather species can be a problem at higher elevations and in temperate zones.

Control measures include use of crop rotation, fallow and resistant cultivars. The economics of the use of nematicides generally precludes their use in cowpea cultivation.

Reniform nematodes

Rotylenchulus spp., the reniform nematodes, are widely distributed throughout the world's warmer areas and cause cowpea yield losses where their population levels are relatively high. Like the root-knot nematodes, they cause economic losses on a wide range of crops and interact with fungi to cause plant disease (Linford and Yap, 1940; Linford et al., 1938, Ayala and Ramirez, 1964; Caveness, 1967a; Jensen, 1972; Siddiqi, 1972; Gupta and Yadav, 1980).

Rotylenchulus reniformis and *R. parvus* attack cowpeas and reduce yields particularly where a crop-rotation scheme contains a series of susceptible crops. *Rotylenchulus reniformis* is more common than *R. parvus*. Very much like root-knot nematodes, the reniform nematodes tend to increase under intense food-crop cultivation and in lighter soils. Cowpea yield reductions of 15 and 20 per cent have been reported (Caveness, 1967a; Heald, 1978).

There may be thousands of males, vermiform females and juveniles in the soil around the cowpea roots. The mature, egg-producing female is sedentary and semi-endoparasitic on the roots, whereas the male is not parasitic. Eggs are laid in a gelatinous matrix outside the root and are fewer than 100. They hatch and develop quickly into second-stage juveniles; with little or no feeding the juveniles moult three times in the soil and become adults, with *R. reniformis* males and females being about equal in numbers. After a few days of feeding on epidermal cells the female invades the root, partially embedding herself in root tissue, and forms a permanent feeding site in the cortex and phloem. The posterior portion of the female outside the root enlarges to the characteristic kidney shape. At a soil temperature of 29°C the life cycle is completed in 19 days.

Reniform nematode attack can be light to severe, depending on the susceptibility of the cultivar and on the density of the nematode population. Common symptoms are unthrifty top growth and root injury indicated by stunting, discoloration and the presence of adult females with an enlarged posterior end, with the gelatinous matrix and a complement of sand grains protruding beyond the epidermis. Numerous males, immature females and juveniles can usually be recovered from the soil.

The host range is less extensive than that of root-knot nematodes but includes artichoke, bean, beet, cabbage, carrot, cauliflower, cucumber, eggplant, lettuce, okra, sweet potato, tomato and water yam.

Rotylenchulus reniformis occurs throughout tropical and subtropical countries, whereas *R. parvus* is widespread in Eastern and Southern Africa and the Imperial Valley of California.

The reniform nematodes can be controlled with crop rotation, as many non-susceptible crops are available. Crops reported resistant to *R. reniformis* include certain soybean cultivars (Rebois et al., 1970), sugarcane, Crotolaria, Pangola, Merker, and Para grasses (Roman, 1964), capsicum, jackbean, velvet bean (Peacock, 1956), *Leucaena glauca, Stenotaphrum secumdatum* and *Cynodon dactylon* (Linford and Yap, 1940). Crops reported resistant to *R. parvus* include cotton, tomato, peach, cabbage, beet, lucerne, carrot, cucumber, celery and cauliflower (Dasgupta and Raski, 1968).

Root-lesion nematodes

Pratylenchus is a large genus of root-lesion nematodes, of which *P. brachyurus, P. penetrans* and *P. zeae* are parasites of cowpea and may cause yield reductions. These are endoparasitic, migratory nematodes, widespread and found on many economic crops (Jensen, 1972; Corbett, 1973, 1976; Fortuner, 1976).

Lesions on the roots begin as elongate brown spots that become darker and more extensive with age. Borders of the lesions are delineated, dead and discoloured epidermal cells. Coalescing lesions eventually involve the whole of the cortical region followed by death of the vascular tissues resulting in loss of the root.

Root pruning by massive root-lesion attack results in a reduced and inadequate root system. The plants suffer from poor top growth and reduced grain yield.

These are obligate plant-parasitic nematodes with a simple life cycle. Eggs are laid in root tissue and in the soil. The juveniles undergo four moults, with the first within the egg, to become adults. Depending on the species, males may be common or rare. The time needed to complete one generation varies greatly depending on the host and especially temperature. *Pratylenchus brachyurus* completes one generation in 4 weeks at 30°C. Populations generally decline during the dry season and reach maximum numbers during the rains when temperatures are lower.

General but non-specific symptoms of distressed plants include chlorosis, stunting, sensitivity to moisture stress, a general lack of vigour and an early death of older leaves. Lesions are produced on large roots, but the fine feeder roots are the preferred feeding sites, often being destroyed. The lesions are caused by nematodes feeding on and killing root cells. Usually large numbers of root-lesion nematodes in all stages of development are found in the cortex in a limited area where feeding kills the cells, resulting in the formation of a lesion that often involves secondary invaders. The nematodes are generally found at the periphery of the damaged tissue and gradually enlarge the lesion by feeding on healthy cells.

Root-lesion nematodes are recorded as parasites of several hundred crops, many of which are commonly grown annual food crops as well as many grasses and trees. Most species are polyphagous and feed on woody perennials as easily as on succulent annuals.

Pratylenchus brachyurus and *P. zeae* are widely distributed in the world's warm regions, while *P. penetrans* is generally found in the temperate areas and at high elevations in the tropics.

The known host ranges of the different species indicate that, in some instances, crop rotations could be an effective means of control. Cotton and groundnuts have been reported as resistant to *P. zeae*. However, the possible occurrence of mixed populations makes it necessary that the species present be accurately identified. Summer fallow in dry climates causes a decline in nematode numbers through exposure to heat and drying and the elimination of the host-plant food source.

Lance nematodes

Four species of lance nematodes, *Hoplolaimus seinhorsti, H. columbus, H. indicus* and *H. pararobustus*, have been reported in association with roots, and soil around roots, of cowpea (Caveness, 1967a; Jensen, 1972; Hogger and Bird, 1976).

Hoplolaimus seinhorsti has been shown to cause economic losses of cowpea (Bridge, 1973a,b), reducing yields particularly where the crop-rotation scheme contains a series of susceptible crops. The endoparasitic migrations of *H. seinhorsti* cause considerable mechanical damage to root cells, resulting in poor top growth and reduced grain yield. Heavily infected roots die, and plants wither.

Hoplolaimus spp. are reported from several countries in association with a wide variety of plants; control measures have not been worked out but would be much the same as for other genera of nematodes.

PART 7

Agronomy

Cowpea Research, Production and Utilization
Edited by S. R. Singh and K. O. Rachie
© 1985 John Wiley & Sons Ltd

21

Optimizing Cultural Practices for Cowpea in Africa

N. Muleba* and H. C. Ezumah†
*IITA-SAFGRAD (International Institute of Tropical Agriculture and Semi-Arid Food Grains Research and Development Project), Ouagadougou, Burkina Faso, Upper Volta and †IITA, Ibadan, Nigeria

Cowpea in Africa is cultivated under diverse soil and climatic conditions and is traditionally grown with cereals such as millet, sorghum and maize (Steele, 1972; Rachie and Roberts, 1974). The bulk of production comes from smallholdings in the semiarid zones of West Africa, particularly in Nigeria (77.7 per cent of world production), Burkina Faso (5.2 per cent) and Senegal (2.1 per cent) (FAO, 1973).

Yield potential is high, averaging 1.5–3 t/ha (Raheja and Hayes, 1975; Warui, 1979), but actual yields are the world's lowest, averaging 0.2–0.3 t/ha. The current agronomic practices such as date of planting, plant populations, maintenance of the soil's physical properties and fertility, weed control and cropping patterns strongly influence yields of cowpeas.

DATE OF PLANTING

Research programmes studying the time of planting should search for a date that does not conflict with the period when staple foods make high labour demands (Summerfield et al., 1974) and that allows pods to mature during dry, bright sunny weather (McDonald, 1970).

Experiments conducted in semiarid zones of West Africa in monomodal or monsoon type climates have shown early planting, as soon as rains become well established in mid to late June, to be associated with high grain yield (IITA–SAFGRAD, 1981, 1982, 1983). Such early planting would, however, conflict with critical periods for planting and weeding cereals. It also would mean that, in the northern guinea and sudan savannas, photoperiod-insensitive cultivars would mature in August and early September under humid, cloudy weather that favours pod rots. Therefore, in these areas, the IITA–SAFGRAD programme recommends planting cowpea in mid-July.

In humid zones, at Ibadan, Nigeria, Rachie and Roberts (1974) recommended planting photoperiod-insensitive cowpeas in late May and late August. They also recommended that photoperiod-sensitive cowpeas not be planted in the first season in bimodal rainfall regions.

PLANT POPULATION

Cowpea is traditionally grown in a mixed-cropping system at wide spacings (total plant population from 10,000 to 20,000 plants/ha); most research, however, has been conducted in monocropping systems. It has shown optimal plant populations and row spacings vary with plant types and dates of planting.

Prostrate, photoperiod-sensitive cowpeas show little increase in yield when planted at populations greater than 22,000 plants/ha (Nangju, 1979a; IITA–SAFGRAD, 1981, 1982), whereas semierect and erect cowpeas, respectively, respond positively to intermediate (50,000–80,000 plants/ha) and high (>100,000 plants/ha) populations (Ezedinma, 1974; Nangju et al., 1975; Fadayomi, 1979a; IITA–SAFGRAD, 1982, 1983). These plant populations correspond to row spacings of 50 × 25 cm or 75 × 20 cm, for semierect indeterminate, high-branching types and to 16 × 34 cm or 17 × 40 cm, for erect, determinate, and low-branching types. At optimal densities most favourable results were obtained when the spacings between rows were decreased and space within rows was increased. At narrow spacings between plants within rows yield was reduced because of interplant competition (Ezedinma, 1974; Nangju, 1979a).

PHYSICAL PROPERTIES OF THE SOIL

In semiarid zones, particularly in sudan savanna alfisols, which are highly prone to water runoff and erosion, the soils have poor structure and a low rate of water infiltration. They are also subject to silting and crusting. Although farmers traditionally plant without tillage, the physical properties of the soil can be improved markedly and cowpea seed yields increased (55–90 per cent) if farmers hoe the land or 100–215 per cent if they tractor plough before planting. Additional increases (up to 36 per cent) are possible, particularly during dry years, if the cowpeas are planted on tied ridges, which considerably reduce water runoff (IITA–SAFGRAD, 1981, 1982, 1983). Also, the use of mulch, not a tradition in much of the savanna, can vastly increase yields, especially in dry years. For example, zero tillage with mulching significantly outyielded tillage by oxen or tractor in 1984, a particularly dry year with scarce and poorly distributed rainfall.

In humid regions also, zero tillage with mulching conserves the soil's healthy physical properties and has produced cowpea yields equal to or better than conventional tillage (Lal, 1976; Aina, 1979; Kamara, 1980a,b). The only exception in published reports as far as we know was in a rotation where cowpea was grown continuously (Lal, 1979).

SOIL FERTILITY

Soils in humid and semiarid West Africa have been reported as deficient in nitrogen and phosphorus and, to some extent, potassium and sulphur (Ezedinma, 1961; Godfrey-Sam-Aggrey, 1973; Ofori, 1973; Kang and Fox, 1975; Osiname, 1978; IITA–SAFGRAD, 1980, 1981, 1982, 1983). Nitrogen deficiency is found mostly in acidic, highly leached soils such as the oxisols in Sierra Leone (Haque and Gbla, 1978; Rhodes, 1978). Haque and Gbla (1978) recommended split application of N: 67 per cent at 15 days after planting and the remaining 33 per cent at flowering. In southwestern Nigeria, Agboola (1979) reported that cowpea response to nitrogen application was good in soils with low contents of organic matter. He recommended that nitrogen be applied in soils containing less than 2 per cent organic matter no later than 2 weeks after planting date. However, recent results from southwestern Nigeria indicate no response at soil organic matter levels ranging from about 1 to 1.75 per cent (B. T. Kang, personal communication; Ezumah, 1984, Table 7). In northern guinea savanna (semiarid zone), in Burkina Faso, we have observed that cowpeas intercropped with maize and grown on oxisols with 1.23 per cent organic matter after 4 years of fallow take advantage of nitrogen fertilizer and outyield unfertilized cowpea in pure stands (IITA–SAFGRAD, 1983).

Phosphorus, although not required in large quantities, is critical to cowpea yield (particularly for improved photoperiod-insensitive cultivars) because of its multiple effects on nutrition. It not only increases seed yields but also nodulation (Ezedinma, 1961; Ofori, 1973; Luse et al., 1975; Coulibaly, 1984). Other workers have reported that P application influences the contents of other nutrients in leaves (Kang and Nangju, 1983) and seed (Omueti and Oyenuga, 1970). At the present level of management in tropical Africa, P is the single most recommended fertilizer (P_2O_5: 20–60 kg/ha) (Rachie and Roberts, 1974).

Other nutrients (K, S and Ca) are of little importance and need only to be supplied where soils are particularly deficient, for instance in highly leached and eroded soils.

WEED CONTROL

Weeds directly damage cowpeas, competing for light, water and nutrients. Moody (1973) and Enyi (1973a) respectively recommended two and three weedings for cowpea plots during the first 1.5 months following planting. Their recommendations were from work in northern Nigeria and central regions of Tanzania but are generally applicable. This time is critical, as the cowpea is establishing groundcover. Pre-emergence herbicides that have been reported to provide better weed control than hand weeding with no damage to cowpeas in humid southern Nigeria are fluorodifen 3.5 kg/ha; sulfalate 8 kg/ha; metolachlor 2–3 kg/ha; metolachlor and metribuzin 1.5 and 0.25 kg/ha (Akobundu, 1978; Olunuga and Akobundu, 1980). Alachlor at 2–3 kg/ha damages cowpea, particularly when applied before emergence in moist soils. In semiarid zones, herbicides that have been reported as ineffective because they either lacked persistence or were phytotoxic to cowpeas include trifluralin, chlorthaldimethyl, nitralin, fluorodifen, chlorpropham, alachlor, linuron and chloramben (Moody, 1973).

Weeds also indirectly damage cowpeas by harbouring insect pests or intercepting insecticides and reducing their effectiveness (Moody and Whitney, 1974).

Parasitic weeds (*Striga gesnerioides* and *Alectra vogelii*) sporadically cause damage to cowpeas in semiarid zones. *Striga gesnerioides* is more prevalent in dry, hot regions, such as the sudan and sahel savannas, and *A. vogelii* is mostly found in relatively dry, cooler regions such as the northern guinea savanna. The occurrence of these weeds is generally associated with continuous cropping of cowpeas.

Experiments conducted at Kamboinse, Burkina Faso, involving cultivars and dates of planting have shown that:

○ The number of days from cowpea planting to first *Striga* emergence varies between 30 and 45 days, with the later emergence being associated more with low soil moisture than with date of planting.

○ The effects of cultivars as well as cultivar and date-of-planting interactions are significant: some cultivars are resistant (immune), some tolerant, and some susceptible to *Striga* damage. The susceptible photoperiod-sensitive cultivars escape damage when planted so that they flower a long time before or slightly after *Striga* first emerges in the plot. Thus, to reduce *Striga* damage where resistant varieties of cowpea are not available, farmers can either plant photoperiod-insensitive cowpeas that flower at about 40–45 days or manipulate the date of planting of photoperiod-sensitive cowpeas such that they flower at an appropriate time (IITA–SAFGRAD, 1982, 1983).

MIXED CROPPING

In semiarid West Africa, cowpea is traditionally intercropped with millet or sorghum. As soon as the rains become well established in early to late June, the cereal is planted; then, in mid to late July, prostrate photoperiod-sensitive cowpeas are planted at wide spacings and at densities inversely proportional to the cereal densities and plant vigour. Under good rain conditions, cowpeas are subject to heavy shading by cereals and yields are low. The reverse is true if rains improve late in the season. Farmers who want high yields of cowpea seed and relatively low yields of cereal grains plant cowpeas and cereals simultaneously in alternate or the same hills. For this purpose, semierect and photoperiod-insensitive cultivars are preferred to prostrate photoperiod-sensitive ones.

Several attempts to improve cowpea performance in the mixed-cropping system have been made. By planting cereals and cowpea at the same time and manipulating cereal row spacings and densities of both crops, researchers have obtained good yields of cowpea (>500 kg/ha) (Haizel, 1974; Adetiloye, 1980; Cunard, 1981; IITA–SAFGRAD, 1983). Isenmilla et al. (1981) reported that yield losses of cowpeas intercropped with maize could be reduced from 68 to 48 per cent by proper choice of cultivar; in their study, New Era, a spreading type, sustained less damage (48 per cent yield reduction) than did Ife Brown (62.2 per cent) and Adzuki (67.7 per cent), which are semierect and erect plants, respectively. Similarly, yields were reduced by only 41 per cent in spreading VITA 5 intercropped with maize, whereas the determinate, early IT 82E-60, sustained 54 per cent yield loss (IITA, unpublished mimeo, 1984).

The detrimental effect of shading on cowpea in association with cereals was demonstrated by Wahua et al. (1981). They showed that the more light transmitted to cowpea the greater were its growth and yield. Similar yield results were obtained from maize varieties intercropped with semideterminate TVx 3236 and indeterminate VITA 5 (IITA, in press). Adetiloye (1980) showed that yield of semierect and semiprostrate cowpea cultivars was reduced by an associated maize intercrop. He also reported that a cowpea cultivar with a climbing growth habit performed satisfactorily in association with maize. However, results reported by Wien and Nangju (1976) showed that the climbing cultivars cause increased lodging in maize and lower maize yields much more than do erect or spreading cowpea cultivars.

In IITA trials of cowpea intercropped with maize that had a wide range of growth habits (short: population 49, intermediate height: TZSRW, and sturdy, spreading, tall: TZPB), leaf-area index for the maize varieties at 8 weeks was respectively 4.2, 5.4 and 5.7; corresponding total (ear) heights were 176 (66), 270 (121) and 273 (117) cm. Higher cowpea yields were realized in intercrops with the short maize variety, but higher gross returns were obtained from the high-yielding, tall maize variety. The portion of gross returns attributed to maize ranged from 70 to 82 per cent.

Where the crop season is long enough, 150–200 days, such as in the northern and southern guinea savannas, cereal–cowpea relay-cropping can be practised. Cowpea, in these systems, is planted toward the end of the crop season so that pods mature during dry, sunny weather. In the northern guinea savanna (120–150 days' growing season), a maize–cowpea relay-cropping system has been developed (IITA–SAFGRAD, 1981, 1982, 1983): cowpeas are planted 4 weeks after maize, with very little maize yield reduction (<10 per cent). Maize cultivars that mature in 100–105 days were found less suitable than early-maturing varieties (90 days); they had a yield advantage over early varieties, but it never exceeded 15 per cent in good years, and they depressed cowpea yield by 30 per cent more than early cultivars. Photoperiod-sensitive cowpea cultivars were more adapted to relay cropping with maize than were photoperiod-insensitive ones. The former flower after maize harvest at the end of September or early October and thus form pods and mature when there is no competition from maize. Good yields for both maize (>3 t/ha) and cowpea (500–1500 kg/ha) have been repeatedly observed in this system. In contrast, photoperiod-insensitive cultivars flower, form and mature pods while under maize cover.

Andrews (1972a) described a relay-cropping system that worked well in southern guinea savanna (180–200 days' growing season). It involved early millet followed by cowpeas or early maize followed by cowpeas in alternate rows with a high-yielding, late-maturing (200 days) dwarf sorghum. Millet followed by cowpeas in relay with sorghum gave double the financial return obtained with sorghum alone; maize followed by cowpea gave more than 60 per cent better return than sorghum alone. According to Andrews (1972b), the ratio of the intercrops made a difference, with the most profitable being one row of sorghum for every two rows of relay crops. This configuration also gave the best yield of cowpeas.

In the humid regions, cowpeas may be profitably intercropped or relay-cropped with root and tuber crops such as cassava (IITA, 1980). The wide maturity gap between cowpea (about 90 days) and cassava (>360 days) and the slow initial growth of cassava make the combination compatible. In studies of this combination, yield advantage varied from 29 to 89 per cent, depending upon cowpea variety and season of growth (Mba, unpublished thesis).

ROTATION SYSTEMS

Cowpea is commonly incorporated in rotation systems in semiarid, humid and subhumid ecologies. The biological nitrogen fixation, estimated to vary from

20 to 25 kg/ha, has been reported as beneficial to subsequent (but not current) cereal crops (Nnadi, 1978; Kang, 1983). Recent reports suggest that a system of double cropping of rice and legumes will prove suitable in Africa, particularly where the early-maturity types can be grown in residual soil moisture (Sinner *et al.*, 1983).

The early maturity of cowpea, which does not exceed 90 days, makes it ideal for the short second season in bimodal rainfall ecologies common in the humid tropics where the pods mature during the dry season.

22

Agronomic Research Advances in Asia

R. K. Pandey* and Anake Topark Ngarm†
*IITA/IRRI Collaborative Grain Legume Improvement Program, International Rice Research Institute, Manila, Philippines; and †Faculty of Agriculture, Khon Kaen University, Thailand

Cowpea is grown in many parts of Asia as a monoculture in rotation with cereals, intercropped with maize, sorghum, pearl millet, cotton, cassava, sugarcane and relayed in standing rice. The cultivation practices differ widely in the region.

It is often grown in areas where other crops either fail or do not perform well. It grows under a wide range of climatic conditions—arid to subhumid— and performs well on acidic soils. In many areas cowpea is grown in sloping land as a cover crop for grazing. In Sri Lanka cowpea is grown in both riceland and hilly areas where slope is 15–20 per cent. In Burma (Rajan, 1977), the Philippines and Thailand, it is grown after rice where soil moisture is limited. In Indonesia it is grown in the highly acidic soils and performs better than other grain legumes.

The results of studies into environmental adaptation at IRRI show that cowpea performs better than mung bean in both extremes of soil moisture. In Jodhpur, an arid region of India, analysis of the climatic pattern for 1901–72 (Ramakrishna and Singh, 1978) showed cowpea was more successful than mung bean, sunflower and groundnut. Gadre and Umrani (1972), working on water availability and crop planning potential based on evapotranspiration values of Jeur and Sholapur district of Maharastra, India, found that only short-duration cowpea and mung bean can be successfully grown in periods of limited water availability.

In another study, Downes et al. (1977) subjected six crops to simulated sandstorms. Because of new lateral growth, cowpea was superior to the other crops in tolerance to sandstorm exposure and had normal seed yield. First-harvest yields decreased with increasing sandstorm velocity, indicating that sandstorm injury delayed maturity.

CROP ROTATION; MIXED/INTERCROPS AND RELAYS

Cowpea is grown in rotation with rice in Kerala, India (Varkey and Jacob, 1979), Bangladesh, Burma (Rajan, 1977), Philippines (Nadal and Carangal, 1977, Indonesia, Thailand and Sri Lanka (Figure 1). It is also grown for grain and fodder in wheat fallow in northern India. In Maharastra, Gujrat, Karnataka and Rajasthan states it is rotated with pearl millet, sorghum, maize and cotton. A maize–wheat–cowpea sequence is also common in some parts of northern India. Rao and Prasad (1978) found cowpea suitable in rotation with upland jute under low soil moisture in west Bengal, India. Yadahalli *et al.* (1973) reported cowpea cultivar C 152 grown in summer after rice gave a seed yield of 2.18 t/ha. In northeast Thailand VITA 3 was grown successfully in paddy fields after rice harvest. Among six food legumes tested after rice, cowpea performed best and gave highest economic return at Yezin, Burma (Rajan, 1977).

In India cowpea is widely grown as an intercrop with sorghum, castor, pearl millet and pigeonpea (Freyman and Venkateswarlu, 1977), and maize (Gangwar and Kalra, 1979). Saxena and Yadav (1979) observed that growing cowpea with pigeonpea had no adverse effect on seed yield of pigeonpea and increased the total production. In Kerala, cowpea is first in area and production among pulse crops and is traditionally grown as an intercrop with coconut, cassava and banana (Sasidhar and Sadanandan, 1976). At Jodhpur, India, intercropping of cowpea with sunflower did not affect the seed yield of

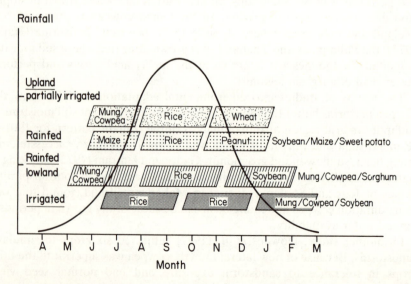

Figure 1. The generalized rainfall and cropping pattern in South and Southeast Asia

sunflower but gave an additional yield (Singh and Singh, 1977). Cowpea is grown in coconut plantations (Karunaratne *et al.*, 1978) in Sri Lanka and in Kerala, India (Nair, 1979). Recently in the studies at IRRI, growing cowpea cultivars with upland rice has shown promise in sustaining productivity in areas where rainfall is erratic.

Cowpea is also a relay crop after rice: seeds are broadcast 10–15 days before rice harvest, and the crop is allowed to grow on residual moisture with little or no input.

GROWING SEASON AND PLANTING TIME

For good crop performance, planting time must be considered. Cowpea is grown in the rainy season (July–October) and the summer (March–June) throughout Asia, and postmonsoon (November–February) in tropical Asia. Temperature during the postmonsoon period favours crop growth, and cultivation depends on the availability of water. When the crop depends entirely on rainfall, planting time may be critical for the successful completion of the life cycle. In the Indian subcontinent cowpea is planted in late June and July with the onset of rain. In northern India, Nepal, Bangladesh and Pakistan, cowpea could be planted after winter in late February, March or even April with irrigation.

In the tropical areas (southern India, Sri Lanka, Philippines, Indonesia and Thailand), most of the crop is planted after rice. During this period rainfall is diminishing, and later in the season soil moisture is limited. In rainfed areas of Southeast Asia, early maturing cowpea could be planted in late April or early May before rice, with harvest at the end of June or early July—after rice is transplanted.

PLANT TYPE

Traditionally, cowpea cultivars with highly indeterminate growth habit were grown in Asia, posing the problem of several harvestings. The development of synchronous maturing and erect cultivars has made this crop more attractive (Singh *et al.*, 1983). In South and Southeast Asia there are two varietal requirements in the different cropping systems:

○ Short-duration (55–60 days), erect cultivars with synchronous maturity, highly determinate growth habit and resistance to diseases and insects. Such cultivars are ideal for the short growing period before rice or wheat fallow. They also perform well after rice in areas where rainfall ends abruptly or soil has low moisture-holding capacity. They are suitable in semiarid and arid environments and have advantages over long-duration crops such as soybean or peanut.

○ Medium maturing (70–80 days' crop duration) cultivars with synchronous podding and pod held above canopy, and with green foliage for animal fodder. These are ideal for the relatively long growing period. They must have insect and disease resistance, tolerance for excessive moisture and drought, which are common features of rice land and upland rainfed areas.

In intensive cropping patterns in irrigated areas, short-duration cowpea cultivars fit in the rice–wheat–cowpea sequence in Pakistan, India, Nepal and Bangladesh. Similarly, short-duration cultivars fit into the intensive rice–rice–cowpea cropping pattern in south India where soil moisture is available for 60–70 days (Varkey and Jacob, 1979).

In the rainfed wetland rice areas, cowpea could be planted before and after rice. Before rice, the short-duration determinate varieties such as IT 82D-889 and IT 82D-9 could be grown in such countries as the Philippines, Indonesia, Thailand, Bangladesh, Sri Lanka and south India for either green pod or dry seed. After harvest the biomass could be incorporated into the soil for raising the rice yields. Cowpea cultivars such as IT 81D-716 and IT 81D-1069 with indeterminate growth habit, 70–80 days to maturity, and high-yield potential are suitable for cultivation in rice fallow in the Philippines, Thailand, Indonesia, Bangladesh, Burma and Sri Lanka. Dual-purpose cultivars such as TVx 2907–02D and IT 81D-1069 (grain and fodder types) will be preferred by small farmers, as animal fodder is often scarce in the dry season in many Southeast Asian countries and cowpea provides better fodder than other grain legumes.

In intercropping systems, short-duration, determinate cultivars fit as an intercrop with maize, sorghum, pearl millet, cassava, cotton, pigeonpea, sugarcane, coconut and rubber. The cowpea crop should mature in 60 days so that its yield will not be reduced by shading. It should also not compete with the major staple food crop for moisture in the later part of the rainy season so that full yield potential of the main crop is realized.

Development of bush types of cowpea at IITA has opened up the possibility of eliminating the expensive and time-consuming practice of providing mechanical support (trellis) for climbing types of cowpea. The new varieties should gain popularity as labour becomes more expensive in the future, as should varieties with pods held above the canopy, which are easier and cheaper to harvest than others.

CULTURAL MANAGEMENT

Cowpea can be grown with zero, low or high tillage. The previous crop, soil condition and weed-control measures determine the effectiveness of type of tillage. In upland fields it can be planted with zero tillage after wheat but may need irrigation for maximal establishment. After maize or sorghum, ploughing and two harrowings may be needed for a good stand.

Following rice, it can be planted in either untilled or well-tilled seedbeds, but tillage has no advantages. Pulverizing and aerating the soil are expensive, time-consuming, and wasteful of residual moisture, and experimental results show that yields at zero or minimal tillage are comparable with those from conventional tillage (Table 1). Zero tillage significantly reduces the risk of crop failure due to drought and represents savings in labour and energy in land preparation. Another study in Pangasinan, Philippines, compared low tillage (furrow planting with no land preparation but supplemented with herbicide spray before seeding), medium tillage (one ploughing and one harrowing), and high tillage (with rototiller) for the establishment of cowpea following rainfed lowland rice. Yield was highest under low tillage (Table 2), probably because soil-moisture conditions were better during early crop growth. Land preparation and, hence, crop establishment in the medium- and high-tillage plots were about a month late because the soil was too wet to till.

Table 1. Grain yield of cowpea as affected by tillage and plant population, IRRI, 1979, dry season*

| Tillage | Grain yield (t/ha) at a plant population of | |
	0.2 million/ha	0.4 million/ha
Conventional	1.0b	1.8a
Minimum	1.6a	1.6a
Zero	1.7a	2.0a

* Values followed by different letters differ significantly (P<0.05).

Table 2. Yield at three tillage levels of upland crops following lowland rice, Manaoag, Pangasinan, Philippines, November 1975–February 1976

| Crop | Yield (t/ha) | | |
	Low tillage	Medium tillage	High tillage
Mung bean	0.66	0.56	0.59
Soybean	0.04	0.16	0.17
Cowpea	0.76	0.43	0.46

Planting method

There are two common methods of planting cowpea: drilling in rows and broadcasting. In many areas where high tillage is practised, seeds are drilled in rows. Recently, the rolling injection planter (RIP) was tested for planting cowpea in rows with zero tillage after rice (IRRI, 1980). In some areas, cowpea is mixed with seeds of other crops and broadcast and then the soil is harrowed. This approach often provides a low plant stand. In rice fields, relay planting is also common where farmers have limited time and resources for land preparation and wish to take advantage of soil moisture. Under such conditions, cowpea seeds are broadcast in standing rice about 10–15 days before rice harvest. Seeds germinate with available soil moisture and can provide a good stand, although a slightly higher seed rate than in drilling is often required.

Cowpea could also be grown in the sorjan system—common in Indonesia—where dryland crops are grown in rows on beds bordered by rice in flooded depressions. During the rainy season two successive dryland crops can be grown on beds with two rice crops in the depressions, and a third upland crop (cowpea or soybean) can be grown on residual moisture in the depression.

Plant density

Optimal plant density is a key to realizing high yields, although cowpea has considerable plasticity in compensating for suboptimal densities. At IRRI, two plant types (strap leaf vs. normal leaf) did not respond positively when plant density was increased from 260,000 to 520,000 plants/ha (IRRI, unpublished), but in north India a plant population of 300,000–400,000 plants/ha (obtained from a seed rate of 20–25 kg/ha) produced the highest yield from determinate early-maturing cultivars. The spreading, full-season cultivars required lower densities (200,000–300,000 plants/ha).

Several studies in South and Southeast Asia have varied plant densities at different row spacings. Spacings of 0.5 m apart between rows with 2 plants/hill at 15 cm proved to be the best combination, resulting in a yield of 924 kg/ha (Braga Paiva et al., 1971). Furthermore, maintaining 2 plants/hill provides some insurance against crop failure caused by disease or insect damage in both wet and dry seasons (Jenkins and Hare, 1957). In Malaysia cowpea varieties, a determinate and indeterminate type, were tested at various plant densities with different row spacings. The results showed an increasing trend in dry seed yields as plant density increased from 67,000 to 200,000 plants/ha (Mohdnoor, 1980).

Fertilizer requirement

The well-nodulated cowpea crop can meet much of its nitrogen requirement, and in most tropical soils cowpea nodulates well with indigenous *Rhizobium*.

Results of several trials showed that artificial inoculation of the soil with elite rhizobia did not increase seed yields from cowpea. Similarly, in many soils, application of basal nitrogen did not improve seed yields (Faroda, 1973).

Phosphorus is critical for cowpea production. Singh and Lamba (1971) reported that, in low-phosphorus soils, application of 40 kg P_2O_5/ha increased cowpea grain yield. Similarly, application of phosphorus improved the grain yield and dry matter of cowpea in studies by Swami and Pal (1974), and seed yields increased from 426 to 551 kg/ha in response to an increase of P_2O_5 (from 20 to 40 kg/ha) (Kumar and Pillai, 1979).

In contrast, potassium plays a limited role in cowpea production, except perhaps in soils where initial exchangeable potassium is low. In most studies reporting fertilizer effects, potassium application did not increase seed yields (Faroda, 1973; Kozak, 1977; Kumar and Pillai, 1979). Experimental evidences so far indicate that, in most Asian soils, exchangeable potassium is adequate for cowpea production.

Generally, micronutrients have not been found to affect the performance of cowpea in most soils, although Bansal and Singh (1975) reported 33.4 per cent increase in yield by foliar spray of $FeSo_4$.

Water relations

Most cowpea crops are rainfed, a small proportion are irrigated, and others utilize residual moisture in the soil after harvest of a rice crop. Crops grown after rice often receive little or no rainfall during the growing season and must be capable of producing grain using only stored soil water. Thus, cowpea is particularly well suited for rice-based systems and may be grown in countries such as Burma in soil prepared immediately after rice harvest (De Wit, 1958). In India, among the crop species tested for high water-use efficiency, cowpea cultivars at four sites performed well (Sinha, 1972).

The ability of crops to extract soil water from a rainfed dryland site was tested at IRRI and reported by Angus et al. (1983). Cowpea extracted water rapidly during its 66-day cycle, but the total water extracted was only 140 mm. At depths greater than 80 cm, cowpea roots proliferated in rainfed plots but were few in an irrigated control (Figure 2). In another IRRI study 10 crop species were compared for soil-moisture extraction and performance after lowland rice. Among the grain legumes, cowpea and pigeonpea set down deeper roots than did soybean and mung bean (Figure 3), and cowpea extracted water from soil up to 1 m deep. During the crop season it maintained higher plant-water status and a cooler canopy temperature than did other crops. Using a line-source sprinkler system, we (Pandey et al., 1984) compared the performance of cowpea, soybean and mung bean and, again, found that cowpea extracted water from a deeper soil profile and maintained a higher plant-water status and cooler canopy than did the other two.

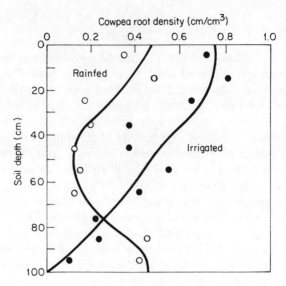

Figure 2. Root density of cowpea in rainfed and irrigated plots (IRRI, 1979)

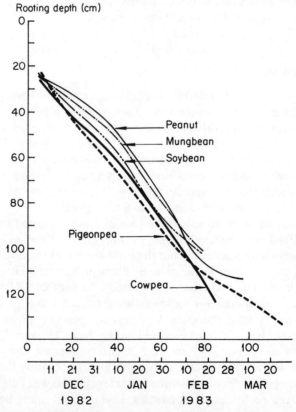

Figure 3. Changes in depth of the rooting zone of rainfed upland crops determined from water extraction patterns

A linear relationship between water use and total dry matter has been reported (Turk and Hall, 1980c; Pandey *et al.*, 1984), and, according to Ahlawat *et al.* (1979) cowpea responds favourably to irrigation in dry environments, with increases in seed yield being reported for cowpea irrigated at 50–75 per cent of available soil moisture. Under favourable moisture supply, however, cowpeas, especially indeterminate cultivars, tend to produce abundant vegetation and reduce the amount of dry matter being partitioned to seed. Water-use efficiency seems to be improved by moderate deficits in soil water, especially during the vegetative stage (Turk and Hall, 1980c).

Although short-duration cultivars complete their life cycle early, their roots remain shallower than those of long-duration cultivars. Thus, long-duration cultivars are able to extract soil moisture from greater depths and produce large seed yields in areas where moisture is limited. Short-duration cultivars have an advantage in areas where the growing period is short with reliable soil moisture for 2 months or so, whereas indeterminate cultivars are probably better than determinate ones in fluctuating rainfall environments where the growing season is relatively long.

Weed control

Cowpea suppresses weed growth (Tables 3 and 4), but yield losses caused by weeds can be 30–65 per cent. The major weeds infesting cowpea plots are *Echinochloa colona*, *Amaranthus spinosus*, *Rottboellia exaltata*, *Cyperus rotundus*, *C. kyllingia* and *Lindernia anagallis*.

When intercropped with maize, cowpea significantly reduced weed dry weight (IRRI, 1978)—a valuable trait in South and Southeast Asia where small farmers generally control weeds either by hand weeding or intercropping. Herbicides are rarely applied, although several pre-emergence herbi-

Table 3.   Effect of time of planting on total weed weight in an unweeded plot 30 days after plant emergence, IRRI, 1979, wet–dry transition period (IRRI, 1980)

Crop	Total weed weight (g/m^2)		
	16 Oct.	30 Oct.	24 Nov.
Mung bean	233.6a	200.4a	51.0bc
Cowpea	134.8a	130.0a	42.6bc
Maize	200.6a	214.4a	43.6bc
Sorghum	173.4a	264.2a	105.2ab

* Treatment means with different letters are significantly different ($P<0.05$).

Table 4.   Effect of weeding regimens and date of planting on grain yield of cowpea planted after dry-seeded wetland rice, 1979, wet–dry transition season (IRRI, 1980)*

Weeding regimen	Grain yield (t/ha)		
	16 Oct.	30 Oct.	24 Nov.
Weeded twice	1.8a	2.0a	2.4a
Weeded once	1.7a	1.7ab	2.5a
No weeding	1.4a	1.5b	2.3a

* In a column, treatment means followed by different letters are significantly different ($P<0.05$).

cides (butachlor, terbutryne and alachlor) work well with cowpeas, as do pendimethalin and butralin (IRRI, 1980). Rate of application is critical and must be adjusted to soil type and organic matter.

Harvesting and threshing

Cowpea is harvested when about 85–90 per cent of the pods are dry. Delay in harvesting often results in shattering, yield loss and poor-quality seeds. After harvest the crop should be piled up, threshed, and allowed to dry 2–3 days. Cowpea as green pods must be harvested at 10–12 days after flowering for the best combination of protein, yield and tenderness. If harvested at 20 days the pods become fibrous (Sambandan et al., 1965), and crude protein and total pod yield are affected. Since flowering and pod formation continue over several weeks, green pods are picked at 3–4-day intervals to maintain tenderness of pods and high yield. Cowpea grown for fodder should be harvested at the stage between flowering and pod formation. Kandaswamy et al. (1976) observed maximal green matter and crude protein at 60 days.

Cowpea Research, Production and Utilization
Edited by S. R. Singh and K. O. Rachie

23

Nitrogen-fixing Symbiosis and Tropical Ecosystems

K. MULONGOY
International Institute of Tropical Agriculture, Ibadan, Nigeria

Like many other legumes, cowpea can symbiose with nodule bacteria (rhizobia) present in most, if not all, tropical soils. The rhizobia possess the nitrogenase complex, an enzyme capable of reducing atmospheric nitrogen (N_2) into compounds assimilable by the host plant. Effective cowpea–*Rhizobium* symbiosis fix more than 150 kg N/ha and supply 80–90 per cent of the host plant N requirement (Eaglesham *et al.*, 1977; Summerfield *et al.*, 1977a). This attribute allows adequate yields in N-deficient soils where non-nodulated crops such as cereals fail. Maximizing N_2 fixation is an economic way to cope with the shortage of expensive nitrogenous fertilizer in tropical countries (FAO, 1976).

In this paper I report on research work, conducted during the past 5 years, to describe the major attributes enabling cowpea rhizobia to survive and fix N_2 effectively under tropical conditions; to identify the most appropriate *Rhizobium* strains for cowpea-growing areas where drought, high temperatures, excessive moisture and acidity limit the productivity of cowpeas; and to assess the potential benefits from using these elite strains in inoculation trials. On the basis of the findings, I propose areas of research where the management of biological N_2 fixation would help improve yields of cowpeas and other crops in tropical farming systems.

CHARACTERIZATION OF THE 'COWPEA' RHIZOBIA

At IITA more than 800 rhizobial isolates were obtained from cowpea cultivars sown in West Africa at three locations selected to provide diversity in soil and climatic factors. Onne, southeastern Nigeria, is in the humid tropics, with acid soils, low levels of calcium, high levels of exchangeable aluminium and relatively many rhizobia (4.3×10^4/g soil). Maradi soils, in Niger, receive little rain and are subject to prolonged drought and extreme temperatures; they have low N, Ca and available phosphorus, and they contain fewer than 4.9×10^2 cowpea rhizobia/g soil. IITA at Ibadan, in southwestern Nigeria, is intermediate, with a cowpea rhizobia population of 3.5×10^3/g soil. Details about the soil physicochemical characteristics have been presented elsewhere (Ahmad et al., 1981b; Mulongoy et al., 1981).

All the cowpeas nodulated with indigenous rhizobia. Many lines, particularly at Maradi, responded positively to N fertilization in terms of fresh shoot weight and thus failed to derive optimal amount of N from symbiotic fixation (Ahmad et al., 1981b; Mulongoy et al., 1981). Rhizobia were collected and characterized on the basis of differences in colony morphology, ability to utilize various carbon sources, tolerance to high levels of salts and high temperatures, chlorosis inducement, intrinsic antibiotic resistance, formation of dark nodules, host-range promiscuity, effectiveness, enzyme-linked immunosorbent assay (ELISA) and polyacrylamide gel electrophoresis (PAGE).

Many characteristics of the West African rhizobia correlated with the colony morphology type in yeast-extract mannitol (YEM) (Vincent, 1970) medium (Table 1). 'Wet' strains form flat colonies, translucent, with copious extracellular polysaccharides. The colonies of 'dry' strains are small (<2 mm diameter), punctiform, with raised elevation and not confluent. Dry-colony forming isolates produce dark nodules (Mulongoy et al., 1981; Eaglesham et al., 1982). PAGE (Stowers et al., in preparation) and ELISA (Ahmad et al., 1981a) data showed that wet and dry strains were fundamentally different, with higher homology between isolates of the same colony type from different locations than between isolates of different colony types from the same location. Wet strains were found more tolerant to 0.5 per cent NaCl

Table 1. Number of rhizobial isolates collected in West Africa, and forming dry or wet colonies on yeast-extract mannitol, after 7–10 days of incubation at 30 °C (Mulongoy et al., 1982)

Nodule origin	Colony type	
	Dry	Wet
Maradi, Niger	287	9
Onne, Nigeria	84	187
IITA, Nigeria	32	238

(Eaglesham and Ayanaba, 1984) and more resistant to antibiotics than dry strains (Sinclair and Eaglesham, 1984). All rhizobia isolated from Maradi grew at 37°C in YEM, and more than 90 per cent of these strains grew also at 40°C, whereas the rhizobia from IITA and Onne, in the humid zone, showed little or no growth at these temperatures (Eaglesham et al., 1981; Eaglesham and Ayanaba, 1984). Ability to tolerate elevated temperatures is a desirable property of native rhizobia and strains to be used as inoculant in the tropics where temperatures during transportation, storage and planting are usually high. These observations suggest that the wet and dry colony morphology can be used to screen large collections of cowpea rhizobia.

ENVIRONMENTAL STRESSES

Tropical soils are constantly subjected to stresses affecting the rhizobia, the host plant, nodulation and symbiotic effectiveness. In West African savannas where most of the world's cowpea is produced, the cowpea–*Rhizobium* symbiosis must possess some tolerance to high temperatures and drought. In the surface layers of the soil in this region, temperatures often exceed 40°C and can reach 60°C during dry periods (Day et al., 1978). The detrimental effects of desiccation and high temperatures on cowpea *Rhizobium* survival in soils, and the differences among *Rhizobium* strains in their susceptibility to these factors were described recently by several investigators (Van Rensburg and Strijdom, 1980; Boonkerd and Weaver, 1982; Osa-Afiana and Alexander, 1982; Hartel and Alexander, 1984); their findings confirmed earlier observations in laboratory media (Eaglesham et al., 1981; Hartel and Alexander, 1984). Cowpea rhizobia from hot, dry areas were more tolerant to high temperatures and drought than were strains from cooler, more humid regions, in both sand and other soils. Rhizobial tolerance of elevated temperatures varies with soil structure. The more sandy the soil is (Mahler and Wollum, 1981), and the higher the water tension (Boonkerd and Weaver, 1982), the more adverse are the effects of high temperature. Nodule formation and development decrease, and the proportion of ineffective and abnormal nodules increases, at elevated temperatures probably because of the changes in root-hair conformation at the time of infection by the rhizobia (Day et al., 1978). Mulching may improve nodule formation, as it limits soil-temperature fluctuations (Lal, 1974). After canopy closure, soil temperatures are unlikely to exceed 35°C and should not limit the formation and function of cowpea nodules.

Acid soils contain high concentrations of hydrogen ions and free Al. They are low in Ca and available P and can be deficient in molybdenum, with phytotoxic levels of manganese. Ayanaba et al. (1983) have developed an agar plate method for rapid screening of *Rhizobium* for tolerance to acid–aluminium stress. Cowpea rhizobia as well as the host plant exhibit various degrees of tolerance to soil acidity (Edwards *et al.*, 1981; Hartel and Alexander, 1984). Cowpea nodulation is generally reduced in acid, aluminium-rich soils where even tolerant strains fail to infect root hairs. However cowpeas grown in acidic, high-aluminium soils produced more nodules and dry matter when they were inoculated with strains of *Rhizobium* identified in liquid media as tolerant of pH 4.5 and 50 μM Al (Keyser *et al.*, 1979). Other factors such as Mn toxicity may also be involved in reducing cowpea nodulation at low pH. Liming can correct the acidity and aluminium stresses.

Legumes have a high-P requirement for nodule development and optimal plant growth. For example, the response of 10 cowpea cultivars to urea (150 kg N/ha) and single superphosphate (30 kg P/ha) was studied in the poor soils of Maradi (soil N and P: 0.04 per cent and 10.6 ppm respectively); shoot weight and grain yield increases were superior in plots fertilized with P (Table 2). Plants were heavily infected with vesicular–arbuscular (VA) mycorrhizal fungi, with great variability among cowpea cultivars. At IITA and Onne (unpublished data), mycorrhizal infections were fewer. Reports have shown that nodulation, N_2 fixation, utilization of rock phosphate and growth of cowpea were improved after inoculation with VA-mycorrhizal fungi (Godse *et al.*, 1978; Bagyaraj and Manjunath, 1980; Islam and Ayanaba, 1981).

When the soils become waterlogged during and after periods of high rainfall, and in lowlands where rotation with rice is the common practice, the potential of a *Rhizobium* strain to survive depends on whether it possesses a dissimilatory nitrate reductase utilizing nitrate as an electron acceptor (Daniel *et al.*, 1982). The enzyme has been detected in cowpea rhizobia, in both free-living and bacteroid form (Zablotowicz and Focht, 1979). The anaero-

Table 2. Effect of urea and single superphosphate on mycorrhizal infection, fresh shoot weight, and grain yield of cowpeas at Maradi, Niger (Mulongoy and Ayanaba, unpublished data)*

Fertilization (kg/ha)		Mycorrhizal infection (%)	Fresh shoot weight (g/10 plants)	Grain yield (g/10 plants)
N	P			
0	0	31	390	63
150	0	44	600	69
0	30	29	690	108
150	30	40	1980	206
LSD at 5%		12	30	41

* Means of 10 cowpea cultivars.

biosis in waterlogged soils reduces plant root growth, nodule formation and development, as well as nitrogen fixation (Minchin and Pate, 1975). Although the root nodules rapidly recover their N_2-fixing ability when moisture excess is corrected (Hong et al., 1977), they may be adversely affected by predatory protozoa, lytic bacteria and phages that are favoured by waterlogged conditions, organic acids, ethanol and ethylene production.

RESPONSE OF FIELD-GROWN COWPEAS TO INOCULATION

Results in pot experiments indicated that inoculation of cowpeas with *Rhizobium* stimulated effective nodulation and increased dry-matter production, seed yield, crude protein and N content (Ezedinma, 1964b; Rotimi, 1972; Deshmukh and Joshi, 1973). However, little or no response of cowpea to inoculation with *Rhizobium* has been reported from limited field experiments (Ezedinma, 1964b; Rotimi, 1972). The lack of a positive response may have resulted from the unnecessary inoculation of cultivars capable of effective nodulation with native rhizobia; the use, as inocula, of strains having poor effectiveness, persistence, competitiveness or nodulating ability; or uncontrolled environmental constraints.

To solve some of these problems, microbiologists at IITA, Nigeria, and Boyce Thompson Institute (BTI), at Cornell University, USA, screened more than 500 strains of rhizobia from a diverse cowpea germ plasm. In 1981 the 13 best isolates selected on the basis of their ability to grow at elevated temperatures and for their superior performance in pot experiments were tested as inoculants in West Africa, at three locations where the soils contain few cowpea rhizobia, and the maximum air temperatures average 40°C. In these field trials the proportion of nodules formed by introduced strains was low in general (IITA, 1982). Up to 95 per cent yield increases were recorded, however, in plots inoculated with strains IRc 252 (wet), IRc 430A (dry) and IRc 500A (wet) from Onne, Maradi and IITA respectively. These rhizobia were tested again in 1982 at Samaru, northern Nigeria; Maradi, Niger; and Kamboinse, Burkina Faso. Cowpea VITA 7, an IITA-recommended variety, was grown along with local checks (Table 3). Nodulation was poor, and the proportion of nodules formed by introduced rhizobia was low, although at Kamboinse, where soil moisture was not limited, plants inoculated with *Rhizobium* IRc 252 had 90 per cent of their nodules formed by this strain (Table 3). Probably most important, inoculation did not increase grain yields. Similar results were obtained, in 1983, by collaborators testing the same rhizobial strains in Latin America and Asia.

The poor survival and competitive ability of introduced strains raise doubts about the relevance of the methods of strain selection in controlled environ-

Table 3.   Effect of N fertilizer and inoculation with *Rhizobium* on cowpea at three locations in West Africa (IITA, 1983)*

	Nigeria Samaru		Niger Maradi		Burkina Faso Kamboinse	
	Vita 7	IAR 48	VITA 7	TN 88-63	VITA 7	SUVITA-2
Nodule dry weight (mg/plant)						
Uninoculated	156	106	88	150	156	156
N fertilized	38a	47	44	88a	25a	38a
IRc 252	69a	72	88	88a	125	144a
IRc 430A	106	106	38	106	88	166a
Proportion of nodules formed by inoculant strain (%)						
IRc 252	41	7	43	10	89	90
IRc 430A	25	15	25	40	59	69
Grain yield of cowpeas (t/ha)						
Uninoculated	1.09	0.61	0.33	1.04	2.00	1.98
N fertilized	0.98	0.64	0.54	1.09	1.98	1.90
IRc 252	1.05	0.50	0.36	0.93	2.00	1.72
IRc 430A	0.83	0.57	0.36	1.18	2.05	1.72

* IRc = *Rhizobium* strain; a = significantly different from uninoculated control at 0.05 level.

ments. Screening of cowpea strains for inoculant production would be more appropriate at the location of inoculant use.

Yield increases have been recorded in other experiments after inoculation of field-grown cowpeas. The inconsistent results among seasons and locations when the same cowpeas and rhizobia were used (Tables 3 and 4) justify extensive work to assess and ensure the survival and performance of inoculant strains in soils.

Also, *Rhizobium* populations should be monitored in the seasons following initial introduction because strains surviving initial stresses may multiply in the presence of the host and eventually dominate the native populations. Marcarian (1982) suggested that the granular inoculant could improve the survival of rhizobia and thus nodulation by introduced strains in the dry and hot areas, but little research has been done to formulate appropriate inoculant forms and methods of placement. At present, with the identification of various suitable carriers for inoculant production and use in Africa, legume inoculation is becoming popular, particularly in Egypt and Eastern Africa.

Table 4. Positive response of field-grown cowpeas to inoculation with *Rhizobium**

Cowpea cultivar	*Rhizobium* strain	Grain yield		Location (reference)
		kg/ha	% increase	
VITA 6	Vi F2	1200	26	Fashola, Nigeria (Mafuka, 1984)
VITA 6	Vi F2	1660	68	IITA, Nigeria (Mafuka, 1984)
TN 88-63	IRc 252	1970	95	Maradi, Niger (IITA, 1983)
TN 88-63	IRc 334A	1400	39	Maradi, Niger (IITA, 1983)
TN 88-63	IRc 400A	1680	66	Maradi, Niger (IITA, 1983)
TN 88-63	IRc 430	1430	42	Maradi, Niger (IITA, 1983)
n.s.	n.s.	1950	16	Trinidad (Mughogho et al., 1982)
PLS 370	AH-6	211	20	Coimbatore, India (Balasubramanian et al., 1980)
PLS 370	CMBS-1	203	15	Coimbatore, India (Balasubramanian et al., 1980)
PLS 370	Multistrain	253	44	Coimbatore, India (Balasubramanian et al., 1980)
Mississippi Silver	CB 756	5950	27	Starkville, USA (Kremer and Peterson, 1983)
Mississippi Silver	CV 030	5900	26	Starkville, USA (Kremer and Peterson, 1983)
Mississippi Silver	CV 041	5670	21	Starkville, USA (Kremer and Peterson, 1983)

* Seed yields, except for Starkville—shelled peas; percentage increase in grain yield over the uninoculated control, significant at 0.05 level; n.s. = not specified.

NITROGEN FIXATION AND CONTRIBUTION IN CROPPING SYSTEMS

A judicious breeding programme conducted in low-N soils to select for high-yielding cowpea varieties could simultaneously select for high N_2-fixing ability.

The 'difference' method described by Eaglesham *et al.* (1982) proved effective, simple and inexpensive in comparison with the acetylene reduction assay and the 'A' value method. It can be recommended for screening large collections of cowpea germ plasm once an adequate non-fixing control has been identified.

At IITA, cowpea VITA 3, an indeterminate variety from Kenya, fixed 124 kg N/ha, almost three times the N_2 fixed by TVu 4552, a determinate cowpea from Nigeria. Eaglesham *et al.* (1982) concluded that late-maturing cowpea cultivars increase the available soil N after seeds are harvested and plant residues returned to the soil and that early varieties, because of their smaller potential for N_2 fixation, were not likely to contribute N. When he assessed N contribution of cowpeas in a poor soil (0.073 per cent N) at IITA, Eaglesham (1982) found a positive balance of 2–52 kg N/ha when a starter dose of ammonium sulphate (25 kg N/ha) was applied to late- and medium-maturing cowpeas. With N fertilizer (100 kg/ha), nodule numbers, weights and activities were reduced, and the amount of N_2 fixed was 50 per cent less. In these conditions cowpeas removed up to 34 kg N/ha from the soil and fertilizer-N pool (Eaglesham, 1982).

All legumes can deplete soil N unless their nodules fix a lot of N_2 and their harvest index for N is low. In West Africa farmers often utilize vegetative parts of cowpeas for cattle fodder. Thus, only the roots and the nodules, containing, at best, about 6 per cent of plant N (Minchin and Summerfield, 1978) account for the residual N. These farming systems need cowpea cultivars with high N_2-fixing ability and little or no dependence on soil N for growth.

In the tropics, where cowpea is predominantly intercropped with cereals, positive interactions between the crops have been documented (Eaglesham *et al.*, 1978; Remison, 1978). In soils low in N, cowpea roots may excrete N for the benefit of the companion crop. Cowpea litter, soluble leaf N, and N from decaying nodules could also account for the N transferred to the cereal.

Subsequent cereal crops also benefit from the N contributed to the soil by cowpeas. Stoop and Van Staveren (1982) found that when sorghum followed cowpea at Kamboinse, yields were 320 kg/ha higher than when sorghum followed millet. Also, Mughogho *et al.* (1982) calculated yield increases of subsequent maize crops resulting from incorporation of cowpea residues to be equivalent to the yields in plots fertilized with 40–80 kg N/ha in Trinidad. In the latter experiment inoculation of cowpea with *Rhizobium* improved the residual effect of the legume on maize yield.

Cowpea Research, Production and Utilization
Edited by S. R. Singh and K. O. Rachie
© 1985 John Wiley & Sons Ltd

24

Selection for Enhanced Nitrogen Fixation in Cowpea

J. C. MILLER, JR. and G. C. J. FERNANDEZ
Department of Horticultural Sciences, Texas A&M University, College Station, Texas, USA

Biological nitrogen fixation in cowpea, as in other legumes, is a process whereby *Rhizobium* spp., in association with the host plant, reduce free nitrogen from the soil air to a form usable by the plant. The role played by the plant in the symbiosis is to furnish carbohydrates that supply energy to the nitrogen-fixing bacteria, while nitrogen compounds produced by the bacteria are used by the plant for growth. Obviously only a small, but sometimes significant, portion of the carbohydrates produced by the host plant is used to supply the energy requirements of the bacteria. Likewise, not all nitrogen fixed by the bacteria is made available to the host plant for its use. Nitrogen may pass into the soil by excretion, by the sloughing off of the roots, especially their nodules, or by the decomposition of the cowpea plant at the end of the season. Thus, crops grown in association with cowpea or succeeding crops may benefit from the fixed nitrogen.

There is general agreement that cowpea yields could be increased through proper plant nutrition, especially nitrogen, if other factors such as disease and insect incidence were controlled through plant resistance or other means. There is, however, disagreement as to precisely how much of the cowpea plant's nitrogen requirement could be met through biological nitrogen fixation alone. Many factors, such as plant genotype, quality and quantity of available rhizobia and numerous environmental factors such as available soil nitrogen, extreme temperatures, low pH, lack or excess of soil moisture, etc., influence the biological nitrogen fixation process. The important point is that biological nitrogen fixation could contribute more to cowpea yield than it is at present if it were being fully exploited. For example, we found that, on low nitrogen soils (<15 kg/ha), the benefits in seed yields derived from rhizobial

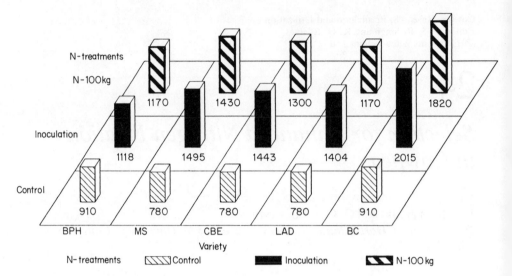

Figure 1.   Effects of inoculation vs 100 kg/ha nitrogen on seed yield (kg/ha) in cowpea
varieties 'Bush Purple Hull' (BPH), 'Mississippi Silver' (MS), 'California Black Eye
No. 5' (CBC), 'Lady' (LAD) and 'Brown Crowder' (BC)

inoculation compared with noninoculation were equivalent to at least 100
kg/ha of fertilizer nitrogen (Figure 1).

THE NITROGEN FIXATION SYMBIOSIS

Several steps need to be successfully completed before effective biological
nitrogen fixation can occur. The process begins on the root surface and follows
a multistage sequence of interdependent steps culminating in the establishment
of an effective N_2-fixing nodule. Briefly stated, this series of events starts with
the bacteria contacting the surface of the root hair, perhaps with lectin-assisted
recognition, curling of the root hair, and penetration of the root by the
bacteria. After entering, the bacteria form infection threads through which
they migrate to the root cortex, the first layer of cells under the epidermis.
Nodules are then formed as the result of plant-cell division and cell
enlargement provoked by the bacteria. The bacteria then increase in the
nodule tissue and develop bacteroids, which have the capacity to fix nitrogen.
    To breed for increased biological nitrogen fixation in cowpea, or any other
legume species for that matter, one needs to have some understanding of the
nature of the symbiosis. The events in the symbiotic process can be grouped
in a number of ways to conform with three major steps that are commonly
agreed upon, namely: preinfection, infection and nodule development and
nodule function (Vincent, 1980) (Table 1). This sequence of events is

Table 1. Analysis of symbiotic sequence in biological nitrogen fixation (Vincent, 1980)

Stage	Abridged description	Phenotypic code
Preinfection		
Multiplication on root surface ('rhizoplane')	Root colonization	Roc
Attachment to root surface	Root adhesion	Roa
Branching of root hairs	Hair branching	Hab
'Marked' curling of root hairs	Hair curling	Hac
Infection and nodule formation		
Formation of infection thread	Infection	Inf
Development of polyploid (disomatic) meristem, nodule development and differentiation	Nodule initiation	Noi
'Intracellular' release of rhizobia from infection thread	Bacterial release	Bar
'Intracellular' multiplication of rhizobia and development of full bacteroid form	Bacteroid development	Bad
Nodule function		
Reduction of N_2 to NH_4^+ (nitrogenase)	Nitrogen fixation	Nif
Complementary biochemical and physiologic functions	Complementary functions	Cof
Persistence of nodule function	Nodule persistence	Nop

complex as well as interrelated and interdependent. Furthermore, attempts by plant breeders to define the genetic, physiologic or environmental control of symbiosis in such general terms as nodule number (invasion, nodule initiation or nodulation) and nitrogenase activity (nodule effectiveness) leave them open to criticism by microbiologists.

Breeders need to recognize that looking at components such as nodule number is examining only one of several components that contribute to infection and nodule formation. Likewise, when looking at nitrogenase activity, they are dealing with only one facet of the broader component of nodule function. At the same time the microbiologist must recognize the difficulty in breeding for (i.e. evaluating, indexing) a trait that cannot be seen without a microscope. Also the way to improve the whole is by improving its component parts, and the parts that are easiest to deal with from a breeding standpoint will probably be the first order of business for the breeder.

IMPROVING THE SYMBIOSIS

Interactions between the host plant and *Rhizobium* are complex, and made even more complicated by the influence of environment. Little can be accomplished by debating whether host plant, microbial symbiont or environment is most important, as cooperation among all three is essential. In fact, for symbiosis at maximal levels of efficiency, all three factors must mesh perfectly.

Rhizobial approach to enhancing biological nitrogen fixation

Historically, the study and enhancement of biological nitrogen fixation have centred on the microsymbiont, with efforts to increase nitrogen fixation efficiency channelled primarily toward the development of improved strains of *Rhizobium*. These improved strains have not always had an impact on cowpea production. The introduction of new strains into soil where indigenous strains are well established is frequently unproductive. The introduced strains are usually not as competitive as the indigenous strains for survival in that soil and to effect good nodulation must often be introduced at uneconomic levels (LaRue, 1980).

It is frequently proposed that superior or improved strains of *Rhizobium* be provided to farmers—an approach that can and does work under certain circumstances. However, the cost of implementing such a proposal can be high, when the establishment of an inoculant factory and distribution of the inoculant to farmers are considered. Additional obstacles to this approach include the fact that rhizobia are quite sensitive to desiccation and heat, and developing countries lacking an adequate transportation network may find distribution of viable inoculant difficult or impossible. An alternative approach is the development of cowpea cultivars capable of enhanced, effective symbiosis with indigenous rhizobia.

Host plant genetics and enhancing biological nitrogen fixation

Historically, those attempting to increase fixation have concentrated on the *Rhizobium*, ignoring the contribution of the host plant. The most notable exception is P. S. Nutman (now retired) of the Rothamsted Experimental Station who, for almost 40 years, investigated the plant genetics of nodule formation in clover and expounded on the need to study the host plant as well as the microsymbiont.

The host plant is involved in determining when and which strains infect the root, the extent of nodulation, and the potential for N_2 fixation (Graham, 1982). In fact, the host can control all steps of the symbiotic sequence (Table 1). It is only speculative and probably unproductive to suggest the number of host genes that contribute to such a complex process, but many must be involved. The identification and manipulation of this plant genetic material is essential if the potential contribution of biological nitrogen fixation is to be realized.

Plant breeders have rarely consciously selected for biological nitrogen fixation, and the N_2-fixing potential of new lines has seldom been evaluated and reported. Only in the past 10 years have they begun to recognize fully the contribution of the host genotype and the potential for its exploitation in maximizing the benefits from symbiosis. However, the research community working on biological fixation now includes a small number of plant physiologists and plant breeders in addition to the microbiology core. There are now a number of programmes throughout the world that are attempting to develop legume cultivars with improved nitrogen-fixation potential.

ENHANCING BIOLOGICAL NITROGEN FIXATION IN COWPEA

The Texas A&M University programme to improve nitrogen fixation in cowpea was established in 1976. The programme has an active microbiology component; however, most of the work has centred on plant physiology and genetics as they relate to nitrogen fixation in cowpea. The programme was initiated with the development of a screening procedure for detecting quantitative differences in nitrogenase activity, nodule number, nodule weight and plant top weight, while keeping rhizobial strains constant. To date, more than 1400 genotypes have been screened, and extremely wide variability for all traits has been found (Zary et al., 1978). However, non-nodulating cowpea genotypes have yet to be identified with this procedure.

Numerous studies have been conducted, including the effect of viruses, mycorrhizal fungi, soil nitrogen (including nitrogen uptake and partition), soil phosphorus, plant density (population), water and temperature stress, and grafting to determine the effect of root stock and scion on nitrogen fixation in cowpea (Zary and Miller, 1980; O'Hair and Miller, 1982). In addition, rather extensive investigations on the genetics of the host plant have been, and are currently being, carried out. For example, investigations on the genetics of quantitative differences in nodule number (nodulation or nodule initiation) and nitrogenase activity (nodule function) resulted in heritability estimates of 0.55 and 0.62, respectively, indicating that progress should be significant in breeding for enhanced biological nitrogen fixation in cowpea by improving these two components (Tables 2 and 3).

Table 2.   Variance components and heritability estimation on log-transformed data in a cross between cowpea varieties Brown Crowder (P_1) and Bush Purple Hull (P_2) for nodules/plant*

Generation	N	σ^2	Three-parameter model		
			Observed means	Expected means	Residual
P_1	20	0.0169	1.292	1.313	−0.021
P_2	20	0.0096	1.590	1.594	−0.004
F_1	20	0.0118	1.520	1.549	−0.029
$F_{1'}$	20	0.0135	1.560	1.549	0.010
BC_1	80	0.0202	1.483	1.431	0.052
BC_2	80	0.0234	1.592	1.571	0.021
F_2	240	0.0300	1.501	1.501	0.000

* Variance components include additive variance 0.0329, dominance variance 0.00239, environmental variance 0.0129, degree of dominance 0.27 and narrow sense heritability 0.55. The chi-square value is 0.34, indicating that the three-parameter model is adequate.

Table 3.   Variance components and heritability estimation on log-transformed data in a cross between cowpea varieties Brown Crowder (P_1) and Bush Purple Hull (P_2) for nitrogenase activity*

Generation	N	σ^2	Three-parameter model		
			Observed means	Expected means	Residual
P_1	20	0.0139	0.744	0.735	−0.009
P_2	20	0.0196	0.285	0.302	−0.017
F_1	20	0.0190	0.648	0.641	−0.007
$F_{1'}$	20	0.0209	0.629	0.641	0.012
BC_1	80	0.0387	0.627	0.688	−0.061
BC_2	80	0.0429	0.542	0.472	0.070
F_2	240	0.0593	0.589	0.580	0.009

* Variance components include additive variance 0.0738, dominance variance 0.0162, environmental variance 0.0183, degree of dominance 0.47 and narrow sense heritability 0.62. The chi-square value is 0.32, indicating that the three-parameter model is adequate.

Cowpea is probably one of the best candidates for plant breeding attention in the area of nitrogen fixation. It could certainly be classified as a poor person's crop, worldwide and, as such, the likelihood that it will be grown under optimal nutritional conditions, especially with expensive fertilizer nitrogen, is negligible. Therefore, attention should be directed and focused on meeting nutritional needs for maximal seed and biomass yield, through full exploitation of biological nitrogen fixation. As it is difficult or impossible to control the environment, attention should be directed to the microbial symbiont and the host plant. The most promising approach would be that of developing host plants with superior biological nitrogen fixation characteristics.

With the identification of biological nitrogen fixation components that can be seen and worked with, variability for these factors indexed and found to be substantial, and encouraging heritability estimates, a breeding strategy was then formulated. However, at least two important and rather unique considerations had to be kept in mind in establishing the methodology. First, superior levels of biological nitrogen fixation cannot be seen by the breeder through casual observation and, secondly, measurement of relative levels of fixation usually requires destruction of the cowpea plant, obviously leading to a non-productive dead end. These problems can be overcome if breeders take cuttings before, or in conjunction with, whatever procedure is used to measure a given nitrogen fixation component. Dealing with large numbers of plants makes this approach cumbersome, time-consuming and, consequently, expensive.

With these limitations and restraints in mind we propose two breeding approaches. The first, and by far the easiest and at the same time least likely to meet with success, simply involves the selection of parents from genotypes that have been identified as possessing high nitrogen-fixation potential. In this case, whether the pedigree, population improvement, or any other method is followed, no monitoring of progeny for nitrogen-fixing potential throughout successive generations is practised. Breeders would depend on the relatively high heritabilities for the various nitrogen-fixation components to ensure that these desirable traits are passed forward. For example, they might follow this approach in a programme for disease resistance. In selecting parents, breeders might choose lines that, in addition to possessing resistance to the disease, also possess high nitrogen-fixation potential.

The second method is a modification of that proposed for common bean by McFerson *et al.* (1982) and involves a continual assessment that is not directly destructive and does not involve cuttings. We describe this approach as a modified backcross–inbred method and believe that it offers a high probability for success. Cowpea lines possessing high N_2-fixation potential (donor parent) are crossed to an adapted, high-yielding cultivar (recurrent parent), and approximately 30 F_1 hybrid seeds are obtained (Figure 2). These seeds are grown in the greenhouse and crossed to the recurrent parent and approximately 100 backcross (BC) seeds are obtained. Each BC seed is a different genotype, as the inheritance of N_2 fixation is quantitative. Therefore, the selfed offspring from each BC seed constitutes a unique family.

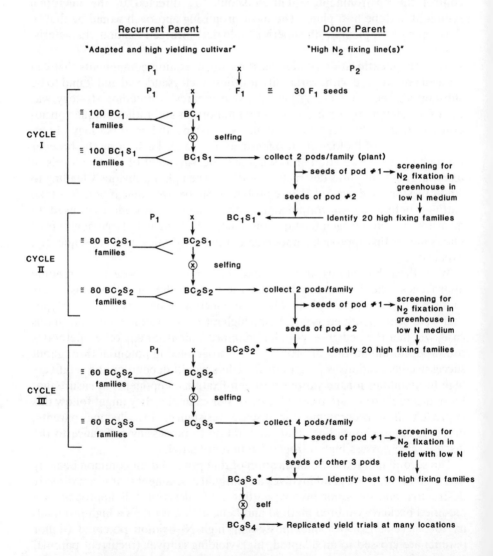

Figure 2.   Modified backcross–inbred method to introduce high N_2 fixing potential into an adapted and high-yielding cultivar

Seeds of the first backcross (BC_1) are grown and selfed (S) in the greenhouse. From each BC_1 family, two pods are collected and the selfed progeny identified as $BC_1 S_1$. Seeds from one pod are grown in a nitrogen-free medium in the greenhouse, and the BC_1 families evaluated for N_2-fixation potential based on the mean performance of the selfed progeny. Seeds of 20 high N_2-fixing families ($BC_1 S_1$) from the second pod are grown in the greenhouse and advanced to the second cycle. Thus, one cycle involves a backcross, selfing, evaluation for N_2 fixation, and identification of high-fixing families.

This cycle is repeated twice, and in cycle II, about 80 $BC_2 S_2$ families are evaluated and 20 $BC_2 S_2$* high fixing families identified. In cycle III, 60 $BC_3 S_3$ families are evaluated in the field, with low nitrogen. Four pods are collected from each family and one pod used for evaluation, while the other three pods are used for seed multiplication. The best 10 high-fixing families ($BC_3 S_3$*) are selected and seeds multiplied in the field by selfing. The $BC_3 S_4$* families are evaluated for yield and N_2-fixation potential in many locations. This method comprises three alternating cycles of BC and S, then one final selfing. Therefore, at the end of this breeding procedure, the selected plants will have achieved greater than 98 per cent homozygosity. This modified backcross–inbred method offers several advantages for the introduction of superior N_2-fixation potential into an adapted cultivar:

○ Only the high N_2-fixing families are carried forward to the final selection; therefore the probability of success is high.

○ Evaluation of a BC family from the mean performance of the selfed progenies facilitates screening for quantitative traits such as N_2-fixation potential.

○ The tedium and expense of taking cuttings and growing them out are eliminated.

○ The destructive C_2H_2–C_2H_4 assay is not a problem since screening is carried out on the selfed progenies from seed of an extra pod harvested for this purpose.

○ The number of crosses required is reduced because with three BC and four S more than 98 per cent homozygosity is attained.

CONCLUSION

In conclusion, biological nitrogen fixation in cowpea depends on a complex interaction between host plant, microbial symbiont and environment. Microbial symbiont as well as host-determined factors influence nodule initiation, development and function. Both partners of the symbiosis are subject to genetic variation; thus, substantial variation exists in type of nodulation, amount of nitrogen fixed and, ultimately, cowpea yield. Evidence is that the genetic manipulation of the host plant offers the greatest potential for improvement of current levels of biological nitrogen fixation in cowpea.

Cowpea Research, Production and Utilization
Edited by S. R. Singh and K. O. Rachie
© 1985 John Wiley & Sons Ltd

25

Weed Management Designed for Smallholdings

J. A. POKU and I. O. AKOBUNDU
International Institute of Tropical Agriculture, Ibadan, Nigeria

Weeds associated with cowpea differ in flora and intensity from one region to another. Their distribution is affected by factors that include cropping systems, method of weed control and cropping frequency. Some of the commonly observed weeds in cowpea fields are *Synedrella nodiflora, Talinum triangulare, Acanthospermum hispidum, Amaranthus* spp., *Commelina benghalensis, Brachiaria* spp., *Digitaria* spp., *Cynodon dactylon, Paspalum* spp. and *Eleusine indica*. Although these and many other dryland weeds may occur in cowpea fields, some weeds are known to be troublesome in cowpea either because they are hard to kill or they seriously interfere with cowpea production. These weeds include wild poinsettia (*Euphorbia heterophylla*), iron weed (*Vernonia galamensis*) and *Striga gesnerioides*.

Research on weed control in cowpea in the tropics is scanty, but the few studies that have been conducted show that uncontrolled weed growth accounts for 40–81 per cent reductions in grain (Moody, 1973; Nangju, 1980; Akobundu, 1982; Lagoke *et al.*, 1982; Poku and Akobundu, 1983). Weeds have also been found to increase insect pest damage in cowpea, probably because the weeds reduce the effectiveness of insecticide applications and provide a favourable environment for the insects (Moody and Whitney, 1974).

This paper examines research on weed control in cowpea with particular reference to the tropics where the bulk of the crop is grown.

WEED INTERFERENCE

The critical period of weed interference in cowpea is influenced by the competing weed species, the cultivar, plant density and environmental factors including light, water, nutrients and allelopathy. In a study conducted in a subhumid climate with bimodal rainfall, the impact of weed interference was reported to be more severe in late-season cowpea than in the early-season crop because of limited available moisture (Ayeni, 1982). Among the four cowpea cultivars examined, the yield reductions due to weeds were 25 per cent for VITA 1, 33 per cent for VITA 5, 46 per cent for ER-1 and 54 per cent for TVx 33-1G (Nangju, 1980). The competitive ability of VITA 1 and VITA 5 was associated with their high leaf area index and canopy height. The nature of weed interference strongly influences the choice of weed-management practices. Weed competition is most critical during the first 20–40 days of cowpea growth (Medrano *et al.*, 1973; Moody, 1973; Fadayomi, 1979b). Recommended production practices should therefore minimize weed interference during this growth phase of the crop. Yield losses are negligible if the crop is hand-weeded twice within 5–6 weeks of crop emergence (Moody, 1973; Enyi, 1975). In other words, farmers can obtain maximal yields by weeding early in the growth cycle of the crop, but they seldom have the time because of other demands.

CONTROL MEASURES

Methods for weed control are cultural, biological and chemical. Each has been shown to minimize weed competition in cowpea, but the choice of the best method for individual farmers depends on factors that include farm size and other socioeconomic considerations.

Cultural methods include those farmers employ to remove growing weeds physically from the crop environment; the energy required may be provided by humans, animals or machinery (Akobundu, in press). Hand-weeding is the most common practice among small-scale producers. Repeated hand-weeding is effective against weeds like itchgrass (*Rottboellia exaltata*) and wild poinsettia. Hand-weeding is, however, expensive and not adapted to large-scale operations in most parts of the tropics.

Crop rotation can be effective in preventing build-up and spread of weeds that are difficult to control with other measures, such as wild poinsettia in cowpea. A rotation involving maize is particularly effective because atrazine or 2,4-D can be used with maize for control of the poinsettia.

Biological control technology has potential in cowpea production; it is the control or suppression of weeds by natural or introduced enemies, or by the manipulation of the weeds, organism, or environment (Anon, 1983). Preliminary investigations were made in Papua New Guinea (Bennett *et al.*, 1970) into the possibilities of biological control of wild poinsettia, a serious weed problem in cowpea, but results have not so far been published. Some biological control studies on wild poinsettia have also been initiated in Brazil. Although classic biological control involving the use of insects for weed control has been used successfully to control prickly pear (*Opuntia* spp.), St. Johnswort (*Hypericum perforatum* L.) and more recently alligator weed (*Alternanthera philoxeroides* (Mart) Griseb) (Crafts and Robbins, 1962; Andres, 1981), it has little relevance in broad-spectrum weed control in cowpea because no insect is known to feed on a cross-section of cowpea weeds without attacking the crop. Bioherbicides, a recent innovation in biological weed control, are now on the market (Templeton, 1982). The fungus *Phytophthora palmivora* Butler is recommended for control of strangler vine (*Morrenia odorata* Lindl) in citrus, and *Colletotrichum gloeosporioides* f. sp. *aeschynomene* (CGA) for control of northern jointvetch (*Aeschynomene virginica* (L) B.S.P.) in rice and soybean. Acifluorfen can be tank-mixed with CGA to control hemp sesbania (*Sesbania exaltata* (Raf) Cory) in soybean without loss of biological activity. Another notable innovation is a granular formulation of *Alternaria macrospora* Zimm. applied to control spurred anoda (*Anoda cristata* Schlecht) in cotton (Walker, 1981).

Chemical weed control combined with other cultural practices may be practical in reducing weed competition, crop losses and labour costs. With increasing demand for cowpea, opportunities exist for large-scale production, which would be highly dependent on chemical weed control. Much of the present research on weed control in cowpea is directed toward finding suitable pre-emergence, post-emergence and pre-plant herbicides. Ideally, the pre-plant or pre-emergence herbicides should maintain relatively weed-free conditions within the first 6 weeks of cowpea growth when the canopy is developing. This is hard because high temperatures and humidity favour fast dissipation of herbicides and rapid, excessive weed growth. Therefore, herbicide applications should be supplemented later by other methods of weed control.

Several herbicides have been identified for weed control in cowpea (Table 1). Excellent control of most annual grasses in cowpea has been obtained with pre-emergence applications of metolachlor, pendimethalin and DCPA (Braithwaite, 1978; Talbert *et al.*, 1979; Akobundu, 1982; Lagoke *et al.*, 1982; Poku and Akobundu, 1983, 1984), and post-emergence applications of fluazifop-butyl and sethoxydim (IITA, 1983a). Itchgrass was effectively controlled with pendimethalin (Akobundu, 1982) and fluazifop-butyl (IITA, 1983a) applied pre-emergence and post-emergence, respectively. Excellent control of *C. dactylon* has also been observed with fluazifop-butyl. Alachlor was phytotoxic to cowpea in the forest transition zone (Moody, 1973; Akobundu, 1982) while found safe on the crop at a higher rate in the southern guinea savanna zone of Nigeria and in India (Singh *et al.*, 1975; Lagoke *et al.*, 1982). The results may reflect varietal response, which has been reported for use of alachlor with cowpea (Akobundu, 1979). Chloramben and a formulated mixture of metobromuron plus metolachlor (Galex®) have provided good control of most annual broadleaf weeds in cowpea when used as

Table 1. Herbicides for weed control in cowpea (Akobundu, 1983b)*

Herbicides	Time of application
Fluchloralin	Pre-plant incorporated
Trifluralin	Pre-plant incorporated
Metolachlor	Pre-emergence
Pendimethalin	Pre-emergence
Alachlor	Pre-emergence
DCPA	Pre-emergence
Chloramben	Pre-emergence
Metobromuron, metolachlor	Pre-emergence
Linuron	Pre-emergence
Prometryne, metolachlor	Pre-emergence
Fluorodifen	Pre-emergence
Fluazifop-butyl	Post-emergence
Sethoxydim	Post-emergence
Paraquat	Post-emergence (as directed)

* Rates of herbicide vary with soil type, rainfall and weed problems.

pre-emergence herbicides (Glaze, 1970; Braithwaite, 1978; Poku and Ako-bundu, 1983, 1984). Incorporating herbicides during land preparation is not likely in the tropics, partly because of erosion hazards and partly because few farmers have the tools for herbicide incorporation.

In the southern guinea zone of Nigeria where *V. galamensis* poses a problem, we (Poku and Akobundu, 1983) found that a tank-mix of chloramben and metolachlor provided satisfactory control. Wild poinsettia is resistant to most of the herbicides used for broadleaf weed control in cowpea. A tank-mix of metribuzin and metolachlor was found to suppress it long enough to minimize yield reduction (IITA, 1977). Scepter, a new experimental herbicide, discovered recently through extensive screening of herbicides for *E. heterophylla* control in cowpea, has provided good control of this weed in cowpea and soybean (Table 2). The effective dose range with no phytotoxicity to cowpea (cv. IT 82E-9) was 100–250 g a.e./ha when applied pre-emergence.

A tank-mix of scepter and metolachlor or pendimethalin should greatly improve chemical weed control in cowpea and minimize related yield losses.

No one weed control method is applicable in all cases because of the wide variability in soil type, climatic conditions, growth habits and life cycles of

Table 2. Effect of herbicides on *Euphorbia heterophylla* control in cowpea (IITA, 1984)*

Herbicide	Rate (kg a.i./ha)	TOA	Crop injury rating (%) (2 WAT)	Weed control rating (%) (3 WAT)	*E. heterophylla* dry weight (g/m^2) (6 WAT)
Scepter	0.10	PE	0	69	32
Scepter	0.15	PE	0	85	7
Scepter	0.25	PE	0	84	4
Scepter	0.30	PP	4	79	10
Acifluorfen	1.12	PE	9	54	94
Acifluorfen	1.56	PP	0	33	121
Metribuzin	0.50	PP	0	41	105
Metribuzin	0.25	PE	13	42	144
Oxyfluorfen	0.50	PP	28	34	125
Oxyfluorfen	0.25	PE	73	56	114
Oxadiazon	1.00	PP	4	33	152
Oxadiazon	0.50	PE	6	47	164
Weed-free check	—	—	0	100	0
Unweeded check	—	—	0	0	147

* TOA = time of application; PE = pre-emergence; PP = applied 7 days before planting; WAT = weeks after treatment.

weeds. Therefore, a combination of two or more weed control methods is likely to prove more effective than any single method in alleviating the build-up of hard-to-kill weeds. The use of low rates of pre-emergence herbicide to provide a weed-free environment during early crop establishment, supplemented by hoe- or inter-row weeding later in the season, is an example of an integrated weed-control system applicable to cowpea production.

REDUCED-TILLAGE SYSTEMS

Reduced-tillage systems, such as strip-tillage and no-tillage, for grain legume production have been reviewed recently (Akobundu, 1983a), and in a study of tillage systems, Nangju (1979) reported that weed problems were more severe in cowpea than in soybean. There was a build-up of *Digitaria horizontalis, C. dactylon* and *E. indica*. Preliminary results with tillage research in cowpea at IITA (Poku and Akobundu, 1983) indicate that most weeds increase in the following crop in each tillage system (Figure 1). In the study, some interactions were apparent, as sedges increased in strip tillage at the expense of broadleaf weeds and grasses whereas the reverse was true in the other two tillage methods. No conclusion can be drawn yet as to specific weed species being associated with a specific tillage method, although control of *Brachiaria deflexa* and *D. horizontalis*, the dominant grasses, was slightly better in the conventional and strip-tillage plots than in the no-tillage plots

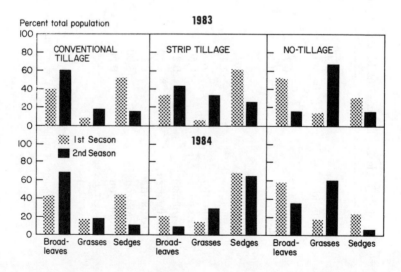

Figure 1.   Composition of weed species in untreated weedy control (IITA, 1983–84.)

Table 3. Effect of tillage method and herbicide treatment on visual weed control rating, 2nd season 1983*

Treatment	Rate (kg/ha)	Weed species (percentage control)									
		Euphorbia heterophylla		*Talinum triangulare*		*Commelina benghalensis*		*Brachiaria deflexa*		*Digitaria horizontalis*	
		3WAT	8WAT	3WAT	8WAT	3WAT	8WAT	3WAT	8WAT	3WAT	8WAT
Conventional tillage											
Metobromuron, metolachlor	2.5	33	41	100	100	96	98	100	100	100	100
Metobromuron, metolachlor	4.0	53	44	98	100	98	100	100	100	100	100
Chloramben, metolachlor	2.0, 2.0	45	61	97	99	97	99	100	98	100	99
Two hoe weedings (3,6 WAP)	—	100	99	100	97	100	100	100	100	100	97
Untreated check	—	0	0	0	0	0	0	0	0	0	0
Strip tillage											
Metobromuron, metolachlor	2.5	30	53	81	99	85	99	83	99	100	98
Metobromuron, metolachlor	4.0	38	54	83	99	100	100	100	100	100	100
Chloramben, metolachlor	2.0, 2.0	29	51	95	100	90	99	100	99	98	98
Two hoe weedings (3,6 WAP)	—	100	97	100	98	100	100	100	99	100	97
Untreated check	—	0	0	0	0	0	0	0	0	0	0
No tillage											
Metobromuron, metolachlor	2.5	60	75	95	98	83	98	92	96	81	94
Metobromuron, metolachlor	4.0	38	75	94	98	95	100	100	98	92	98
Chloramben, metolachlor	2.0, 2.0	31	59	97	98	100	100	100	99	97	98
Two hoe weedings (3,6 WAP)	—	92	93	98	98	96	98	98	100	95	95
Untreated check	—	0	0	0	0	0	0	0	0	0	0

* WAP = weeks after planting; WAT = weeks after treatment.

(Table 3). Selective control of grasses in cowpea in reduced tillage systems should be good with recent commercialization of fluazifop-butyl and sethoxydim. A no-tillage package for grain legume production in the tropics has been recommended recently (Akobundu, 1983b) and should include post-emergence application of fluazifop-butyl and sethoxydim for selective control of annual and perennial grasses.

STALE SEEDBED

Problem weeds such as wild poinsettia in cowpea may be reduced by use of the stale-seedbed technique, a weed-control strategy in which weed seedlings emerging after land preparation are killed with a contact herbicide before the actual planting of the crop. Where blind cultivation is used in place of the stale-seedbed technique, soil disturbance should be as shallow as possible or buried weed seeds will be brought to the surface.

Cowpea Research, Production and Utilization
Edited by S. R. Singh and K. O. Rachie
© 1985 John Wiley & Sons Ltd

26

Cowpea Striga *Research*

V. D. AGGARWAL
IITA–SAFGRAD (International Institute of Tropical Agriculture and the Semi-Arid Food Grains Research and Development Project), Ouagadougou, Burkina Faso, Upper Volta

The genus *Striga* is distributed in many parts of the world and is known to have approximately 50 recognized species that are parasitic in nature (USDA, 1957). The species that invades cowpea and is economically important in Africa is known as *S. gesnerioides*. The main cowpea-growing areas of West Africa are seriously infected with it, resulting in large yield losses. In comparison with sorghum and millet *Striga* (*S. hermonthica* and *S. asiatica*), it has been researched little. In addition, most of the work done has been included as a small portion in studies with the main cereal *Striga* species, so the scope of previous research efforts is difficult to pinpoint. The objective of this paper, therefore, is to review, collate and update the available research information on cowpea *Striga*.

TAXONOMY AND MORPHOLOGY

Of the known species of *Striga*, 23 occur in Africa (Musselman and Parker, 1981), and only a handful are considered agriculturally important. Wettstein (1893) divided the genus into two broad taxonomic sections, the Pentapleurae and the Polypleurae, characterized by a 5-ribbed calyx tube and a 5–10-ribbed calyx tube, respectively. Although this classification has been observed at times to be imprecise (Saldanha, 1963), the species *S. gesnerioides* has been taxonomically placed into Pentapleurae.

This species is widely considered to be the most advanced in the genus (Ayensu, in press) because it contains less chlorophyll than do other species and has a complex haustorium (Ba, 1983; Ayensu, in press). It generally has a single, large, well-developed haustorium, whereas other *Striga* species have a diffuse system of several small haustoria (Ozenda and Capdepon, 1972; Ayensu, in press). This species, for these reasons, seems to be closely related to Orobanchaceae, a family containing non-chlorophyllous species (Tiagi, 1956; Cronquist, 1968).

The aerial portion of *S. gesnerioides* is typically 15–20 cm high. It is composed of primary floral axes, extensively branched at the base giving rise to numerous secondary floral axes. The floral axes bear scale-shaped leaves, ranging from 0.5 to 0.7 cm long. They are green, thick and covered with stiff white hairs. The inflorescences are simple, spike-shaped racemes. Single flowers are axilled by a single thick, hairy bract. The flowers are typically up to 2 cm long and slightly irregular, as is typical of the Scrophylariaceae. The flowers are blue, pink, purple or creamy white. The species is strongly autogamous. The androecium is composed of five stamens, the most external having been aborted. The pollen is sticky and clumps together inside the corolla. The gynoecium contains a superior ovary with axile placentation. The fruit, an oblong capsule, contains approximately 400–500 wrinkled grains. These grains are 0.20–0.35 mm long and 0.15 mm wide. A single plant can produce an estimated 50,000–500,000 seeds, strongly suggesting dispersion by wind or water (Musselman and Parker, 1981; Musselman *et al.*, 1981; Ramaiah *et al.*, 1983; Ba, in press).

The underground portion of the plant can be divided into three sections: the stem, the roots and the parasitic system (haustorium). The subterranean portion of the stem, and the leaves that it bears, are white. Secondary roots, depending on the host, may or may not emerge from the base of young submerged leaves. The haustorium is formed when the *Striga* seedling attaches itself to the host root. A haustorium developed from the first contact is called a primary haustorium (Tiagi, 1956) and continues to develop, becoming tuber-shaped. According to Ba (in press), it may reach the size of a cola nut and can weight 150 g. If secondary roots come in contact with the host roots, secondary haustoria develop.

Ba (in press) has done the most complete work to date on the anatomic features of the haustorium of *S. gesnerioides*. According to him, there are two distinct parts of the haustorium: the endophyte and vascular core. The endophyte consists of intrusive cells that invade the host. These cells have been found frequently to be preceded by a dark fluid substance that seems to dissolve the host's cell walls. The vascular core contains phloem tissue—an indication that this species translocates compounds from the host root through its haustorium. To date, phloem tissue has been found in fewer than 10 of the known 3000 parasitic phanerograms.

HOST SPECIFICITY

Striga gesnerioides is a variable species with specific physiologic strains (Saldanha, 1963; Musselman and Parker, 1981). It has a wide host range of predominantly broadleaf species, including members of Solanaceae, Convolvulaceae, Leguminosae, Euphorbiaceae and Gramineae (Parker and Reid, 1979). Parker and Reid (1979) reported that, in West Africa, a species that is much-branched, green-stemmed, white and blue-flowered has been observed on cowpeas, but on no other species, not even groundnut. They observed another form, different from the one parasitizing cowpea, growing on *Tephrosia pedicellata* (Leguminosae), *Jacquemontia tammifolia* and *Merremia tridentata* (both Convolvulaceae) but never on cowpea. Herbaugh *et al.* (1980) found a strain of *S. gesnerioides* parasitizing sweet potato, in addition to *J. tammifolia*. However, Musselman and Parker (1981), in pot experiments, found it to be extremely host-specific, emerging only on the two *Indigofera* species they used in their studies. Even the strain attacking *Nicotiana tobaccum* in Zimbabwe, which was close in appearance to the cowpea type and was stimulated by cowpea to germinate, was not found parasitizing cowpea, although cowpea served as a trap crop (Wild, 1948).

The literature indicates that host-specific forms of *S. gesnerioides* do exist, but the mechanism governing host specificity is not yet fully understood. Parker and Reid (1979) considered two possibilities: that morphology of the weed is influenced by the host on which it grows, and that a combination of special requirements together with resistance barriers (Musselman and Parker, 1981) plays a great role in host specificity. According to Parker and Reid (1979), probably at least two or three strains within the species had a much narrower host range than did the species as a whole.

GERMINATION

According to Parker (1983a), the germination of *Striga* seeds depends on dormancy, preconditioning at a suitable temperature and exposure to a suitable germination stimulant, also at a suitable temperature, which may differ from that for preconditioning. Parker (1983a) used the phrase 'after ripening' to refer to the need of *Striga* seeds to lie dormant for a certain time (up to 3 months for some *Striga* spp.). In the semiarid tropics *S. gesnerioides* produces seeds at the end of the rainy season, and the dry season provides a natural period of dormancy. I could find no information describing optimal dormancy for *S. gesnerioides*.

'Preconditioning' (or pretreatment) refers to 'the requirement of the seed to imbibe for a period of time before it is exposed to a stimulant' (Parker, in press). Reid and Parker (1979) reported that *S. gesnerioides* required longer preconditioning than did other *Striga* spp. They also showed that more seeds germinated at 33°C than at 22°C; however, they admitted that the 'after-ripening' period ('few months') that their samples underwent was too short to allow maximal germination. They reported that a 'second dormancy' or 'wet dormancy' was possible if the preconditioning period were too long.

The final requirement for *Striga* germination, and seemingly the most important, is the exposure to a suitable germination stimulant. Parker (in press) stated: 'germination is normally triggered by substances leaking from the roots of potential host plants'. The one natural substance identified as a stimulant of *S. asiatica* is known as strigol (Cook *et al.*, 1972), but its activity on *S. gesnerioides* has not been reported. Several synthetic analogues of strigol have been produced (Johnson *et al.*, 1976); two, namely GR 7 and GR 24, were tested with *S. gesnerioides* but did not stimulate germination (Parker, 1983a). Other chemicals have been used for germination studies with other *Striga* species (Parker, in press), but only one, ethylene, has been tested with *S. gesnerioides*. Parker (1983a) reported that *S. gesnerioides* 'to some extent' was stimulated by ethylene to germinate. This finding, if confirmed, opens up an avenue for control of *S. gesnerioides*. To date, no stimulant has been isolated from cowpea.

PHYSIOLOGY AND DEVELOPMENT

As stated by Parker and Reid (1979), any complete host of a particular species of *Striga* must provide the stimulant that induces *Striga* germination and the environment that allows successful attachment and development. Based on the work of a number of authors studying *Striga* spp. of sorghum and millet, Parker (in press) stated that, upon germination, the *Striga* glues itself to the nearest object with haustoria attaching to almost any inert surface. Although the chemical nature of the 'glue' was not known, an adhesive substance was found in the hairlike protrusion on the haustorium. Penetration of the host, according to Parker, involves the intrusive cells' separating the host cells with enzymes rather than crushing or penetrating them.

Pectinases were found to be involved. In *S. gesnerioides*, Ba (1983) used a series of bioassays of haustorial material and detected phosphatase activity in what he termed the 'contact zone'. He concluded that since phosphatases are known to catalyse the breakdown of certain compounds, these enzymes play a role in the penetration and breakdown of host tissue by the parasite. Peroxidase was also found in regions where the parasite tissue was under greatest stress. These enzymes, he also concluded, are related to the active transport of metabolites.

To achieve effective parasitism the haustorium has to establish contact with the host stele, which entails penetration through the endodermis. Factors in the endodermis that discourage penetration may be responsible for varietal resistance, but no research on this possibility has been carried out.

Certain ecologic factors play a role in the development of the species—namely light, water and nutrient uptake. I could find no specific information about the effects of such factors on *S. gesnerioides*, although Parker (in press) hypothesized that light positively influenced the development of *Striga*.

CONTROL MEASURES

Although *S. gesnerioides* has been known for a long time in the Old World (Africa, Arabian peninsula and India), it was first reported in the New World (in Florida) only in 1979. It is widespread in the semiarid areas of Africa, and Obilana (1983) observed it spreading rapidly southward to the northern guinea savanna in Nigeria, from its traditional habitat in the sahel and sudan savannas. My colleagues and I (Aggarwal *et al.*, 1984) expect it to increase further with increased area and frequency of cowpea cultivation. This, in fact, is already happening in some places, particularly in Nigeria: Emechebe (1981b) reporting yield losses as high as 100 per cent. Likewise, in Burkina Faso, *Striga* greatly reduced yields of susceptible varieties compared with a resistant one. The yield losses were particularly high when *Striga* emerged before flowering occurred (IITA–SAFGRAD, 1982, 1983).

According to Ogborn (in press), 'control techniques can be classified into "crop production" methods, which protect the current susceptible host crop and must produce a profitable increase in yield, and "infestation reduction" or "eradication" methods which reduce the infestation of *Striga* seed in the soil in the absence of susceptible hosts'. There has already been considerable work done on various control methods for *Striga* species, although *S. gesnerioides* has received relatively little attention.

Use of resistant varieties is the most convenient and the simplest solution to controlling *Striga*, and two varieties of cowpea, SUVITA-2 and 58-57, have been found to be resistant in Burkina Faso (Aggarwal *et al.*, 1984). However, SUVITA-2, when tested in Mali, Niger and Nigeria, was found to be susceptible, an indication of different strains (IITA–SAFGRAD, 1983). Parker (personal communication, 1984) has confirmed that the varieties are resistant to the strain found in Burkina Faso and are susceptible to strains from Niger and Nigeria.

In studies involving SUVITA-2, resistance was found to be inherited and controlled simply by one or two genes (Aggarwal *et al.*, 1984), and work is in progress to incorporate the resistance in promising cowpea varieties.

Cultural practices that limit *Striga* infestation include such methods as roguing, trap cropping and rotation. The main objective is to decrease the number of *Striga* seeds in the soil. The benefits are gradual and cumulative so the practices are long term (Parker, 1983b).

Olunuga and Akobundu (1980) demonstrated a marked increase in yields from sorghum that is weeded weekly and kept free of *Striga*, but Ogborn (in press) found this practice to be ineffective because of intensive labour requirements and re-emergence of the *Striga*.

Cowpea was used as a trap crop to control *S. gesnerioides* in tobacco in Zimbabwe (Wild, 1948), but I could find no reference to use of a trap crop to control *Striga* on cowpea. Any susceptible cowpea variety can serve as a trap crop and be destroyed before *Striga* flowers. Use of rotation, shifting date and depth of planting, applying nitrogen fertilizer have been reported to be helpful in controlling *Striga* in cereal crops (Pieterse and Pesch, 1983) and are likely to be useful for its control in cowpea as well.

Chemical control by artificially stimulating seeds to germinate has been relatively well studied, and the methods can be grouped into abortive germinators and suicide germinators. Among these methods, use of ethylene gas holds promise (Parker, 1983a) and has been studied extensively as a control against other species of *Striga* (Eplee, 1973, 1979). Synthetic germination stimulants GR 5, GR 7, GR 9 and GR 24, may also prove useful. Various other compounds, such as Ethepon®, have been reported to be effective against *Striga* species (Chancellor *et al.*, 1971; Ogborn, 1979); yet they were not reported specifically for *S. gesnerioides*. Soil fumigants, for example methylbromide, have also proved successful against *Striga* (Eplee, 1969, in press). However, all these synthetic compounds have one major drawback: they are expensive and sometimes need to be dispersed with special equipment that puts them out of reach of a smallholder farmer.

Herbicides such as 2,4-D have been examined and found effective in the control of *Striga* on cereals, but their efficacy against cowpea *Striga* has not been reported. As *S. gesnerioides* possesses less chlorophyll than the other species of *Striga*, it may tolerate herbicides that inhibit photosynthesis (Ogborn, in press). In addition, *S. gesnerioides* grows underground for a relatively long period and is protected from foliar applications because the cowpea crop is generally well established before *Striga* emerges.

To date, all of the work to find predatory insects and fungi to act as a biological control has been done with sorghum and millet *Striga* species (Greathead, in press) and has been extremely limited.

Cowpea Research, Production and Utilization
Edited by S. R. Singh and K. O. Rachie
© 1985 John Wiley & Sons Ltd

27

Farming-systems Approach: Research on Cowpeas and Extension

M. ASHRAF
International Institute of Tropical Agriculture, Ibadan, Nigeria

This paper describes the use of a farming-systems approach for transferring new high-yielding cowpea varieties to the smallholders in the subhumid tropics of Nigeria. The methods that emerged from a 3-year on-farm research–extension experience consist of a procedure used for grouping small farmers into various homogeneous recommendation domains; assessing opportunities and leverages for improving the productivity of the existing cropping systems; and the on-farm testing of short-season cowpea varieties that farmers were keen to accept. The collaborative on-farm adaptive research programme was undertaken with the agronomy and extension sections of the Bida Agricultural Development Project (BADP).

BADP covers about 17,000 km^2 in southern Niger State in Nigeria and lies in the southern guinea savanna zone. The topography is gently undulating, and the underlying base is sandstone. One of the major determinants of the cropping system is the network of rivers, especially the Niger and its tributaries and the Kaduna. These rivers are bordered by large, swampy plains—the lowlands—that flood during the rainy season and then gradually dry out. The area is used to grow rice.

Part of the project area is outside the rivers' reach—the uplands—and is used for rainfed farming. Soils in the uplands are generally sandy and acidic with low levels of organic matter and low cation-exchange capacity. They are highly permeable and liable to erosion, especially on steeper slopes. Soils in the lowlands are loamy and can support longer periods of cropping without need of fallow. Rainfall is the major constraint: it is seasonal, monomodal and variable from year to year. Seedbed preparation, germination and the early stages of crop growth are entirely dependent on the amount and frequency of precipitation, since the soil, especially the surface, has no stored moisture at the beginning of the rainy season. On average (1961–83), annual rainfall (1182 mm) exceeds evapotranspiration from 11 May until 10 October (152 days) (Kowal and Knabe, 1972). However, rainfall can only be relied on to exceed evapotranspiration from 1 July to 30 September (91 days). The period between early May and mid-June presents considerable risk to farmers because the moisture available is often not sufficient for sustained crop growth.

An estimated 9 per cent of the total project area is cultivated (BADP, 1983), with concentrations of farm land around the lowlands and Bida, the major urban centre. Cultivation in the lowlands is semipermanent to permanent (following Ruthenberg's classification, 1980), whereas in the uplands, bush fallow or shifting cultivation is the norm. During the dry season only cassava, sugarcane and vegetables are grown on the lowlands for consumption during the hungry period before the main harvest.

The project serves an estimated 65,000 farming families, with an average 6.5 persons; farms range from 1 ha to 3 ha. Farming is the full-time business of the rural population and is the primary source of income. Rice, sorghum and dried fish are the major surplus commodities in the area, and a local marketing system has arisen for these items. About 500–600 t of cowpea beans are brought into the area every year from the north to meet local demand, so improving cowpea production was a focus on the project. As a first step, local cropping systems were analysed.

TARGET CROPPING SYSTEMS

The obvious differences between the systems used by farmers in the lowlands and those in the uplands led the researchers initially to define two target groups; however, later analysis of project agronomic data (BADP, 1983) revealed four cropping systems: one in the lowlands and three in the uplands—lowland rice-based, upland yam-based, upland cassava-based and upland cereal-based (Table 1).

The major determinants of the four cropping systems were soil types (including the soil fertility) and the availability of moisture during the wet and dry seasons. Much of the land within the project areas is in the top section of the catena, where soils are sandy, highly eroded, leached and infertile. The fields are planted to sorghum, millets, egusi melon, Bambarra nut and groundnut, which are relatively well adapted to low-fertility soils.

For the purpose of the project, farmers having land in both the lowlands and the uplands have been regarded as practising the rice-based cropping system (Ashraf et al., 1984).

The dominant crop by area in all four target domains is sorghum, but within their respective domains, rice, yam and cassava are the major cash crops,

Table 1.   Percent area grown to different crops and mixtures by target domain in Bida
project area, Nigeria, 1982

| Crop enterprise | Cropping system domain | | | | |
	Rice-based	Yam-based	Cassava-based	Cereal-based	Overall
Rice	25	*	—	3	7
Yam	9	12	—	4	8
Yam, millets	—	5	—	—	2
Cassava	1	1	44	2	4
Sorghum	8	27	15	18	19
Egusi melon	3	1	—	3	2
Sorghum, melon	40	16	8	17	22
Sorghum, maize	5	14	3	6	9
Sorghum, millet	*	3	—	4	2
Sorghum, cowpea	*	4	—	—	2
Sorghum, groundnut	1	1	3	12	4
Sorghum, millet, melon	2	1	—	12	4
Cereals, others	1	8	—	3	5
Tubers, others	1	3	24	5	4
Others	4	4	3	11	6

* Denotes value less than 0.5%.

producing higher yields than other crops and failing less often. Crop failures are highest in the upland cereal-based system, which has traditionally included cowpea. The researchers found that farmers were growing a local indeterminate type of cowpea on about 2 per cent of the land in mixtures with sorghum.

The rice-based system accounted for 25 per cent of the farmers sampled. In late April or May they plant upland fields with sorghum, usually in mixtures with millet, maize or egusi melon. They grow rice in two ecologic environments: on the flood plains and in inland valleys where seepage from the surrounding uplands accounts for much of the water. The dry season is a period of slack labour for most farmers in the rice-based system, since only a small part of the lowlands is utilized for production of dry-season crops on residual moisture and few opportunities exist for off-farm employment (Figure 1).

In the upland systems 25 per cent of the farmers cultivate yam, 10 per cent cassava and 40 per cent solely cereals. For both the yam- and cassava-based systems, labour input per farm is high at 365 and 239 work-days, respectively. Peak labour demand is at the start of the rains during the planting of guinea corn and after the rains at harvesting (Figure 1). Potential conflicts in the

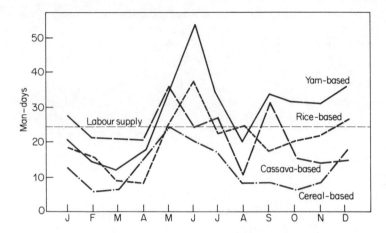

Figure 1.   Average household labour input (work-days) for agricultural activities by cropping system in BADP, Nigeria, 1983

allocation of scarce labour for establishment of yam and sorghum are avoided because most of the yam is planted during the dry season. Within the cassava-based system (Figure 1), the planting of cassava occurs after the establishment of the sorghum crop because cassava is relatively drought resistant, with a growth cycle of more than 12 months.

The upland cereal-based system is the poorest domain: crop yields are all lower than in the other domains and farms are smaller. The first half of the rainy season is usually busy, with farmers' planting and weeding, whereas during the second half many of these farmers sell their labour to the richer farmers who practise rice- and yam-based systems (Figure 1).

The farm productivity, measured in terms of farm income and returns to farm inputs, varies considerably among the four target domains. The systems based on root crops yield greater quantities of food and higher incomes. The rice-based system, which has access to good land and water resources, is comparatively less productive, because cropping intensity is lower in the lowland fields. Farmers' capital costs are for seed, small quantities of fertilizer and primitive hand tools; they do not have access to animal or mechanical power. Thus, the return to capital is negligible. Nearly 85 per cent of the farm costs are for labour inputs (mostly family members) and returns to labour input are close to the rural wage rate. After one accounts for costs of labour and capital, return to land is negative for the rice- and cereal-based systems and positive for the yam- and cassava-based systems (Table 2).

Table 2. Average farm cost and income budgets for target domains in Bida project area, Nigeria, 1982*

Item	Target domain				
	Rice-based	Yam-based	Cassava-based	Cereal-based	Overall
Average farm size (ha)	2.02	3.21	2.1	1.32	2.07
Total labour input					
Work-days	248	365	239	147	250
Cost ($)	1674	2464	1613	992	1686
Seed cost ($)	153	285	66	86	147
Fertilizer input cost ($)	16	3	18	9	12
Farm tools annual cost ($)	92	79	79	77	82
Land rent ($)	27	43	28	18	29
Total farm cost ($)	1962	2874	1804	1182	1956
Total farm income ($)	1802	3842	2427	1053	2281
Net farm income ($)	−160	968	623	−129	326
Measures of efficiency					
Return to land ($/ha)	−66	315	310	−84	171
Return to capital ($)	0.4	3.6	4.8	0.3	2.3
Return to labour, management ($/work day)	6.1	9.4	9.4	5.9	8.0

* Labour is valued at the current wage rate of 5 naira (or $6.75)/work-day; fertilizer cost is calculated at the subsidized price of 2 naira (or $2.70)/50 kg bag; costs of farm tools (two large and three small hoes, two cutlasses, one axe, three sickles for the rice system only, three baskets and two to five sacks) are charged at 25% interest rate; land rent is assumed to be 10 naira (or $13.5)/ha.

INTRODUCTION OF IMPROVED COWPEA VARIETIES

From the analysis of local farming systems there emerged four leverage points for the introduction of short-season, high-yielding cowpea varieties. (Varieties of cassava and rice were also introduced as part of the project— BADP, 1983.) They were:

○ Cowpeas are a favoured food for the rural population. An average family consumed 2–4 kg grains weekly and spent an equivalent of $250/year on the purchase of beans. Small farmers with limited cash resources were eager to produce their subsistence needs on their farms. Furthermore, the demand for cowpea consumption was elastic to income changes, thus indicating a secure position of cowpeas for human consumption (Sinner *et al.*, 1983).

○ In most of the lowlands, moisture coming from seepage kept the fields wet for a considerable time during the dry period. The idle fields and available moisture provided potential for a short-season catch crop in the dry period.

○ Because of the low productivity of the coarse cereal-based and rice-based cropping systems, farmers practising these systems were eager to try innovations to improve their food supply and farm income.

○ An analysis of the labour-use calendar for the four cropping systems revealed certain slack periods that were opportune times for the introduction of farm innovations. For the farmers practising the cereal-based systems, labour was plentiful throughout the year. For the farmers practising rice-based and cassava-based systems, slack periods emerged around and after the month of August. However, for the yam-based system, most farmers remained busy throughout the year (Ashraf *et al.*, 1984).

Among the various options for improvement of the productivity of the systems, short-season cowpeas were the best because they could serve as a catch crop in the inland swamps of the lowlands after paddy and in the cassava- and cereal-based systems during the last 2 months of the rainy season. To test the applicability of cowpeas for the different cropping systems the researchers designed an on-farm research programme consisting of agronomic experimentation with planting dates, toposequential planting, intercropping, multicropping, complemented by socioeconomic surveys.

The cowpea varieties TVx 3236 and IT 82E-60 during the main seasons of 1982–84 under local management conditions produced reasonably high yields. TVx 3236 yielded more than 1200 kg when sprayed three times for pest control; IT 82E-60 yielded less, but both the yield levels and high returns to spraying were quite attractive to the farmers (IITA, 1982).

The planting-date trials of sole cowpeas and cowpea–cassava intercrops revealed that the crop matures successfully when planted in August, and that it can be grown twice in one season by the farmers belonging to the cereal-based cropping system.

The success of short-season cowpea trials conducted on residual moisture during the dry seasons of 1982–83 and 1983–84 was of major consequence. Both TVx 3236 and IT 82E-60 varieties matured successfully and produced economic yields (IITA, 1983b). The three methods of tillage—zero, strip and conventional hand-hoe—were all successful for the germination of seed and good establishment of the crop in paddy fields. Thus, cowpea did not necessarily mean large labour inputs and could be planted on a timely basis.

The dry-season crop on the middle and bottom parts of the inland swamps (fadamas) matured successfully, with economic yields (600–700 kg/ha with two sprayings) both on the experimental and on the extension-verification plots (Table 3).

For the rice-based system dry-season cowpea offered a major improvement (Table 4). With a marginal cost increase of 8 per cent, the overall farm productivity increased by more than 25 per cent. The return to capital has increased by 275 per cent, and productivity of farm labour has increased by 13 per cent. Similarly, for the cereal- and cassava-based systems, a substantial improvement in farm productivity was achieved.

Because of the project's focus on four homogeneous target groups the agronomists found it easy to design appropriate trials and locate them with farmers belonging to different cropping systems. Within 3 years a substantial number of farmers have picked up the extension advice and have begun to grow short-season cowpeas. Many of these early adopters have begun to adapt the short-season crop to their specific socioeconomic conditions. Farmers have started to invest in insecticide sprayers and chemicals as well as in seeds of the new varieties of cowpea.

The project monitoring and evaluation unit has estimated a sharp rise in the numbers of farmers growing cowpeas: compared with the 7 per cent growing cowpeas pre-1983, the percentage was about 25 during the 1984 main season. Even though the subsidy on the cost of sprayers and insecticides was

Table 3. Average yield (kg/ha) of IT 82E-60 and TVx 3236 cowpeas by lowland type and topographic position in Bida, Nigeria, 1983–84

Variety/ topographic position	Fadama	River valley	Flood plain	Overall
IT 82E-60				
Middle	571	750	—	597
Bottom	620	672	—	646
Average	578	711	—	608
TVx 3236				
Middle	707	750	—	712
Bottom	536	—	830	634
Average	669	750	830	691

* These yields were obtained with two insecticide sprayings and are an average of 20 verification sites: 9 for IT 82E-60 and 11 for TVx 3236 or 16 in fadama, 3 in river valley and 1 in flood plain (15 located on the middle catena and 5 on the valley bottom).

Table 4. Estimated gains of farm productivity with dry-season cowpeas in the rice-based cropping system

Item	Rice-based system		
	Benchmark	Dry-season cowpeas	Percentage increase
Farm costs ($/ha)			
Labour	1674	1775	6
Capital	261	314	20
Land rent	27	27	0
Farm returns ($/ha)			
Gross	1802	2263	26
Net after land, capital	1514	1775	17
Net after labour, land rent	101	461	356
Net after labour, capital	−133	174	α
Net of all costs	−160	147	α
Returns per unit of			
Land ($/ha)	−66	86	α
Capital ($)	0.4	1.5	275
Labour, management ($/work-day)	6.1	6.9	13

* The calculations were based on an average farm of 2.02 ha cropped land, and dry-season labour costs, because the time is a slack period, were valued at 50% of the normal wage rate ($6.76/work-day).

removed, the project's entire stock of 100 Electrodyn® sprayers and 300 cymbush bozzles was sold, and additional orders for 100 Electrodyn sprayers and 1000 cymbush bozzles were placed by the farmers. One reason that farmers have so quickly started investing their valuable cash in the Electrodyn sprayer is the possibility of growing more than one cowpea crop in a year.

During 1984–85 three crops of cowpeas were grown. The first upland crop was planted in May and harvested in July; the second upland crop planted in August and harvested in October; and the third lowland crop will be planted in November–December and harvested 2–2.5 months later. The three-crop cycle has the advantage to the farmers of improving the food-and-cash flows; reducing on-farm storage and, consequently, losses; and making good-quality seeds available on a timely basis. Because of the expansion of cowpea in all the cropping systems, the farmers practising the different systems have begun to interact more often than before.

CONCLUSIONS

Even though small farmers in the humid and subhumid tropics grow different crops in various mixtures, they can be grouped into target groups based on the principal crop, the methods, labour inputs, etc. The groupings allow one to detect leverage points for the introduction of high-yielding crop varieties. This approach enabled the extension service to disseminate information and bring about a rapid adoption of a cowpea-production package.

PART 8

Nutrition

Cowpea Research, Production and Utilization
Edited by S. R. Singh and K. O. Rachie
© 1985 John Wiley & Sons Ltd

28

Nutritive Value of Cowpea

RICARDO BRESSANI
Instituto de Nutrición de Centro América y Panamá (INCAP), Guatemala, Guatemala

The diet of most people in developing countries is based on processed cereal grains such as maize, sorghum and rice, on roots such as cassava, and on fruit such as plantain. Other than the starchy roots and tubers, these foods, because they are eaten in large quantities, provide considerable protein, but the quality of the protein leaves much to be desired, particularly for children and pregnant and lactating women. Food legumes, because of their high protein content, in general constitute the natural protein supplement to staple diets, and cowpeas, in Africa at least, represent the legume of choice for many such populations.

The discussion of the nutritional quality of cowpeas will be developed therefore in the context of diet; that is, as a food component consumed together with such staples as cereal grain and starchy foods. It covers variability in the chemical composition of cowpea, antiphysiologic factors, the supplemental effect of cowpeas when eaten with staple foods, processing and its effects on nutritional quality of cowpeas and some recommendations for improvement.

CHEMICAL COMPOSITION

The chemical composition of cowpeas (Table 1) is similar to that of most edible legumes. It contains about 24 per cent protein, 62 per cent soluble carbohydrates and small amounts of other nutrients (Elías, 1964). Thus, most of its nutritional value is provided by proteins and carbohydrates.

Protein

Variability in protein content has been reported to be from 23 to 30 per cent and is influenced by genotypes as well as by environmental factors (Bliss,

Table 1.   Nutrient content of eight cultivars of cowpea

Nutrient	Range	Average
Protein (g/100 g)	24.1 –25.4	24.8 ± 0.48
Ether extract (g/100 g)	1.1 – 3.0	1.9 ± 0.62
Crude fibre (g/100 g)	5.0 – 6.9	6.3 ± 0.64
Ash (g/100 g)	3.4 – 3.9	3.6 ± 0.17
Carbohydrate (g/100 g)	60.8 –66.4	63.6
Thiamin (mg/100 g)	0.41– 0.99	0.74 ± 0.22
Riboflavin (mg/100 g)	0.29– 0.76	0.42 ± 0.14
Niacin (mg/100 g)	2.15– 3.23	2.81 ± 0.26

1975). Because of differences between cultivars, selection for higher amounts may be possible. Lysine content (Table 2) is relatively high, making cowpea an excellent improver of the protein quality of cereal grains. However, cowpea protein, as is the case for other food legumes, is deficient in sulphur-containing amino acids. The deficiency is definitely important when the diet is based on root crops or starchy foods, and, even in diets based on cereal grains, increases in sulphur-containing amino acids in cowpea protein would be nutritionally beneficial.

Protein quality, which reflects total sulphur amino acids, tends to be slightly higher in cowpea than in *Phaseolus vulgaris* (2.30 *vs*. 1.42 in samples of the

Table 2.   Essential amino acid content of cowpea (mg/g N)

Amino acid	Range in eight cultivars	Average
Arginine	433–572	500
Histidine	169–236	213
Isoleucine	305–333	318
Leucine	434–543	484
Lysine	467–497	486
Methionine	74– 82	79
Cystine	26– 38	32
Phenylalanine	251–290	263
Tyrosine	113–137	124
Threonine	242–281	251
Tryptophan	58– 82	68
Valine	252–368	314

Table 3.   Effect of sulfur amino acid supplementation to cowpea on its protein quality

Cowpea	Biological value (%)	Net protein utilization
Meal	58.17 ± 2.31	50.60 ± 1.83
Meal + cystine	80.26 ± 1.87	72.74 ± 0.94
Meal + cystine + methionine	94.61 ± 1.26	82.12 ± 1.07
Meal + methionine	95.84 ± 1.45	81.46 ± 0.87
Albumìn	101.72 ± 2.54	99.52 ± 1.85

two legumes processed identically) (Elías, 1964; Bressani *et al.*, 1977a) but still needs to be improved. Addition of methionine to cowpea protein (Table 3) increases the protein quality significantly, demonstrating the nutritional value of increasing sulphur amino acid content in cowpea protein (Boulter *et al.*, 1975). Similar results have been shown by others and represent a common finding for all food legumes (Sherwood *et al.*, 1954; Bressani, 1975).

Carbohydrate

Because the protein content in legumes is higher than in other vegetables, the other major component, carbohydrate, has been somewhat neglected (Srini-vasa, 1976; Reddy *et al.*, 1984). Total carbohydrate varies from 56 to 68 per cent, with starch contributing from 32 to 48 per cent. Data from other food legumes suggest that more than 50 per cent of the starch is in the form of amylose. The carbohydrate fraction is relatively rich in total sugars, and cowpeas have their share (about 6 per cent of total chemical composition) of these oligosaccharides, well known for their flatulence effects.

The amount of amylose in the starch influences starch solubility, lipid binding and many functional properties, such as swelling and solubility, water absorption, gelatinization and pasting, that affect cooking practices and acceptability.

Because cowpeas and other legumes are not considered carbohydrate sources or energy foods, studies on the nutritional value of the starch they contain are not readily available. The relatively high amylose content has been shown *in vitro* to cause slow digestibility (Srinivasa, 1976). However, baking, cooking, roasting and germination increase carbohydrate digestibility *in vitro* and may facilitate it *in vivo* to some extent (Srinivasa, 1976; Geervani *et al.*, 1981; Reddy *et al.*, 1984).

In a study using rats as experimental animals, to measure carbohydrate digestibility of cowpeas (Cabezas *et al.*, 1982), the percentage energy digestibility varied from 87.2 to 89.2 and was little influenced by processing.

Bomb calorimetry showed the digestible energy of cowpeas to be 3.67 in raw, 3.93 in cooked and 3.75 kcal/g in extruded cowpeas, values not statistically different. Supplementing the diet with methionine did not affect the digestible energy of cowpea, so the small differences between processed samples may reflect the small amounts of cowpeas present in the diet, representing only about 30 per cent of cowpea carbohydrate. A more rigorous test was the metabolizable energy (1.81, 2.56 and 2.76 kcal/g of raw, cooked and extruded samples respectively), which shows processed cowpeas to have a higher value than raw grain (Cabezas *et al.*, 1982).

Other factors of nutritional interest

Cowpeas contain antiphysiologic substances such as lectins and trypsin inhibitors as well as polyphenols or condensed tannins, which have recently been receiving attention, since they decrease protein digestibility and reduce protein quality (Bressani and Elías, 1979). The values for improved cowpea varieties are much lower than values found in *P. vulgaris* (Price *et al.*, 1980), which may be a reason for their higher protein quality. However, breeders in their efforts to produce insect-resistant varieties have sometimes increased the levels of trypsin inhibitors (Gatehouse and Boulter, 1983). Both tannic acid and catechin are quite variable among samples, the former varying from 0.31 to 1.63 per cent in 10 IITA selections and the latter from an undetectable level to 143.6 mg/100 mℓ.

One aspect that should be studied further in cowpeas, as is being done with common beans, is the protein quality of the cooking broth. Weight gain has been shown to increase in subjects who eat legumes (both cowpeas and common beans) without the cooking broth, and protein quality is considerably increased without broth (Elías *et al.*, 1976). The findings were more marked for common beans than for cowpeas. Further studies have shown the factors responsible to be located in the seed coat: its removal improves the protein quality of both the cooked cotyledons and the cooking broth. The cooking broth of whole cooked cowpeas and common beans may be responsible for the diarrhoea reported in children upon consumption of these foods.

NUTRITIVE VALUE

One of the most important nutritional characteristics of food legumes, including cowpeas, is that they complement cereal grains (Table 4) (Bressani and Scrimshaw, 1961; Cabezas *et al.*, 1982). For example, cowpea–maize mixtures provide the highest quality of protein at the weight ratio of 45 parts maize to 15 parts cowpea and 27 parts maize to 21 parts cowpea. The sorghum–cowpea combination is best between 63 parts sorghum to 9 parts

Table 4.   Complementary effect of cowpea protein to maize and sorghum

Diet (g)		Maize		Sorghum	
Cereal	Cowpea	Weight gain (g)	Protein equivalency ratio	Weight gain (g)	Protein efficiency ratio
90.0	0	54	1.22	21	0.97
71.9	6.1	—	—	33	1.43
67.5	7.6	78	1.59	—	—
63.0	9.1	—	—	40	1.48
53.9	12.2	—	—	42	1.51
45.0	15.0	102	1.84	43	1.55
36.0	18.2	—	—	33	1.40
27.0	21.3	104	1.82	—	—
18.0	24.3	—	—	34	1.27
0	30.0	78	1.41	25	1.13

cowpea and 45 parts sorghum to 15 parts cowpea. For practical purposes and as a recommendation for human diets, a weight ratio of 45 parts cereal grain to 15 parts cowpea could be used. A food product for weaning small children made of 75 per cent cereal grain and 25 per cent cowpea would be about 13 per cent good-quality protein.

Protein quality is synergistically improved in cereal–legume mixes because of the lysine contributed by the cowpea and the methionine contributed by the cereal. Studies of maize–beans mixtures have shown that the quality of the food is highest when the food legume contains high levels of sulphur-containing amino acids (Bressani and Elías, 1980).

In studies by my colleagues and me, processing increased the quality of cowpea–maize and cowpea–cassava mixtures. Much of the improvement reflects the destruction of inhibitory substances in cowpea (João *et al.*, 1980a,b); however, some studies suggest an increase in the efficiency of carbohydrate utilization by extrusion cooking.

Cowpea protein digestibility

One of the problems with the protein of food legumes, including cowpea, is their relatively low digestibility, a problem that deserves more attention. In experimental diets in animals (Bressani *et al.*, 1979), all the food legumes had significantly lower digestibility than did casein.

Cowpea protein gave an apparent protein digestibility of 72 per cent, whereas red bean was lowest with a value of 60 per cent. No information is yet available to explain the low values in legumes, but the regression equations between nitrogen intake from legume and faecal nitrogen output indicate that as intake of protein increases, the nitrogen in faeces also increases. Although the condensed polyphenols in common beans have a role in increasing faecal nitrogen output, thus decreasing protein digestibility, these compounds account for only about 10 per cent of the low values (Bressani et al., 1982). Some studies suggest that a soluble nitrogen fraction, with a low digestibility, is found in variable amounts in the cooked legumes (Bressani et al., 1977, 1982).

EFFECTS OF PROCESSING

Food legumes are always processed in one way or another before consumption. The most common method is some form of cooking which, if not carried out under controlled conditions, may decrease their nutritional value and, therefore, their supplementary effects to cereal grains and other foods (Onayemi et al., 1976; Zamora, 1979a,b; Geervani and Theophilus, 1980; Hamid et al., 1984). Cooking cowpeas in water under 15 lb (ca. 7 kg) pressure at 121°C for 15, 30 and 45 min decreased trypsin inhibitors from 6.7/mℓ in the raw product to 2.8, 2.5 and 3.0/mℓ, respectively (Elías et al., 1976). Protein digestibility in all three cooked products was similar but weight gain and protein quality decreased progressively with increased cooking time. Weight gain from cowpeas cooked for 15 min was twice that from the product cooked for 45 min (34 vs. 17 g). Other studies have also shown that soaking before cooking decreases nutritive value. Thermal processing, if not overdone, in general increases the quality (Table 5) (Elías et al., 1976). Toasting definitive-

Table 5.   Effect of different heat treatments on the protein quality of cowpeas

Treatment	Protein efficiency ratio	Digestibility (%)	Available lysine (g/16 g N)
Raw	1.21	73.2	6.4
Autoclave, 15 lb, 15 min, 121°C	1.33	77.4	5.7
Atmospheric pressure, 45 min, 90–95°C	1.34	—	6.3
Toasting, 30 min, 210°C	1.29	76.1	4.8
Toasting, 30 min, 240°C	0.36	65.0	2.9
Extrusion	1.73	80.2	5.6

ly reduces protein quality because of its destruction of available lysine. Finally, extrusion cooking shows increments in protein quality probably as a result of increased availability of the carbohydrates and increased susceptibility of the protein to enzymatic action.

Removing the carbohydrate fraction from cowpea produces a protein isolate that can be compared with cowpea meal (Table 6). At two different pH values, removing most of the carbohydrate improved protein quality compared with that of cowpea meal. Thus, the presence of carbohydrates in cowpeas affects the quality of protein and merits further study (Molina *et al.*, 1976).

Table 6.   Protein quality of protein isolates from cowpea

Product	Average weight gain (g)	Protein efficiency ratio
Cowpea meal	57	1.43 ± 0.42
Protein isolate (pH 9.0)	56	1.84 ± 0.56
Protein isolate (pH 6.8)	57	1.56 ± 0.31
Casein	74	2.42 ± 0.56

IMPROVEMENTS

From the nutritional point of view two factors merit improvement and should be considered in breeding programmes. One is the sulphur amino acid content. Results from studies indicate that in long-day varieties methionine content averaged 164 mg/g N in 11 cultivars and, although values for cystine were not reported, this value would probably be about 67 mg/g N (Bliss, 1972). Thus, total sulphur amino acids would be about 234 mg/g N. A second is improvement in protein digestibility. In fact, a higher sulphur amino acid content and higher protein content would be useless if protein digestibility were not increased as well. The role of the seed coat and of the carbohydrate in cowpeas also needs to be better understood, as does the effect of different methods of processing. Acceptability and cooking-quality characteristics must be improved and problems associated with storage need to be resolved. These are problems pertaining to all food legumes, and now that breakthroughs in yield have been obtained, more emphasis must be given to other factors in the food chain.

Cowpea Research, Production and Utilization
Edited by S. R. Singh and K. O. Rachie
© 1985 John Wiley & Sons Ltd

29

Functionality of Cowpea Meal and Flour in Selected Foods

KAY H. MCWATTERS
University of Georgia, Experiment, USA

For several years we at the Department of Food Science in the University of Georgia have been involved in studies to determine the functionality or performance of cowpea meal and flour in several foods. We consider that diversifying the use of cowpeas will benefit the producers and consumers in the long term. Successful use of cowpea meal or flour largely depends upon its compatibility as a food ingredient and the quality of end products. Functionality studies in model systems have provided insights into the behaviour of cowpea products as they deal with properties such as oil and water absorption, gelation, foaming, emulsification and dispersibility. The results indicate possible food uses for cowpeas but need to be coupled with performance studies in actual foods.

I would like to review some results we have obtained from using cowpea meal and flour in several foods—some more successful than others. The foods included ground beef patties, several bakery products and a traditional West African food.

Looking first at the ground meat system, we replaced part of the meat with cowpea meal because of potential cost savings and increased product yield. We used a meal made from Dixiecream cowpeas to replace 5, 10 and 15 per cent of the meat in patties (McWatters, 1977; McWatters and Heaton, 1979) made from minced beef. The meal was steamed for 30 min at 100°C before use. Moist heat treatment is frequently applied to strong-flavoured protein products such as soybeans to improve flavour and, in some instances, to improve nutritional quality. We found that the percentage of water and fat retained by patties increased as the level of cowpea meal increased; this was reflected in lower cooking losses for extended patties than for all-beef patties (Table 1).

Table 1. Characteristics of ground beef patties extended with steamed cowpea meal
(McWatters, 1977)

Pattie formula		Percentage retention		
Percentage beef	Percentage cowpea	Water	Fat	Cooking loss (%)
100	0	66.1	43.8	34.8
95	5	72.5	46.2	28.5
90	10	76.2	51.6	23.8
85	15	81.5	59.8	18.0

We also found that extended patties had lower compression and shear values than did all-beef products (Table 2). The force required to compress and shear patties decreased as the cowpea level increased. Overall sensory quality was adversely affected when cowpea meal was used at levels greater than 5 per cent (Table 2) because the beany aroma and flavour were noticeable. The beneficial effects of adding legume ingredients to a food cannot be at the expense of sensory quality if a product is to be acceptable to consumers. More successful results may have been obtained with the ground meat product if a textured or fibre form of cowpeas had been available. These forms of plant proteins are more commonly used in extended meat products than are meal or flour.

Another potential use for cowpea flour is to make bread, which enjoys worldwide acceptability. One approach to studies of this type involves replacing wheat flour with an alternative flour to determine the maximal level at which the substitute can be used without adversely affecting breadmaking

Table 2. Characteristics of ground beef patties extended with steamed cowpea meal
(McWatters, 1977)

Pattie formula				
Percentage beef	Percentage cowpea	Compression (N/g)	Shear (N/g)	Sensory score
100	0	1.29	1.86	7.7 (good)
95	5	1.01	1.84	7.1 (good)
90	10	0.88	1.27	6.0 (fair)
85	15	0.68	0.81	5.1 (borderline)

quality. Since little information existed concerning the use of cowpea flour in quick breads, as opposed to yeast breads, we investigated the effects of using the flour as a milk-protein substitute rather than as a wheat-flour substitute (McWatters, 1980). The quick bread selected for study was biscuits, which are chemically leavened by the action of baking powder. This type of biscuit is an unsweetened bread usually eaten with meals. A sufficient quantity of cowpea flour was used to supply protein equivalent to that provided by whole milk. We encountered no problems in preparing and handling the doughs that contained cowpea flour.

In addition to sensory quality, physical characteristics such as external and internal colour and density are important (Table 3). Colour differences were more apparent in the crust than in the crumb. The crusts of the biscuits made with cowpea flour products were lighter, having higher L values, slightly less red (lower a values), and slightly less yellow (lower b values) than were those incorporating milk. Very little difference in the colour of reference and test

Table 3. Characteristics of biscuits containing heated and unheated cowpea flour (McWatters, 1980)

	Crust colour			Crumb colour			Density (wt/vol)
	L	a	b	L	a	b	
Reference	65.9	8.8	26.7	82.4	−0.7	12.9	0.602
Dixiecream—heated	74.3	5.2	23.2	81.8	0.0	15.1	0.535
Dixiecream—unheated	73.6	5.4	24.6	82.1	−0.1	14.9	0.537

products was evident in the biscuit crumb. The cowpea-flour products were also less dense than reference biscuits. Steaming the flour had little effect on these physical characteristics.

The cowpea flour had no significant ($P<0.05$) effect on appearance and colour (Table 4); however, aroma, texture, and flavour scores for the cowpea-flour products were significantly lower than for the reference. The aroma and flavour of the cowpea-flour products were described as beany and strong. Steaming the flour improved biscuit flavour but not to the level of acceptability of the reference.

We also tested the cowpea flour in cookies, which are popular in the United States, particularly with children. In these tests legume flour was substituted for some of the wheat flour (McWatters, 1978). Each of the flours used in the study contained about 1 per cent fat. On a wet-weight basis, as used in cookie preparation, peanut flour contained 51 per cent protein, soybean flour 46 per

Table 4.   Sensory scores* of cowpea biscuits containing heated and unheated cowpea flour (McWatters, 1980)

	Appearance	Colour	Aroma	Texture	Flavour
Reference	7.6	7.7	7.9a	8.1a	7.5a
Dixiecream—heated	7.3	7.1	5.9b	6.8b	6.4b
Dixiecream—unheated	7.3	7.3	5.2b	7.1b	5.5c

* Scale of 9 to 1 where 9 = excellent, 5 = borderline, 1 = very poor. In columns, values not followed by the same letter differ significantly ($P<0.05$).

cent and cowpea flour 21 per cent. For comparison, soft wheat flour contains about 10 per cent protein. Carbohydrate content was 36 per cent for peanut flour, 38 per cent for soybean flour and 62 per cent for cowpea flour. There were fewer changes in colour, top-grain character and degree of spread with the use of cowpea flour than with the other legume flours (Table 5), and there were no problems in mixing or handling the cowpea-flour doughs. Thus, processors could produce this type of cookie without modifying formulas, equipment or baking procedures. Each increment of cowpea flour increased the protein content of cookies by about 0.5 per cent. Sensory quality attributes were not adversely affected except at the 30 per cent replacement level where a slightly beany flavour was detected.

Another type of bakery product that is highly favoured in some parts of the world is the doughnut. Made from a sweet dough or thick batter leavened by the action of yeast or a chemical agent such as baking powder, doughnuts are deep-fat fried at a temperature of 375°F (190°C). Our first attempts at preparing doughnuts utilized peanut and cowpea meals as wheat-flour replacements and were fairly successful (McWatters, 1982a). The only major adverse characteristic of the legume products was that they absorbed more oil during frying than did the all-wheat flour products.

We conducted a follow-up study to evaluate the effects of using legume flour instead of meal (McWatters, 1982b). Again, the cowpea flour was made from the Dixiecream cultivar, a small cream-type seed with very little dark colour at the hilum. Consequently the intact seed can be milled into flour. Ten per cent of the wheat flour was replaced with peanut or cowpea flours. The legume-supplemented doughnuts were prepared with and without soybean flour, which is frequently added to doughnut formulations to control fat absorption during frying.

There were no problems with mechanical cutting and dispensing of the cowpea batters, and the degree of browning was similar for reference and test products. The grain of these doughnuts was decidedly less open and less coarse than the grain of those made with meal. The moisture and oil content

Table 5. Characteristics of sugar cookies prepared with cowpea flour (McWatters, 1978)*

Wheat flour replacement (%)	Spread ratio (diameter/height)	Protein (%)	Sensory scores				
			Appearance	Colour	Aroma	Texture	Flavour
0	5.56	4.16d	9.0	9.0a	9.0a	9.0a	9.0a
10	6.11	4.67c	8.3	7.7b	8.2bc	8.5ab	8.4ab
20	6.10	5.10b	9.0	9.0a	8.5ab	8.2b	7.7bc
30	6.89	5.60a	8.4	8.5a	7.8c	8.1b	7.3c

* Scale of 9 to 1 where 9 = excellent, 5 = borderline, 1 = very poor. In columns, values not followed by the same letter differ significantly (P<0.05).

Table 6. Sensory scores* of doughnuts prepared with cowpea and soybean flours (McWatters, 1982b)

	Appearance	Colour	Aroma	Texture	Flavour
Wheat-flour reference	7.7	7.8	7.9	7.7	7.9
10% cowpea flour	7.9	8.1	7.2	7.6	6.8
10% cowpea flour, 3% soy flour	7.8	7.9	7.2	7.5	7.1

* Scale of 9 to 1 where 9 = excellent, 5 = borderline, 1 = very poor. None of the values differed significantly.

of the legume-supplemented doughnuts closely resembled that of reference doughnuts and was a definite improvement over the previous study. The addition of soy flour provided no added benefit in controlling fat absorption during frying. The cowpea flour products were quite acceptable in all sensory attributes and compared favourably with the reference products (Table 6).

We are presently involved in a project sponsored by the Bean/Cowpea Collaborative Research Support Program and the United States Agency for International Development. The title of the project is 'Appropriate Technology for Cowpea Preservation and Processing and a Study of Its Socio-Economic Impact on Rural Populations in Nigeria'. In it, our department collaborates with three departments at the University of Nigeria at Nsukka—Food Science and Technology, Home Science and Nutrition, and Sociology/Anthropology.

A major constraint that limits cowpea use in West Africa is the time and effort involved in preparing it. For uses in which cowpea paste is the principal ingredient, removing the seed coat from blackeye-type peas is desirable. Seed-coat removal produces a light-coloured pea that is free of black specks. Dry seeds are usually soaked in water to loosen the seed coat, decorticated by hand—the wetted peas being rubbed or stirred in a mortar until the seed coats loosen and can be floated off in water. The seeds are then ground to paste either on a stone, in a mortar or in a blender if available. The paste can be steamed or deep-fat fried. The fried form is known by several names, but we refer to it as Akara.

A major focus of this project has been the development of techniques to produce meal or flour that consumers can use simply by adding water. This would eliminate the soaking, dehulling and grinding steps and would provide cowpeas in a form convenient to use.

We have found that the production of good-quality Akara depends upon several factors, a key one being the volume and viscosity of the foam produced when cowpea paste is whipped. This is largely influenced by conditions under which peas have been stored and pretreated for mechanical dehulling, the particle size distribution of the milled product (McWatters, 1983), the amount of water used to hydrate the meal and, to a lesser extent, the time the meal is hydrated.

Cowpea Research, Production and Utilization
Edited by S. R. Singh and K. O. Rachie
© 1985 John Wiley & Sons Ltd

30

Novel Foods from Cowpeas by Extrusion Cooking

R. Dixon Phillips, M. B. Kennedy, E. A. Baker, M. S. Chhinnan and V. N. M. Rao
University of Georgia, Experiment, Georgia, USA

As part of a collaborative project between the University of Georgia and the University of Nigeria, Nsukka, novel foods are being produced by the extrusion cooking of cowpea meal, and their textural and nutritional properties are being evaluated. These products are by no means ready to be marketed. Rather, they are at best the ancestors of real foods. Both texture and flavour still need much work before acceptable foods result.

Cowpeas were chosen as the source of nutrients because of their relatively high protein content and their amino acid profile, which is both superior to and complementary to that of cereal grains (Dovlo *et al.* 1976).

Why develop novel foods when a wide range of popular foods is already being made from cowpeas in West Africa and elsewhere? The first, and main, reason is the constraints to the use of cowpeas in their major production areas, and the second is the opportunity to expand cowpea use. The constraints—including storage losses and the laborious processes used for converting the whole grains to popular food items—are a major focus of our project (funded under the US government's initiative, the Collaborative Research Support Program—CRSP) (Anonymous 1984).

While the production of novel foods does not necessarily grow out of improved storage schemes and production of ready-to-use cowpea meal, it is a logical extension of these activities. Such products may be made more convenient, more stable and more nutritious than traditional dishes, thus expanding the use of cowpeas and increasing the total market for the raw commodity rather than competing with traditional uses.

Extrusion cooking is an efficient way of converting starchy or proteinaceous raw materials into finished foods or into intermediates that require only

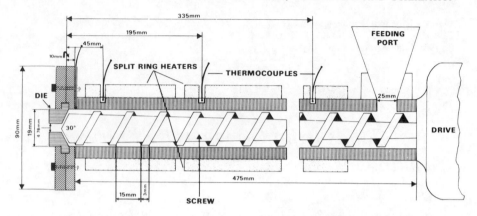

Figure 1. Extrusion cooker used in study

minimal further processing (Figure 1). The process is carried out in an extruder, which consists of a conveying screw rotating in a closely fitting barrel. The barrel is usually jacketed to allow heat to be applied or removed. The screw may taper so as to compress the dough as it is conveyed down the barrel. The raw material is introduced at the feed section of the barrel and is quickly cooked by a combination of externally applied, and mechanically generated, heat. The high temperatures and great shear forces in the extruder barrel result in the gelatinization of starches and denaturation of proteins without appreciable loss of nutrients, producing materials with various textural characteristics. Extrusion is versatile, and is used widely in food processing in developed countries. For example, commercial processors use extrusion cooking to produce snack foods, ready-to-eat cereals, meat analogues, pet foods, etc. Also, because of the diversity of size and design of extruders, they are expected to find increasing application in developing countries at the village and small town as well as the city level (Harper, 1981).

EXPERIMENT

The cowpeas used in this study (Mississippi Silver) are somewhat different from most cultivars found in West Africa. Rather than having a wrinkled, tightly adhering seed coat, they have a smooth, brittle, loosely adhering seed coat that can easily be removed by a rotary mill and peanut sheller.

The percentage of moisture content of decorticated seed was adjusted to 20, 30 and 40 in the study, and the seeds were chopped to a coarse meal before extrusion. The meal was extruded at three barrel temperatures 150°C, 175°C and 200°C—a total of nine products. The screw speed was held at the maximum rate of 180 rpm, and the feed port was kept full during the operation. Each extrusion run was carried out three times. Extrusion was carried out on a small, pilot-scale machine, with a 5:1 compression-ratio screw to achieve maximal shear. The materials with 20 per cent moisture were introduced into the feeding port by a vibrating hopper, while materials with greater moisture had to be forced into the port by staff using a lever and plunger. After emerging from the die the extrudate was allowed to cool and lose moisture for 5 min before being placed in plastic bags and stored at $-20\,°C$.

Small amounts of extrudate were thawed, and the texture was evaluated by rheological tests for shear and tensile properties (Instron Universal testing machine). Other samples were evaluated for moisture content and bulk density. Large lots of extrudate were freeze-dried and ground before nutritional evaluation (Kennedy 1982).

RESULTS

Final moisture content of the extruded products (Table 1) was affected by initial moisture and, to a lesser degree, by barrel temperature. In fact, the latter seemed to exert a consistent effect only on the feed with 30 per cent moisture. Products from materials extruded at 20 per cent moisture contained about 10 per cent moisture—equivalent to that in most extruded snacks. The 30–35 per cent moisture of the products from 40 per cent moisture feeds is similar to that of breads. Bulk density of the extrudate (Table 1) was rather uniform except for the products formed at 20 per cent moisture–150 °C, 20 per cent moisture–175 °C, and 30 per cent moisture–200 °C. The first of these extrudates twisted upon itself as it exited the die to form a very dense material. The second had a highly expanded, snack-like structure, while the last split at the extruder die formed a curled ribbon, the pieces of which

Table 1. Product moisture (M, %), bulk density (D, g/cm^3) of cowpea extrudates

| Barrel temperature (°C) | Initial feed moisture (%) | | | | | |
| | 20 | | 30 | | 40 | |
	M	D	M	D	M	D
200	10.0	0.33	16.0	0.25	28.8	0.32
175	11.4	0.23	18.0	0.33	27.5	0.31
150	11.1	0.44	23.0	0.33	34.4	0.31

packed less closely than the more regularly shaped products. All were subjected to extensive rheological evaluation to elucidate their textural properties, although only the results of the Warner-Bratzler shear test are reported here.

Samples of extrudate were placed in the Warner-Bratzler device and slowly sheared while the force-deformation curve was recorded. Several rheological parameters may be calculated from these data; an example is stress at failure. Shear stress is defined as the amount of force per cross-sectional area necessary to shear a segment of extrudate in two. It is related to the force required to chew a food, and its results can be summarized in a response surface diagram (Figure 2). The surface is generated by data computer-fitted to a mathematical model. The model, then, predicts values of stress for the entire range of dependent variables. The predicted values in our study were in

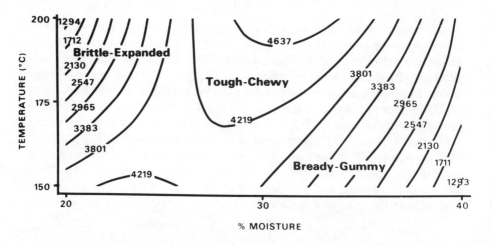

Figure 2.   Texture of extrudates as a function of initial moisture and barrel temperature (Warner-Bratzler shear stress) (kPa)

good agreement with the observed mean values. The toughest products occurred along a temperature–moisture 'ridge' extending from 20 to 25 per cent feed moisture and 150°C barrel temperature to about 30 per cent moisture and 200°C temperature. These products had textures that resembled the spectrum from licorice to leather. At 20 per cent initial moisture, as extrusion temperature was increased, the products became expanded and friable and finally fissured and fragile. The texture of all products from materials with 40 per cent moisture was soft and gummy.

While the sensory characteristics of food are the major factors in acceptability, nutritional quality reflects their true value to consumers. In this investigation we confined our study to protein quality. We were interested in how the nutritional quality of extruded cowpeas compared with that of raw meal and meal cooked in traditional ways. Two approaches to determining nutritional quality were utilized. One involved feeding cowpea products to groups of five weanling white rats, in diets containing levels of cowpea that provided 10 per cent of dietary protein, 8 per cent oil, 3.5 per cent mineral mix and 2.2 per cent vitamin mix. The remainder was corn starch. We calculated nutritional quality from the relationship between change in the animals' body weight and the amount of protein consumed, using the protein efficiency ratio (PER).

PER is defined as the weight gain divided by amount of protein consumed by animals receiving 10 per cent protein diets. The test showed that the quality of raw cowpea meal was unusually high compared with other legumes, but that it was improved by most cooking techniques. One exception was a slurry of meal that was cooked to a paste, then drum-dried. The relatively high temperature and slow drying probably damaged the protein. All the extrudates studied had higher PERs than did raw meal, and while they did not differ significantly from each other, values tended to decrease at lower initial moistures and higher barrel temperatures. The PER of Akara, a traditional deep-fried cowpea paste, made in this case from rehydrated meal, was similar to that of extrudates.

The second approach to measuring protein quality, an *in-vitro* digestibility technique, was developed to permit rapid assessment of quality changes by

Table 2. PER and *in-vitro* digestibility of some cowpea products

Sample	Mean PER*	Corrected PER	*In-vitro* digestibility
Raw meal	2.04a	1.44	77.8a
Steamed/drum-dried meal	2.31ab	1.63	81.2b
Akara	2.68c	1.89	82.8c
Casein	3.54d	2.50	89.8b
Extrudate moisture (%); temperature (°C)			
30; 200	2.57bc	1.81	83.3cd
20; 175	2.65c	1.86	84.4fg
20; 150	2.76c	1.95	84.8g
40; 175	2.80c	1.97	84.3efg
30; 150	2.80c	1.97	83.7de

* Means followed by the same letter do not differ significantly ($P<0.05$).

food processors. Based on an empirical relationship between pH drop in the digestate and apparent digestibility in the rat, percentage digestibility = $234.84 - 22.56x$ where x is the pH of a digestate comprising sample protein and mixtures of proteases after 20 min digestion.

The results generally agreed with the PER values, extending from low digestibilities for raw and drum-dried meal through intermediate values for Akara and extrudates, to high values for casein. These findings indicate that differences in PER correspond with differences in protein digestibility. This hypothesis is supported by the small, apparently random alteration in amino acid profile by processing (Figure 3) (Baker 1983).

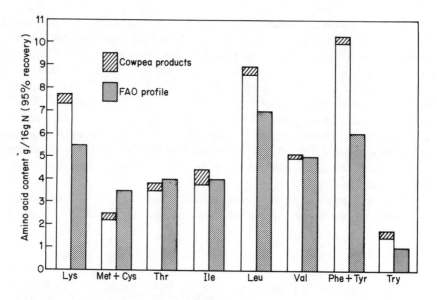

Figure 3.   Essential amino acid profile of cowpea products *vs* FAO profile

SUMMARY AND CONCLUSIONS

We have shown that prototype foods, having nutritional quality equal to that of traditional products, can be produced by extrusion cooking of cowpea meal. Although further work is needed to produce items with acceptable texture and flavour, the way is open to produce foods with nutritional values higher than traditional items. For example, one could enhance protein quality by incorporating complementary amino acids, either in the free form or within other ingredients that could be combined with cowpea meal before extrusion. Work had been undertaken, in our laboratory, to test such combinations, including cowpea and sorghum and cowpea–sorghum–peanut meals. Again, further efforts will be required to optimize the texture of such products. Vitamin and mineral fortification is also theoretically possible, although the effects on the stability and extrusion of the product would also have to be investigated.

References

Abadassi, J. 1984. Inheritance of resistance to brown blotch, *Septoria* leaf spot and scab diseases in cowpea. Abomey-Calavi, Benin, Université nationale du Benin (Diplome d'ingenieur agronome).

Abeygunawardena, D. V. W., and Perera, S. M. D. 1964. Virus diseases affecting cowpea in Ceylon. *Tropical Agriculture Mag. Ceylon Agricultural Society*, **120**, 181–204.

Aboul-Ata, A. E., Allen, D. J., Thottappilly, G., and Rossel, H. W. 1982. Variation in rate of seed transmission of cowpea aphid-borne mosaic virus in cowpea. *Tropical Grain Legume Bulletin* **25**, 2–7.

Aboul-Eid, H. Z., and Ghorab, A. I. 1974. Pathological effects of *Heterodera cajani* on cowpea. *Plant Disease Reporter* **58**, 1130–1133.

Abul-Nasr, S. 1956. Polyhedrosis-virus disease on the cotton leaf-worm *Prodenia litura* F. *Bulletin of the Entomological Society of Egypt* **41**, 591–620.

Abul-Nasr, S., and Assem, M. 1966. Some ecological aspects concerning the bean fly, *Melanagromyza phaseoli* (Tryon.). *Bulletin of the Entomological Society of Egypt* **50**, 163–172.

Acharya, S., and Singh, L. N. 1979. Genotype–environment interaction for quality attributes in cowpeas. *Forage Research* **5**, 147–150.

Aciola, A. 1971. Nova praga do feijoeiro no estado do Ceará—*Chalcodermus* sp. (Col.: Curculionidae). *Biologico* **37** (1), 17.

Adams, M. W. 1967. Basis of yield component compensation in crop plants with special reference to the field bean, *Phaseolus vulgaris*. *Crop Science* **7**, 505–510.

Adams, M. W., and Pipoly, J. J. 1980. Biological structure, classification and distribution of economic legumes. In Summerfield, R. J., and Bunting, A. H., eds, *Advances in Legume Science*, London, Her Majesty's Stationery Office, 1–16.

Adetiloye, P. O. 1980. Growth development and yield in sole and intercropped cowpea [*Vigna unguiculata* (L.) Walp.] and maize (*Zea mays* L.). Nsukka, Nigeria, University of Nigeria. 385 pp. (Ph.D. thesis).

Agble, F. 1972. Seed size heterosis in cowpeas [*Vigna unguiculata* (L.) Walp.]. *Ghana Journal of Science* **12**, 30–33.

Agboola, A. A. 1978. Influence of soil organic matter on cowpea's response to N fertilizer. *Agronomy Journal* **70** (1), 25–28.

Aggarwal, V. D., Muleba, N., Drabo, I., Souma, J., and Mbewe, M. 1984. Inheritance of *Striga gesnerioides* resistance in cowpea. In Parker, C., Mussel-man, L. J., Polhill, R. M., and Wilson, A. K., eds, *Proceedings Third International Symposium on Parasitic Weeds, Aleppo, Syria*, 7–9 May 1984, Aleppo, Syria, ICARDA (International Center for Agricultural Research in the Dry Areas), 143–147.

Agpad-Verzola, E., and Cortado, R. V. 1969. Preliminary experiment in the control of *Heliothis armigera* Hubner attacking pods. *Philippine Journal of Plant Industry* **34**, 187–191.

Agrawal, H. W. 1964. Identification of cowpea mosaic virus isolates. *Mededelingen LandbHoogesch. Wageningen* **64** (5), 1–53.

Agrawal, N. S., and Pandey, N. D. 1961. Bionomics of *Melanagromyza phaseoli* Coq. *Indian Journal of Entomology* **23**, 293–298.

Ahlawat, I. P. S., Saraf, C. S., and Singh, A. 1979. Response of spring cowpea to irrigation and phosphorus application. *Indian Journal of Agronomy* **24** (2), 237–239.

Ahmad, M. 1978. Whitefly (*Bemisia tabaci*) transmission of a yellow mosaic disease of cowpea (*Vigna unguiculata*). *Plant Disease Reporter* **62** (3), 224–226.

Ahmad, M. H., Eaglesham, A. R. J., and Hassouna, S. 1981a. Examining serological diversity of "cowpea" rhizobia by the ELISA technique. *Archives of Microbiology* **130**, 281–287.

Ahmad, M. H., Eaglesham, A. R. J., Hassouna, S., Seaman, B., Ayanaba, A., Mulongoy, K., and Pulver, E. L. 1981b. Examining the potential for inoculant use with cowpeas in West African soils. *Tropical Agriculture (Trinidad)* **58**, 325–335.

Ahmed, S. T., Magoon, M. L., and Mehra, K. L. 1972. Screening of a world collection of cowpea for field resistance to leaf spot and virus diseases. *SABRAO Newsletter* **4**, 103–106.

Ahn, C. S., and Hartmann, R. W. 1978. Interspecific hybridization between mung bean, *Vigna radiata* (L.) Wilczek and adzuki bean, *V. angularis* (Willd.) Ohwi & Ohashi. *Journal of the American Society of Horticultural Science* **103**, 3–6.

Aidar, H., and Kluthcouski, J. 1983. Multiple cropping systems in Brazil. Paper prepared for the international workshop on integrated pest control in tropical grain legumes, Goiânia, Brazil, 4–9 April 1983.

Aina, P. O. 1979. Tillage, seedbed configuration and mulching: effects on soil physical properties and responses of cassava, cowpeas, and maize. *Ife Journal of Agriculture* **1**(1), 26–34.

Akinola, J. O., and Davies, J. H. 1978. Effects of sowing date on forage and seed production of 14 varieties of cowpea (*Vigna unguiculata*). *Experimental Agriculture* **14**, 197–203.

Akobundu, I. O. 1978. Chemical weed control in cowpea and soybean in southern Nigeria. In 3 ème symposium sur le desherbage de cultures tropicales, Dakar, 1978. Paris, *Columna*, **2**, 475–482.

Akobundu, I. O. 1979. An evaluation of selected cowpea cultivars for herbicide tolerance. *Proceedings of the Weed Science Society of Nigeria*, **9**, 69–74.

Akobundu, I. O. 1982. Weed control in cowpea in the humid tropics. *Weed Science* **30**, 331–334.

Akobundu, I. O. 1983a. Tillage in relation to cultural weed control in grain legumes. Paper prepared for the international workshop on integrated pest control for grain legumes, EMBRAPA, Goiâna, Goiás, Brazil, 3–9 April 1983.

Akobundu, I. O. 1983b. Weed management in relation to herbicide use and selectivity in grain legumes. Paper prepared for the international workshop on integrated pest control for grain legumes, EMBRAPA, Goiânia, Goiás, Brazil, 3–9 April 1983.

Akobundu, I. O. in press. Advancing weed control strategies for developing countries. In Commonwealth Agricultural Bureaux (CAB), Advancing Agricultural Production in Africa, Farnham Royal, UK, CAB (in press).

Alabi, O. 1981. Fungicidal control of seed-borne infection of *Colletotrichum capsici* (Syd.) Butler & Bisby, the causal agent of cowpea brown blotch. Paper prepared for the 11th annual conference of the Nigerian Society for Plant Protection, University of Nigeria, Nsukka, February 1981.

Alconero, R., and Santeago, A. 1973. *Phaseolus lathyroides* as a reservoir of cowpea mosaic virus in Puerto Rico. *Phytopathology* **63**, 120–123.

Ali, M. A., Trabulsi, I. Y., and Abd Elsamea, M. E. 1981. Antagonistic interaction between *Meloidogyne incognita* and *Rhizobium leguminasarum* on cowpea. *Plant Disease* **65**, 432–435.

Allen, D. J. 1977. The international cowpea disease nursery program: towards multiple virus resistance. *Tropical Grain Legume Bulletin* **8**, 28–32.

Allen, D. J. 1979. New disease records from grain legumes in tropical Africa. *FAO Plant Protection Bulletin* **27**, 134–136.

Allen, D. J. 1980. Identification of resistance to cowpea mottle virus. *Tropical Agriculture (Trinidad)* **57** (4), 325–332.

Allen, D. J. 1981. Two bacterial diseases of cowpea in East Africa; a reply. *Tropical Pest Management* **27**, 144.

Allen, D. J. 1983. *The Pathology of Tropical Food Legumes. Disease resistance in crop improvement.* Chichester, UK, John Wiley and Sons.

Allen, D. J., Anno-Nyako, F. O., Ochieng, R. S., and Ratinum, M. 1981a. Beetle transmission of cowpea mottle and southern bean mosaic viruses in West Africa. *Tropical Agriculture (Trinidad)* **58** (2), 171–175.

Allen, D. J., Emechebe, A. M., and Ndimande, B. 1981b. Identification of resistance to diseases of the African savannas. *Tropical Agriculture (Trinidad)* **58**, 267–274.

Allen, D. J., Nebane, C. L. N., and Raji, J. A. 1981c. Screening for resistance to bacterial blight in cowpea. *Tropical Pest Management* **27**, 218–224.

Allen, D. J., Thottappilly, G., and Rossel, H. W. 1982. Cowpea mottle virus: field resistance and seed transmission in virus-tolerant cowpea. *Annals of Applied Biology* **100**, 331–336.

Allen, R. 1980. *How to Save the World. Strategy for world conservation.* London, International Union for Conservation of Nature and Natural Resources.

Altieri, M. A., Francis, C. A., van Schoonhoven, A., and Doll, J. D. 1978. A review of insects prevalent in maize (*Zea mays* E.) and bean (*Phaseolus vulgaris* L.) polycultural systems. *Field Crops Research*, 33–49.

Amin, K. S., Baldev, B., and Williams, F. J. 1976. Epidemiology of *Cercospora cruenta* on cowpea. *FAO Plant Protection Bulletin* **24**, 43–44.

Amoako-Atta, B., Omoio, E. O., and Kidega, E. K. 1983. Influence of maize, cowpea and sorghum intercropping systems on stem–pod borer infestations. *Insect Sciences Application* **4**, 47–57.

Amosu, J. O. 1974. Reaction of cowpea to root knot nematodes *M. incognita* in western Nigeria. *Nigerian Agricultural Journal* **11**, 165–169.

Amosu, J. O., and Franckowiak, J. D. 1974. Inheritance of resistance to root-knot nematode in cowpea. *Plant Disease Reporter* **58**, 361–363.

Anderson, C. W. 1955a. *Vigna* and *Crotalaria* viruses in Florida. 2: notations concerning cowpea mosaic virus (*Marmor vignae*). *Plant Disease Reporter* **39**, 349–352.

Anderson, C. W. 1955b. *Vigna* and *Crotalaria* viruses in Florida. 1: preliminary report on a strain of cucumber mosaic virus obtained from cowpea plants. *Plant Disease Reporter* **39**, 346–348.

Anderson, C. W. 1957. Seed transmission of three viruses in cowpea. *Phytopathology* **47**, 55 (abstract).

Andres, L. A. 1981. Biological control of naturalized and native plants. Conflicting interests. In Papavizas, G. C., ed., *Biological Control in Crop Production*, Totowa, NJ, Allanheld, Osmun and Co., 341–349.

Andrews, D. J. 1972a. Intercropping with guineacorn—a biological cooperative? Part 1. *Samaru Agricultural Newsletter* **14** (2), 20–22.

Andrews, D. J. 1972b. Intercropping with guineacorn—a biological cooperative? Part 2. *Samaru Agricultural Newsletter* **14** (3), 40–42.

Angadi, S. P., Subramani, A., and Kulkarni, R. S. 1978. Genetic variability for some quantitative traits in cowpea. *Agriculture Research Journal of Kerala* **16**, 60–62.

Angus, J. F., Hasegana, S., Hsiao, T. C., Liboon, S. P., and Zandstra, H. G. 1983. The water balance of post-monsoonal dryland crops. *Journal of Agricultural Science Cambridge* **101**, 699–710.

Anjos, J. R. N., and Lin, M. T. 1984. Bud blight of soybeans caused by cowpea severe mosaic virus in central Brazil. *Plant Disease* **68**, 405–407.

Anno-Nyako, F. O. 1983. Identification and partial characterization of a mild mottle disease in soybean, *Glycine max* (L.) Merril in Nigeria. Kumasi, Ghana, University of Science and Technology (Ph.D. thesis).

Anno-Nyako, F. O., Vetten, H. J., Allen, D. J., and Thottappilly, G. 1983. The relation between cowpea golden mosaic and its vector, *Bemisia tabaci*. *Annals of Applied Biology* **102**, 319–323.

Anonymous. 1955. Overseas news. *Commonwealth Phytopathological News* **1**, 41–45.

Anonymous. 1957. Colombia, Ministerio de Agricultura, Oficina de Investigaciones Especiales. Sexto informe anual mayo 1 de 1955 a mayo 1 de 1956. *Rev. Nov. Agric., Bogotá*, **51**, 1–89.

Anonymous. 1983. Terms, definitions and abbreviations used in WSSA publications. In Weed Science Society of America (WSSA), *Herbicide Handbook*, Champaign, IL, WSSA, xviii–xxiv.

Anonymous. 1984. Collaborative research in the international agricultural research and development network: a case study progress report of the Bean/Cowpea Collaborative Research Support Program (CRSP). East Lansing, MI, Michigan State University, 97.

Apparao, S., and Reddy, B. M. 1976. Crumpled petal mutants in black gram and cowpea. *Indian Journal of Genetics and Plant Breeding* **35**, 391–394.

Araújo, E., and Moreno, R. 1980. Progresso de doenças foliares de feijão macassar [*Vigna unguiculata* (L.) Walp.] em diferentes sistemas de cultivos. *Fitopatología Brasileira* **5**, 31–38.

Araújo, J. P. 1983. Five year review of the Brazilian cooperative research program in cowpea. Goiânia, Goiás, Brazil, EMBRAPA/CNPAF, 179-74000.

Araújo, J. P. P. de, Rios, G. P., Watt, E. E., Neves, B. P. das, Fageria, N. K., Oliveira, I. P., Guimarães, C. M., Filho, A. S. in press. Cultura do caupi, *Vigna unguiculata* (L.) Walp., descrição e recomendaçãoes técnicas de cultivo. Goiânia, Goiás, Brazil, EMBRAPA/CNPAF, Circ. Técnica 17.

Araújo, J. P. P. de, Watt, E. E., Rios, G. P., Neves, B. P. das, Kluthcouski, J., and

Guimarães, C. M. 1980. Situação do caupi no Brasil. Produção, problemas e pesquisa. Goiânia, Goiás, Brazil, EMBRAPA/CNPAF, Internal report.

Armstrong, G. M., and Armstrong, J. K. 1950. Biological races of *Fusarium* causing wilt of cowpeas and soybeans. *Phytopathology* **40**, 181–193.

Aryeetey, A. N., and Laing, E. 1973. Inheritance of yield components and their correlation with yield in cowpea [*Vigna unguiculata* (L.) Walp.]. *Euphytica* **22**, 386–392.

Ashraf, M., Balogun, P., *et al.* 1984. A case study of on-farm adaptive research at Bida ADP, Nigeria. Paper prepared for the farming system symposium, Kansas State University of Agriculture and Applied Science, Manhattan, 7–10 October 1984.

Assis, L. D. de. 1976. Grau de incidência do gorgulho do feijão-de-corda *Callosobruchus maculatus* (Fabr.) no municipio de Caucaia, Ceará, Brasil. *Gencia Agronomicé* **6** (1–2), 35–36.

Atiri, G. I. 1982. Virus–vector–host relationship of cowpea aphid-borne mosaic virus in cowpea [*Vigna unguiculata* (L.) Walp.]. Ibadan, University of Ibadan (Ph.D. thesis).

Atiri, G. I., Ekpo, E. J. A., and Thottappilly, G. 1984. The effect of aphid resistance in cowpea on infestation and development of *Aphis craccivora* and the transmission of cowpea aphid-borne mosaic virus. *Annals of Applied Biology* **104**, 339–346.

Atkins, C. A., Peoples, M., and Pate, J. S. 1982. Studying the postanthesis performance of cowpea. Ibadan, IITA, Internal progress report.

Ayala, A., and Ramirez, C. T. 1964. Host range distribution and biology of the reniform nematode *Rotylenchulus reniformis* with special reference to Puerto Rico. *Journal of Agriculture of the University of Puerto Rico* **48**, 140–161.

Ayanaba, A., Asanuma, S., and Munns, D. N. 1983. An agar plate method for rapid screening of *Rhizobium* for tolerance to acid–aluminum stress. *Soil Science Society of America Journal* **47**, 256–258.

Ayeni, A. O. 1982. Responses of maize (*Zea mays* L.), cowpea [*Vigna unguiculata* (L.) Walp.] and maize/cowpea intercrop to weed interference in a subhumid tropical environment. Ithaca, NY, Cornell University (Ph.D. thesis).

Ayensu, E. S. in press. Taxonomy, systematics and ecology of *Striga*. In ICSU/UNESCO (International Council of Scientific Unions/United Nations Educational, Scientific and Cultural Organization), Proceedings of a Workshop on the Biology and Control of *Striga*, Dakar, Senegal, November 1983, Paris, France, ICSU.

Ayoade, K. A. 1969. Insecticide control of the pod borer, *Maruca testulalis* Geyer, on Westbred cowpea (*Vigna* sp.). *Bulletin of the Entomological Society of Nigeria* **2**, 23–33.

Ayoub, S. M. 1961. *Pratylenchus zeae* found on corn, milo, and three suspected new hosts in California. *Plant Disease Reporter* **45**, 940.

Ba, A. T. 1983. Evidence of enzyme activities in the haustorium of *Striga gesnerioides* (Scrophulariaceae). In ICRISAT (International Crops Research Institute for the Semi-Arid Tropics), Proceedings of the Second International *Striga* Workshop, 5–8 October 1981, IDRC/ICRISAT, Ouagadougou, Upper Volta, Patancheru, India, ICRISAT, 39–42.

Ba, A. T. in press. Morphology, anatomy and ultrastructure of some parasitic species of the genus *Striga* (Scrophulariaceae). In ICSU/UNESCO (International Council of Scientific Unions/United Nations Educational, Scientific and Cultural Organization), Proceedings of a Workshop on the Biology and Control of *Striga*, Dakar, Senegal, November 1983, Paris, France, ICSU.

BADP (Bida Agricultural Development Project). 1983. Project mid-term implementation report, vol. 2. Lagos, Nigeria, Ministry of Agriculture.

Badra, T., and Yousif, G. M. 1979. Comparative effects of potassium levels on growth and mineral composition of intact and nematized cowpea and sour orange seedlings. *Nematologia Mediterranea* **7**, 21–27.

Bagyaraj, D. J., and Manjunath, A. 1980. Response of crop plants to VA mycorrhizal inoculation in an unsterile Indian soil. *New Phytologist* **85**, 33–36.

Baker, E. A. 1983. Protein quality of raw extruded and drum-dried cowpea meals as determined by in vivo and in vitro methods. Athens, GA, University of Georgia (M.Sc. thesis).

Baker, K. F., Snyder, W. C., and Hansen, H. N. 1940. Some hosts of *Verticillium* in California. *Plant Disease Reporter* **24**, 424–425.

Balasubramanian, A., Prabakaran, J., and Sundaram, S. 1980. Influence of single and multistrain rhizobial inoculants on cowpea. *Madras Agricultural Journal* **67**, 538–540.

Balasubramanian, P., and Sivakumar, C. V. 1982. Action of phosfen (Tributyl-2,4-dichlorobenzylphosphonium chloride) on *Meloidogyne incognita*. *Indian Journal of Nematology* **12**, 285–290.

Ballon, R. B., and York, T. L. 1959. Crossing the common and scarlet bean (*Phaseolus* spp.) with *Vigna* species. *Philippine Agriculturalist* **42**, 454–455.

Bancroft, J. B. 1971. Cowpea chlorotic mottle virus. Kew, UK, CMI (Commonwealth Mycological Institute), CMI/AAB Descriptions of plant viruses 49.

Bancroft, J. D., Hiebert, E., Rees, M. W., and Markham, R. 1968. Properties of cowpea chlorotic mottle virus, its protein and nucleic acid. *Virology* **34**, 224–239.

Bansal, K. N., and Singh, H. G. 1975. Interrelationship between sulfur and iron in the prevention of iron chlorosis in cowpea. *Soil Science* **120** (1), 20–24.

Bapna, C. S., and Joshi, S. N. 1973. A study on variability following hybridization in *Vigna sinensis* (L.) Savi. *Madras Agricultural Journal* **60**, 1369–1372.

Bapna, C. S., Joshi, S. N., and Kabaria, M. M. 1972. Correlation studies on yield and agronomic characters in cowpea. *Indian Journal of Agronomy* **17**, 321–324.

Barrios-G., A., and Ortega-Y., S. 1975. Tuy: nuevo cultivar de frijol (*Vigna unguiculata* L.). *Agronomía Tropical* **25** (2), 103–106.

Barton, D. W., Butler, L., Jenkins, I. A., Rick, C. M., and Young, P. A. 1955. Rules for nomenclature in tomato genetics. *Journal of Heredity* **46**, 22–26.

Bastos, J. A. M. 1969. Substâncias orgânicas como atraentes para a postura do gorgulho, *Callosobruchus analis* Fabr., no feijão de corda, *V. sinensis* Endl. *Pesquisa Agropecuaria Brasileira* **4**, 127–128.

Bastos, J. A. M. 1973a. Avaliação dos prejuízos causados pelo gorgulho, *Callosobruchus maculatus*, em amostras de feijão de corda, *V. sinensis*, colchidas em Fortaleza, Ceará. *Pesquisa Agropecuaria Brasileira* **8**, 131–132.

Bastos, J. A. M. 1973b. Influência do tamanho das larvas do manhoso, *Chalcodermus bimaculatus* F., na emergência de adultos. *Pesquisa Agropecuaria do Nordeste, SUDENE, Recife* **5** (1), 45–47.

Bastos, J. A. M. 1974a. Controle de gorgulho de feijão-de-corda, *Callosobruchus maculatus* (Fabr., 1792), com brometo de metila. *Turrialba* **24** (2), 230–232.

Bastos, J. A. M. 1974b. Controle do manhoso, *Chalcodermus bimaculatus* Fied., no campo, com inseticidas orgânicos sintéticos. *Fitossanidade* **1** (1), 7–9.

Bastos, J. A. M. 1974c. Influência das diferentes fases de desenvolvimento de feijão-de-corda, *V. sinensis* Endl., na preferência do manhoso adulto, *Chalcodermus bimaculatus* Fiedler. *Fitossanidade* **1** (1), 2–3.

Bastos, J. A. M. 1974d. Período pupal do manhoso, *Chalcodermus bimaculatus* Fiedler, 1936, a diversas temperaturas. *Fitossanidade* **1** (1), 3–5.

Bastos, J. A. M. 1974e. Profundidade de penetração de larvas do manhoso, *Chalcodermus bimaculatus* Fiedler, em solos arenosos. *Fitossanidade* **1** (1), 1–2.

Bastos, J. A. M. 1981. *Principais pragas das culturas e seus controles.* Livraria Nobel S.A., São Paulo, SP.

Bastos, J. A. M., and Assunção, M. V. 1975. Influência de diferentes tipos de embalagens na ação do phostoxin contra o gorgulho do feijão-de-corda, *Callosobruchus maculatus* Fabr. *Ciencia Agronomía* **5** (1–2), 7–11.

Bates, G. R. 1957. Botany and plant pathology. In Ministry of Agriculture, Report of Ministry of Agriculture for Rhodesia and Nyasaland, 1955–56, 79–86.

Baudet, J. C. 1978. Prodrome d'une classification generique des Papilionaceae. Phaseoleae. *Bulletin du Jardin botanique national de Belgique* **48**, 183–220.

Baudet, J. C., and Maréchal, R. 1976. Signification taxonomique de la presence de poils uncinules chez certains genres de *Phaseolus* et d'Hedysareae (Papilionaceae). *Bulletin du Jardin botanique national de Belgique* **46**, 419–426.

Baudoin, J. P. 1981a. L'amelioration du haricot de Lima (*Phaseolus lunatus* L.) en vue de l'intensification de sa culture en régions tropicales de basse altitude. Gembloux, Faculté des sciences agronomiques de Gembloux. (Thèse de doctorat).

Baudoin, J. P. 1981b. Observations sur quelques hybrides interspecifiques avec *Phaseolus lunatus* L. *Bulletin des recherches agronomiques des Gembloux* **16** (4), 273–286.

Bawden, F. C. 1956. Reversible host induced changes in a strain of tobacco mosaic virus. *Nature* **177** (8), 302–304.

BCPC (British Crop Protection Council). 1983. *The pesticide manual.* London, BCPC.

Behncken, G. M., and Maleevsky, L. 1977. Detection of cowpea aphid-borne mosaic virus in Queensland. *Australian Journal of Experimental Agriculture and Animal Husbandry* **17**, 674–678.

Bellington, R. V. I. 1971. Cowpea diseases in the western zone. Ikiriguru Research Notes, Tanzania **49**, 1–2.

Beningo, D. A., and Favali-Hedayat, M. A. 1977. Investigations on previously unreported or noteworthy plant viruses and virus diseases in the Philippines. *FAO Plant Protection Bulletin* **25** (2), 78–84.

Beniwal, S. P. S., and Nene, Y. L. 1983. Viral diseases of major grain legumes in India. Paper prepared for the International Working Group on legume viruses meeting, Queensland, Australia, 11–12 August.

Bennett, F. D., Yaseen, N., and Cruttwell, R. E. 1970. Biological control of weeds (for Papua New Guinea). In CIBC (Commonwealth Institute of Biological Control), Report of the Commonwealth Institute of Biological Control, Curepe, Italy, CIBC, 92–93.

Bentham, G., and Hooker, F. J. D. 1865. Leguminosae. In *Genera Plantarum* **1**, 434–600.

Bernado, E. N. 1969. Effects of six host plants on the biology of black bean aphid, *Aphis craccivora* Koch. *Philippine Entomologist* **1**, 287–292.

Bhaskaraiah, K. B., Shivashankar, G., and Virupakshappa, K. 1980. Hybrid vigor in cowpea. *Indian Journal of Genetics and Plant Breeding* **40**, 334–337.

Bhowal, J. G. 1976. Inheritance of pod length, pod breadth and seed size in a cross between cowpea and catjang bean. *Libyan Journal of Science* **6**, 17–21.

Bliss, F. A. 1975. Cowpeas in Nigeria. In Milner, M., ed., *Proceedings of a Symposium on Nutritional Improvement of Food Legumes by Breeding*, 3–5 July, 1972, New York, NY, United Nations Protein Advisory Group, 151–158.

Bliss, F. A., and Robertson, D. G. 1971. Genetics of host reaction in cowpeas to cowpea yellow mosaic virus and cowpea mottle virus. *Crop Science* **11** (2), 258–262.

Bliss, F. A., Barker, L. N., Franckowiak, J. D., and Hall, T. C. 1973. Genetic and environmental variation of seed yield, yield components, and seed protein quantity and quality of cowpea. *Crop Science* **13**, 656–660.

Bock, K. R. 1971. Notes on East African plant virus diseases. 1: cowpea mosaic virus. *East African Agriculture and Forestry Journal* **37**, 60–62.

Bock, K. R. 1973. East African strains of cowpea aphid-borne mosaic virus. *Annals of Applied Biology* **74**, 75–83.

Bock, K. R., and Conti, M. 1974. Cowpea aphid-borne mosaic virus. Kew, UK, CMI (Commonwealth Mycological Institute), CMI/AAB Descriptions of plant viruses 134.

Boettinger, C. F., and Bowers, J. L. 1975. Hot water treatment, a possible control for seed-borne bacterial blight in southern pea. *HortScience* **10**, 148.

Booker, R. H. 1965. Pests of cowpeas and their control in northern Nigeria. *Bulletin of Entomological Research* **55**, 663–672.

Boonkerd, N., and Weaver, R. W. 1982. Survival of cowpea rhizobia in soil as affected by soil temperature and moisture. *Applied and Environmental Microbiology* **43**, 585–589.

Booth, C. 1971. *The grains* Fusarium. Kew, UK, Commonwealth Mycological Institute.

Booth, C. 1973a. *Hoplolaimus seinhorsti* an endoparasitic nematode of cowpea in Nigeria. *Plant Disease Reporter* **57**, 798–799.

Booth, C. 1973b. Nematodes as pests of yams in Nigeria. *Mededelingen Fakulteit Laundbouwwelenschappen* **38**, 841–852.

Bordia, P. C., Yadavendra, J. P., and Kumar, S. 1973. Genetic variability and correlation studies in cowpea (*Vigna sinensis* L. Savi ex Hask.). *Rajasthan Journal of Agricultural Science* **4**, 39–44.

Boswell, K. F., and Gibbs, A. J. 1983. Viruses of legumes: descriptions and keys from virus identification and data exchange. Canberra, Australian National University.

Boulter, D., Evans, I. M., Thompson, A., and Yarwood, A. 1975. The amino acid composition of *Vigna unguiculata* meal in relation to nutrition. In Milner, M., ed., *Proceedings of a Symposium on Nutritional Improvement of Food Legumes by Breeding*, 3–5 July, 1972, New York, NY, United Nations Protein Advisory Group.

Bozarth, R. F., and Shoyinka, S. A. 1979. Cowpea mottle virus. Kew, UK, CMI (Commonwealth Mycological Institute), CMI/AAB Descriptions of plant viruses 212.

Braga Paiva, J., Almeida, F. C. G., and Albuquerque, J. J. L. de. 1971. Spacing and density of cowpeas in Ceará. *Ciencia Agronomia* **1** (1), 3–6.

Braithwaite, R. A. I. 1978. Chemical weed control in Bodie bean cowpea in Trinidad. *PANS* **24** (2), 177–180.

Brantley, B. B., and Kuhn, C. W. 1970. Inheritance of resistance to southern bean mosaic virus in southern pea (*Vigna sinensis*). *Proceedings of the American Society of Horticultural Science* **95**, 155–158.

Bressani, R. 1975. Legumes in human diets and how they might be improved. In Milner, M., ed., *Proceedings of a Symposium on Nutritional Improvements of Food Legumes by Breeding*, 3–5 July, 1972, New York, NY, United Nations Protein Advisory Group.

Bressani, R., and Elías, L. G. 1979. The nutritional role of polyphenols in beans. In IDRC (International Development Research Centre), *Polyphenols in Cereals and Legumes*, Ottawa, Canada, IDRC, IDRC-145e.

Bressani, R., and Elías, L. G., 1980. Nutritional value of legume crops for humans and animals. In Summerfield, R. J., and Bunting, A. H., eds., *Advances in Legume Science*, London, UK. Her Majesty's Stationery Office, 135–155.

Bressani, R., and Scrimshaw, N. S. 1961. The development of INCAP vegetable mixtures, 1: basic animal studies. In National Academy of Sciences (NAS) *Meeting protein needs of infants and preschool children*, Washington, DC, NAS, publication 843.

Bressani, R., Elías, L. G., Huezo, M. T., and Braham, J. E. 1977a. Estudios sobre la producción de harinas precocidas de frijol ej caupé, solos y combinados mediatite cocción–deshidratación. *Archivos Latinoamericanos de Nutrición* **27**, 2, 247–260.

Bressani, R., Elías, L. G., and Molina, M. R. 1977b. Estudios sobre la digestibilidad de la proteína de varias especies de leguminosas. *Archivos Latinoamericanos de Nutrición* **27**, 215–231.

Bressani, R., Navarrete, D. A., Hernández, E., Gutiérrez, O., Vargas, E., and Elías, L. G. 1982. Studies on the protein digestibility of common beans (*Phaseolus vulgaris*) in adult human subjects. Paper prepared for a conference on nutrition, Oestrich-Winkel, Germany, 6–8 September 1982.

Bridge, J. 1972. Plant parasitic nematodes of irrigated crops in the northern states of Nigeria. Samaru, Nigeria, Institute for Agricultural Research, Samaru miscellaneous paper 42.

Brittingham, W. H. 1946. A key to the horticultural groups of varieties of southern pea, *Vigna sinensis*. *Proceedings of the American Society for Horticultural Science* **48**, 478–480.

Brittingham, W. H. 1950. The inheritance of date of pod maturity, pod length, seed shape, and seed size in the southern pea, *Vigna sinensis*. *Proceedings of the American Society of Horticultural Science* **56**, 381–388.

Brown, L. G., and Dukes, P. D. 1982. Evaluating for resistance to white leaf spot incited by *Aristatoma oeconomicum* (Ellis and Tracy) Tehon in southern pea [*Vigna unguiculata* (L.) Walp.]. *Phytopathology* **72**, 999.

Brown, O. D. R. 1971. The influence of the plant parasitic nematode *Helicotylenchus dihystera* (Cobb) on the growth and nitrogen fixation in the southern pea *Vigna sinensis* (L.) Endl. College Station, Texas A&M University. (Ph.D. thesis).

Brown, O. D. R. 1972. The influence of the plant parasitic nematode *Helicotylenchus dihystera* (Cobb) on the growth and nitrogen fixation in the southern pea *Vigna sinensis* (L.). Endl. *Dissertation Abstract International* **32B**, 5563.

Brunt, A. A. 1983. Cowpea mild mottle virus. In Boswell, K. F., and Gibbs, A. J., eds, *Viruses of Legumes: descriptions and keys from virus identification and data exchange*, Canberra, Australian National University, 53–54.

Brunt, A. A., and Kenten, R. H. 1973. Cowpea mild mottle, a newly recognised virus infecting cowpea (*Vigna unguiculata*) in Ghana. *Annals of Applied Biology* **74**, 67–74.

Brunt, A. A., and Kenten, R. H. 1974. Cowpea mild mottle virus. Kew, CMI (Commonwealth Mycological Institute), CMI/AAB Descriptions of plant viruses 140.

Buckmire, K. U. 1978. Pests of grain legumes and their control in the Commonwealth Caribbean. In Singh, S. R., Van Emden, H. G., and Taylor, T. A., eds, *Pests of Grain Legumes: ecology and control*, New York, NY, Academic Press, 179–184.

Bunting, A. H. 1971. Productivity and profit or is your vegetative phase really necessary? *Annals of Applied Biology* **53**, 351–362.

Burkhill, I. H. 1953. Habits of man and the origins of the cultivated plants of the old world. *Proceedings of the Linnean Society, London* **164**, 12–42.

Burkholder, W. H. 1930. The bacterial diseases of the bean: a comparative study. Ithaca, NY, Cornell University Agricultural Experiment Station, Memoir 127.

Burkholder, W. H. 1944. *Xanthomonas vignicola* sp. nov. pathogenic on cowpeas and beans. *Phytopathology* **34**, 430–432.

Cabezas, M. T., Cuévas, B., Murillo, B., Elías, L. G., and Bressani, R. 1982a. Evaluación nutritional de la sustitución de la harina de soyà y sorgo per harina de frijol caupí crudo (*Vigna sinensis*). *Archivos Latinoamericanos de Nutrición*, **32**, 559–578.

Cabezas, M. T., García, J., Murillo, B., Elías, L. G., and Bressani, R. 1982b. Valon nutritivo del frijol caupí crudo y procesado. *Archivos Latinoamericanos de Nutrición* **32**, 543–558.

Caner, J., Silberschmidt, K., and Flores, E. 1969. Ocorrência do virus do mosaico da *Vigna* no Estado de São Paulo. *Biologico (São Paulo)* **35** (1), 13–16.

Canerday, T. D., and Chalfant, R. B. 1969. An arrestant and feeding stimulant for the cowpea curculio. *Chalcodermus aeneus*. *Journal of the Georgia Entomological Society* **4**, 59–64.

Capinpin, J. M. 1935. A genetic study of certain characters in varietal hybrids of cowpea. *Philippine Journal of Science* **57**, 149–164.

Capinpin, J. M., and Irabagon, T. A. 1950. A genetic study of pod and seed characters in *Vigna. Philippine Agriculturalist* **33**, 263–277.

Capoor, S. P., and Varma, P. M. 1956. Studies on mosaic disease of *Vigna cylindrica* Skeels. *Indian Journal of Agricultural Science* **26**, 95–103.

Capoor, S. P., Varma, P. M., and Uppal, B. N. 1947. A mosaic disease of *Vigna catjang* Walp. *Current Science* **16**, 151.

Cardoso, J. E., and Mesquita, E. L. 1981. Ocorrência de mela do feijoeiro em germoplasmas de caupi no Acre. Cinc. Těcnico, UEPAE/Acre.

Carneiro, J. da S. 1983. Reconhecimento e controle das principais pragas de campo e de grãos armazenados de culturas temporárias no Amazonas. Manaus, Amazonas, EMBRAPA/UEPAE, Circ. Técnica 7.

Castro, Z. B. de, Cavalcante, R. D., Santos, O. M., and Cavalcante, S. 1975. Ocorrência da *Diabrotica speciosa* (Germar, 1824) em diversas culturas no estado do Ceará. *Fitossanidade* **1** (2), 51–52.

Cavalcante, M. L. S., Cavalcante, R. D., and Castro, Z. B. 1975. Cigarrinha verde (*Empoasca* sp.) praga do feijão macassar (*Vigna sinensis* Endl.) no Ceará. *Fitossanidade* **1** (3), 83–84.

Cavalcante, M. L. S., Pedrosa, F. N. T., and Araújo, F. E. 1974. *Pantomorus glaucus* (Perty, 1830), praga de diversas culturas no Estado do Ceará. *Fitossanidade* **1** (1), 22.

Cavalcante, R. D., Melo, Q. S., Cavalcante, M. L. S., and Chagas, F. A. 1979. *Promecops* sp. (Coleoptera, Curculionidae), praga do feijoeiro macassar no Ceará. *Fitossanidade* **3** (1–2), 60.

Caveness, F. E. 1967a. End of tour progress report on the nematology project. Lagos, Nigeria, Ministry of Agriculture and Natural Resources, western region.

Caveness, F. E. 1967b. Shadehouse host ranges of some Nigerian nematodes. *Plant Disease Reporter* **51**, 330–337.

Caveness, F. E. 1968. A survey of plant-parasitic nematodes in Nigeria. Shafter, CA, USDA (United States Department of Agriculture).

Caveness, F. E. 1971. The distribution of plant-parasitic nematodes in Nigeria. *Journal of the Association for the Advancement of Agricultural Sciences in Africa* **1**, 35–40.

Caveness, F. E. 1972. Changes in plant parasitic nematode populations on newly cleared land. *Nematropica* **2**, 1–2, 15–16 (abstract).

Caveness, F. E. 1973. Nematode attack on cowpeas in Nigeria. In IITA, Proceedings of the First Grain Legume Improvement Workshop, Ibadan, Nigeria, 29 October–2 November 1973. Ibadan, IITA.

Chalfant, R. B. 1972. *Cowpea curculio*: control in southern Georgia. *Journal of Economic Entomology* **66**, 727–729.

Chalfant, R. B. 1976. Chemical control of insect pests of the southern pea in Georgia. *University of Georgia Research Bulletin* 179.

Chalfant, R. B., and Canerday, T. D. 1972. Feeding and oviposition of the cowpea curculio and laboratory screening of southern pea varieties for insect resistance. *Journal of the Georgia Entomological Society* **7**, 272–277.

Chalfant, R. B., and Johnson, A. W. 1972. Field evaluation of pesticides applied to the soil for control of insects and nematodes affecting southern peas in Georgia. *Journal of Economic Entomology* **65**, 1711–1713.

Chalfant, R. B., and Young, J. R. 1982. Chemigation or application of insecticide through overhead sprinkler irrigation systems to manage insect pests affecting vegetable and agronomic crops. *Journal of Economic Entomology* **75**, 237–241.

Chalfant, R. B., and Young, J. R. 1984. Management of insect pests of broccoli, cowpeas, spinach, tomatoes and peanuts and reduction of watermelon virus with chemigation by insecticides in oils. *Journal of Economic Entomology* **77**, 1323–1326.

Chalfant, R. B., Mullenix, B., and Nilakhe, S. S. 1982. Southernpeas: interrelationships among growth stage, insecticide applications and yield in Georgia. *Journal of Economic Entomology* **75**, 405–409.

Chalfant, R. B., Suber, E. F., Canerday, T. D. 1972. Resistance of southern peas to the cowpea curculio in the field. *Journal of Economic Entomology* **65**, 1679–1682.

Chambliss, O. L. 1974. Green seedcoat: a mutant in southernpea of value to processing industry. *HortScience* **9**, 126.

Chambliss, O. L. 1979. 'Freezegreen' southernpea. *HortScience* **14**, 193.

Chambliss, O. L., and Rymal, K. S. 1980. Pod-wall toughness as a measure of cowpea curculio resistance in southernpeas. *HortScience* **15**, 418.

Chambliss, O. L., and Rymal, K. S. 1982. Response of cowpea curculios to volatile extracts from southernpea pods. *HortScience* **17**, 152.

Chambliss, O. L., Hossain, A. M., Johnson, Jr., W. C., and Rymal, K. S. 1982. Inheritance of pod-wall strength in southernpea (cowpea). *HortScience* **17**, 503.

Chancellor, R. J., Parker, C., and Teferedegn, T. 1971. Stimulation of dormant weed seed germination by 2-chloroethyl-phosphoric acid. *Pesticide Science* **2**, 35–37.

Chandola, R. P., Trehan, K. B., and Bagrecha, L. R. 1970. Varietal resistance to *Bruchus* sp. in cowpea (*Vigna sinensis*) under storage conditions. *Current Science* **38**, 370–371.

Chandrappa, H. M., Manjunath, A., Vishwanatha, S. R., and Shivashankar, G. 1974. Promising introductions in cowpea and their genetic variability. *Current Research (Bangalore, India)* **3**, 123–125.

Chang, C. A. 1983. Rugose mosaic of asparagus bean caused by dual infection with cucumber mosaic virus and blackeye cowpea mosaic virus. *Plant Protection Bulletin Taiwan* **25** (3), 177–190.

Chant, S. R. 1959. Viruses of cowpea, *Vigna unguiculata* L. (Walp.) in Nigeria. *Annals of Applied Biology* **47** (3), 565–573.

Chant, S. R. 1960. The effect of infection with tobacco mosaic and cowpea yellow mosaic viruses on the growth rate and yield of cowpea in Nigeria. *Empire Journal of Experimental Agriculture* **28**, 114–120.

Chaturvedi, G. S., Aggarwal, P. K., and Sinha, S. K. 1980. Growth and yield of determinate and indeterminate cowpeas in dryland agriculture. *Journal of Agricultural Science Cambridge* **94**, 137–144.

Chauhan, G. S., and Joshi, R. K. 1980. Path analysis in cowpea. *Tropical Grain Legume Bulletin* **20**, 5–8.

Chen, N. C., Baker, L. R., and Honma, S. 1983. Interspecific crossability among four species of *Vigna* food legumes. *Euphytica* **32**, 925–937.

Chenulu, V. V., Sachchidananda, J., and Mehta, S. C. 1968. Studies on a mosaic disease of cowpea from India. *Phytopathologische Zeitschrift* **63**, 381–387.

Chopra, S. K., and Singh, L. N. 1977. Correlation and path coefficient analysis in fodder cowpea. *Forage Research* **3**, 97–101.

Chowdhary, R. K. 1983. A bold seeded dwarf mutant of cowpea. *Tropical Grain Legume Bulletin* **27**, 20–22.

Christie, J. R. 1952. Some new nematode species of critical importance to Florida growers. *Proceedings of the Soil Science Society of Florida* **12**, 30–39.

Cisse, N., Thiaw, S., and Sene, A. 1984. Project C.R.S.P. nièbé essais varietaux 1983. Dakar, Senegal, Institut senegalais de recherches agricoles.

Clayberg, C. O., Butler, L., Kerr, E. A., Rick, C. M., and Robinson, R. W. 1966. Third list of known genes in the tomato with revised linkage map and additional rules. *Journal of Heredity* **57**, 189–196.

Clayberg, C. O., Butler, L., Kerr, E. A., Rick, C. M., and Robinson, R. W. 1970. Rules for nomenclature in tomato genetics. *Tomato Genetics Cooperative Report* **20**, 3–5.

Clayberg, C. O., Butler, L., Rick, C. M., and Young, P. A. 1960. Second list of known genes in the tomato including supplementary rules for nomenclature. *Journal of Heredity* **51**, 167–174.

Coffee, R. A. 1979. Electrodyn energy—a new approach to pesticide application. *Proceedings British Crop Protection Conference—pests and diseases*.

Cook, C. E., Whichard, L. P., Wall, M. E., Egley, G. H., Coggan, P., Luthan, P. A., and Mcphail, A. T. 1972. Germination stimulants. 2: the structure of strigol—potent seed germination stimulant for witchweed (*Striga lutea* Lour.). *Journal of the American Chemical Society* **94**, 6198–6199.

COPR (Centre for Overseas Pest Research). 1981. *Pest control in tropical grain legumes*. London, UK, COPR.

Corbett, D. C. M. 1973. *Pratylenchus penetrans*. St. Albans, UK, CIH (Commonwealth Institute of Helminthology), *Descriptions of Plant-Parasitic Nematodes* **2** (25).

Corbett, D. C. M. 1976. *Pratylenchus brachyurus*. St. Albans, UK, CIH (Commonwealth Institute of Helminthology), *Descriptions of Plant-Parasitic Nematodes* **6** (89).

Costa, C. L., Lin, M. T., Kitajima, E. W., Santos, A. A., Mesquita, C. M., and Freire, F. R. F. 1978. *Cerotoma arcuata*, um crisomelideo vector do mosaico da *Vigna* no Brasil. *Fitopatologia Brasileira* **3**, 81.

Costa, C. L., Lin, M. T., and Sperandio, C. A. 1981. Besouros crisomelídeos vectors do serotipo IV do "cowpea severe mosaic virus", isolado de feijoeiro. *Fitopatologia Brasileira* **6**, 523.

Coulibaly, M. 1984. Response de cultivars locaux de nièbé a la fumure phosphatée. Ouagadougou, Burkina Faso, Université d'Ouagadougou, Institut superieur politechnique. (These d'ingenieur du développement rural.)

Covell, S. 1984. Interspecific and intraspecific variation in base temperature for selected grain legume species and the influence of alternating temperatures on germination in chick-pea. Reading, University of Reading, UK. (B.Sc. thesis.)

Cowpea Varietal Nomenclature Study Group. 1978. Cowpea variety descriptions, citations, and synonyms. *Newsletter of the Association of Official Seed Analysts* **52**, 50–72.

Crafts, A. S., and Robbins, W. W. 1962. *Weed Control*, 3rd edn, New York, NY, USA, McGraw-Hill Book Co., Inc.

Cronquist, A. 1968. *The Evolution and Classification of Flowering Plants*. London, UK, Thomas Nelson and Sons Ltd.

Cunard, A. C. 1981. Intercropping trials with millet and cowpeas in the Sahel. In ASA (American Society of Agronomy), *Agronomy Abstracts*, 73rd annual meeting, Madison, WI, USA, ASA, 40.

Cuthbert, F. P., Jr, and Chambliss, O. L. 1972. Sources of resistance to cowpea in *Vigna sinensis* and related species. *Journal of Economic Entomology* **65**, 542–545.

Cuthbert, F. P., Jr, and Fery, R. L. 1975a. CR 17-1-13, CR 18-13-1, CR 22-2-21, cowpea curculio resistant southern pea germplasm. *HortScience* **10**, 628.

Cuthbert, F. P., Jr, and Fery, R. L. 1975b. Relationship between cowpea curculio injury and *Choanephora* pod rot of southern peas. *Journal of Economic Entomology* **68**, 105–106.

Cuthbert, F. P., Jr, and Fery, R. L. 1979. Value of plant resistance for reducing cowpea curculio damage to the southern pea [*Vigna unguiculata* (L.) Walp.]. *Journal of the American Society of Horticulture* **104**, 199–201.

Cuthbert, F. P., Jr, Fery, R. L., and Chambliss, O. L. 1974. Breeding for resistance to cowpea curculio in southern peas. *HortScience* **9**, 69–70.

Cutler, J. M. 1979. Adaptation of cowpea to semi-arid environments in West Africa. Ibadan, Nigeria, IITA. (Unpublished report to the Grain Legume Improvement Program.)

Dale, W. T. 1949. Observation on a virus disease of cowpea in Trinidad. *Annals of Applied Biology* **36**, 327–333.

Dale, W. T. 1953. The transmission of plant viruses by biting insects with particular reference to cowpea mosaic. *Annals of Applied Biology* **40**, 384–392.

Dancette, C., and Hall, A. E. 1979. Agroclimatology applied to water management in the sudanian and sahelian zones of Africa. In Hall, A. E., Cannell, G. H., and Lawton, H. W., eds, *Agriculture in Semi-Arid Environments, Ecological Studies*, Heidelberg, Berlin, New York, Springer-Verlag, 34, 98–118.

Dangi, O. P., and Paroda, R. S. 1974. Correlation and path coefficient analyses in fodder (*Vigna sinensis*). *Experimental Agriculture* **10**, 23–31.

Daniel, R. M., Limmer, A. W., Steele, K. W., and Smith, I. M. 1982. Anaerobic growth, nitrate reduction and denitrification in 46 *Rhizobium* strains. *Journal of General Microbiology* **128**, 1811–1815.

Daoust, R. A., Fernandes, P. M., Magalhães, B. P., and Yokoyama, M. 1984. Pathogenicity of *Beauveria bassiana* applied to cowpea foliage and curcubitacid tubers, *Cayaponia* sp., to adult *Diabrotica speciosa* and *Cerotoma* sp. (Col.: Chrysomelidae) in Brazil. Paper presented for the 17th annual meeting of the Society for Invertebrate Pathology, University of California, Davis, CA, 40.

Daoust, R. A., Roberts, D. W., and Soper, R. S. 1983a. Fungal diseases of cowpea pests in north, northeast and central west Brazil. Paper prepared for the 16th annual meeting of the Society for Invertebrate Pathology, Cornell University, Ithaca, NY, 42.

Daoust, R. A., Roberts, D. W., and Soper, R. S. 1983b. The enzootic and epizootic occurrence of diseases in insect species associated with cowpeas in central, north and northeast Brazil. Annual Report, Bean Improvement Cooperative, 26, 86–87.

Das, S. N., Sahn, H., and Ramana, K. V. 1970. *Indian Phytopathological Society Bulletin* **6**, 4351.

Dasgupta, D. R., and Raski, D. J. 1968. The biology of *Rotylenchulus parvus*. *Nematologica* **14**, 429–440.

Dasgupta, D. R., and Seshadri, A. R. 1971. Races of the reniform nematode *Rotylenchulus reniformis*. Linford and Olivera 1949. *Indian Journal of Nematology* **1**, 21–24.

Day, J. M., Roughley, R. J., Eaglesham, A. R. J., Dye, M., and White, S. P. 1978. Effect of high soil temperatures on nodulation of cowpea, *Vigna unguiculata*. *Annals of Applied Biology* **88**, 466–481.

Debrot, C. E., and de Rojas, C. E. B. 1967. El virus del mosaico del frijol, *Vigna sinensis* Endl. en Venezuela. *Agronomía Tropical* **17**, 3–16.

Decker, H. 1969. *Phytonematologie. Biologie und Bekaempfung pflanzenparasitaerer Nematoden*. Phytonematologie Berlin. VEB Deutscher Landwirtschaftsverlag.

De Jager, C. P. 1979. Cowpea severe mosaic virus. Kew, UK, CMI (Commonwealth Mycological Institute), CMI/AAB Descriptions of plant viruses 209.

De Jager, C. P., and Wesseling, J. B. M. 1981. Spontaneous mutations in cowpea mosaic virus overcloning resistance to hypersensitivity in cowpea. *Physiology of Plant Pathology* **19**, 347–358.

Demski, J. W., Alexander, A. T., Stefani, M. A., and Kuhn, C. W. 1983. Natural infection, disease reactions and epidemiological implications of peanut mottle virus in cowpea. *Plant Disease* **67**, 267–269.

Deshmukh, M. G., and Joshi, R. N. 1973. Effect of rhizobial inoculation on the extraction of protein from the leaves of cowpea (*Vigna sinensis*) Savi ex Hassk. *Indian Journal of Agricultural Science* **43**, 539–542.

Dessauer, D. W., and Hannah, L. C. 1978. Genetic characterization of two high-methionine cowpea lines. *Tropical Grain Legume Bulletin* **13–14**, 9–11.

De Wit, C. T. 1958. Second crop growth during the dry season in Lower Burma. *Netherlands Journal of Agricultural Science* **6** (4), 249–255.

De Zeeuw, D. J., and Ballard, J. C. 1959. Inheritance in cowpea of resistance to tobacco ringspot virus. *Phytopathology* **49**, 332–334.

De Zeeuw, D. J., and Crum, R. A. 1963. Inheritance of resistance to tobacco ringspot and cucumber mosaic viruses in black cowpea crosses. *Phytopathology* **53**, 337–340.

Dhanorkar, B. K., and Daware, D. G. 1980. Differences in number of aphids found on lines of cowpea in a replicated trial. *Tropical Grain Legume Bulletin* **19**, 3004.

Dhery, M., Germani, G., and Giard A. 1975. Resultats de traitements nematocides contre la chlorose et le rabougrissement de l'arachide en Haute Volta. *Cahiers ORSTOM serie Biologie Nematologie* **10**, 161–167.

Diab, K. A. 1968. Occurrence of *Heterodera glycines* from the Golden Island Giza U.A.R. *Nematologica* **14**, 148.

Díaz-Ruíz, J. R. 1976. Properties and electron microscopy of a *Vigna* virus belonging to the cucumovirus group. *Microbiología Española* **29**, 9–18.

Dina, S. O. 1976. Effects of monocrotophos on insect damage and yield of cowpea (*Vigna unguiculata*) in southern Nigeria. *Experimental Agriculture* **13**, 155–159.

Dobereiner, J. 1974. Legumes in Latin American agriculture. *Proceedings of the Indian National Science Academy B* **40**, 768–794.

Dobie, P. 1981. The use of resistant varieties of cowpeas (*Vigna unguiculata*) to reduce diseases due to postharvest attack by *Callosobruchus maculatus*. In Labeyrie, V., ed., *Series Entomologica*, 19, the Hague, Dr. W. Junk Publishers.

Doku, E. V. 1970. Variability in local and exotic varieties of cowpea [*Vigna unguiculata* (L.) Walp.] in Ghana. *Ghana Journal of Agricultural Science* **3**, 139–143.

Donald, C. M., and Hamblin, J. 1976. The biological yield and harvest index of cereals as agronomic and plant breeding criteria. *Advances in Agronomy* **28**, 361–405.

Donli, P. O. 1983. Host range and survival of *Sphaceloma* sp., the pathogen of cowpea scab disease and its interaction with cowpea lines in northern Nigeria. Zaria, Nigeria, Ahmadu Bello University. (M.Sc. thesis.)

Dovlo, F. E., Williams, C. E., and Zoaka, L. 1976. Cowpeas: home preparation and use in West Africa. Ottawa, Canada, IDRC (International Development Research Centre).

Downes, J. D., Fryrear, D. W., Wilson, R. L., and Sabota, C. M. 1977. Influence of wind erosion on growing plants. *Transactions of the American Society of Agricultural Engineers* **20** (5), 885–889.

Duke, J. A. 1980. *Vigna unguiculata* (L.) Walp. In *Handbook of Legumes of World Economic Importance*, New York, NY, USA, Plenum Press Ltd, 302–306.

Dumbre, A. D., Deshmukh, R. B., and Padhey, A. P. 1982. Association of grain yield with other economic characters in cowpea. *Journal of the Maharashtra Agricultural University* **7** (2), 154.

Duncan, L. W., and Ferris, H. 1983. Validation of a model for prediction of host damage by two nematode species. *Journal of Nematology* **15**, 227–234.

Dundas, B. 1939. Host range and strains of the powdery mildew (*Erysiphe polygoni*) of bean and cowpea. *Phytopathology* **29**, 824.

Dutta, D. N., and Lal, R. 1980. Residues and residual toxicity of monocrotophos on cowpea. *Journal of Research, Assam Agricultural University* **1** (2), 177–181.

Eaglesham, A. R. J. 1982. Assessing the nitrogen contribution of cowpea (*Vigna unguiculata*) in monoculture and intercropped. In Graham, P. H., and Harris, S. C., eds, *Biological Nitrogen Fixation Technology for Tropical Agriculture*, CIAT, 9–13 March 1981, Cali, Colombia, CIAT (Centro Internacional de Agricultura Tropical), 641–646.

Eaglesham, A. R. J., and Ayanaba, A. 1984. Tropical stress ecology of rhizobia, root nodulation and legume nitrogen fixation. In Subba Rao, N. S., ed., *Current Developments in Biological Nitrogen Fixation*, New Delhi, IBH, 1–35.

Eaglesham, A. R. J., Ahmad, M. H., Hassouna, S., and Goldman, B. J. 1982a. Cowpea rhizobia producing dark nodules: use in competition studies. *Applied and Environmental Microbiology* **44**, 611–618.

Eaglesham, A. R. J., Ayanaba, A., Ranga Rao, V., and Eskew, D. L. 1981a. Improving the nitrogen nutrition of maize by intercropping with cowpea. *Soil Biology and Biochemistry* **13**, 169–171.

Eaglesham, A. R. J., Ayanaba, A., Ranga Rao, V., and Eskew, D. L. 1982b. Mineral N effects on cowpea and soybean crops in Nigerian soil. 2: amounts of N fixed and accrual to the soil. *Plant and Soil* **68**, 183–192.

Eaglesham, A. R. J., Minchin, F. R., Summerfield, R. J., Dart, P. J., Huxley, P. A., and Day, J. M. 1977. Nitrogen nutrition of cowpea (*Vigna unguiculata*). 3: distribution of nitrogen within effectively nodulated plants. *Experimental Agriculture* **13**, 369–380.

Eaglesham, A. R. J., Seaman, B., Ahmad, H., Hassouna, S., Ayanaba, A., and Mulongoy, K. 1981b. High temperature tolerant 'cowpea' rhizobia. In Gibson, A., and Newton, W. E., eds, *Current Perspectives in Nitrogen Fixation*, Canberra, Australian Academy of Science, 436.

Ebong, V. V. 1971a. Cowpeas: recommendations on cowpea variety 'Dinner'. Ibadan, Nigeria, Federal Department of Agriculture, Information Paper.

Ebong, V. V. 1971b. Strategy for cowpea improvement in Nigeria. *Samaru Journal of Agricultural Research* **13** (2), 25–27.

Edward, J. C., and Misra, S. L. 1968. *Heterodera vigni* n.sp. and second stage larvae of *Heterodera* spp. in Uttar Pradesh India. *Allahabad Farmer* **42**, 155–159.

Edwards, D. G., Kang, B. T., and Danso, S. K. A. 1981. Differential response to six cowpea [*Vigna unguiculata* (L.) Walp.] cultivars to liming in an ultisol. *Plant and Soil* **59**, 61–73.

Ekpo, E. J. A. 1978. Effect of *Xanthomonas vignicola* on the yield of cowpea (*Vigna unguiculata*). *African Journal of Agricultural Science* **5** (1), 67–69.

Ekpo, E. J. A. in press. Comparative virulence of *Xanthomonas vignicola* (bacterial blight) and *X.* sp. bacterial pustule in cowpea (*Vigna unguiculata*). *Nigerian Journal of Agricultural Science* (quoted by Allen, 1983).

El-Din, A. S. G., El-Hammady, M., Mazyad, H. M., El Amrety, A. A., and Morsy, F. T. 1980. Isolation and identification of cucumber mosaic virus from cowpea in Egypt. *Agricultural Research Review* **58** (2), 231–240.

Elías, L. G., Colindres, R., and Bressani, R. 1964. The nutritive value of eight varieties of cowpea (*Vigna sinensis*). *Journal of Food Science* **29**, 118–122.

Elías, L. G., Cristales, F. R., Bressani, R., and Miranda, H. 1976. Composición química y valor nutritivo de algunas leguminosas de grano. *Turrialba* **26**, 375–380.

Elías, L. G., Hernández, M., and Bressani, R. 1976. The nutritive value of precooked legume flours processed by different methods. *Nutritional Reports International* **14**, 385–403.

Elston, J., and Bunting, A. H. 1980. Water relations of legume crops. In Summerfield, R. J., and Bunting, A. H., eds, *Advances in Legume Science*, London, UK, Her Majesty's Stationery Office, 37–42.

Elston, J., Dennett, M. D., and Breeze, V. G. 1975. *Heuristic models of growth and yield in grain legumes*. Topic 6. Rome, Italy, FAO (Food and Agriculture Organization), IAR/75/2.

Embabi, M. S., Hamdi, Y. A., and Tawfik, M. S. in press. Synergistic interaction between *Meloidogyne* spp. root-nodule bacteria on cowpea. *Egyptian Journal of Microbiology Egypt* (in press).

EMBRAPA (Empresa Brasileira de Pesquisa Agropecuaria) Anoil. 1984. *Destaque dos resultados de pesquisa Brasilia*, EMBRAPA.

Emechebe, A. M. 1975. *Some Aspects of Crop Diseases in Uganda*. Kampala, Uganda, Makerere University.

Emechebe, A. M. 1980. Scab disease of cowpea (*Vigna unguiculata*) caused by a species of the fungus *Sphaceloma*. *Annals of Applied Biology* **96** (1), 11–16.

Emechebe, A. M. 1981a. Brown blotch of cowpea in northern Nigeria. *Samaru Journal of Agricultural Research* **1** (1), 20–26.

Emechebe, A. M. 1981b. Cowpea pathology: notes on cowpeas and grain legume improvement program. Paper prepared for a cropping scheme meeting, Institute for Agricultural Research, Samaru, Zaria, Nigeria.

Emechebe, A. M. 1981c. Fungicidal control of fungal diseases of cowpea (*Vigna unguiculata*) in northern Nigeria. Paper prepared for the 11th annual conference of Nigerian Society for Plant Protection, University of Nigeria, Nsukka, February 1981.

Emechebe, A. M., and McDonald, D. 1979. Seed-borne pathogenic fungi and bacteria of cowpea in northern Nigeria. *PANS* **25** (4), 401–404.

Ennis, T. H., and Chambliss, O. L. 1976. Pods resist insect penetration in curculio resistant southern peas. *Highlights of Agricultural Research* **23**, 8. Auburn, AL, USA, Auburn University.

Enyi, B. A. C. 1973a. An analysis of the effect of weed competition on growth and yield attributes in sorghum (*Sorghum vulgare*), cowpeas (*Vigna unguiculata*) and green gram (*Vigna aureus*). *Journal of Agricultural Science* **81** (3), 449–453.

Enyi, B. A. C. 1973b. A spacing/time of planting trial with cowpea [*Vigna unguiculata* (L.) Walp.]. *Ghana Journal of Agricultural Science* **13**, 78–85.

Enyi, B. A. C. 1975. Agronomic practices and yield of crops. Paper prepared for the food crops conference, Papua New Guinea, 159–162.

Eplee, R. E. 1969. Plastic film cover for large field fumigation. Paper prepared for the 22nd annual meeting of the Southern Weed Science Society (US), 405–406.

Eplee, R. E. 1973. Development in the control of *Striga lutea* in the United States. Paper prepared for the 1st symposium on parasitic weeds, Malta, 257–259.

Eplee, R. E. 1979. The *Striga* eradication program in the United States of America. In Musselman, L. J., Worsham, A. D., and Eplee, R. E., eds, *Proceedings 2nd International Symposium on Parasitic Weeds*, 16–19 July 1979, North Carolina State University, Raleigh, NC, 269–272.

Eplee, R. E. in press. Chemical control of *Striga*. In ICSU/UNESCO (International Council of Scientific Unions/United Nations Educational, Scientific and Cultural Organization), *Proceedings of the Workshop on the Biology and Control of* Striga, Dakar, Senegal, ICSU.

Epps, J. M. 1969. Nine varieties of southern peas resistant to the soybean cyst nematode. *Plant Disease Reporter* **53**, 245.

Epps, J. M., and Chambers, A. Y. 1962. Nematode inhibits nodules on soybean. *Crop Soils* **15**, 18.

Erskine, W., and Khan, T. N. 1977. Genotype, genotype × environment and environmental effects on grain yield and related characters of cowpea [*Vigna unguiculata* (L.) Walp.]. *Australian Journal of Agricultural Research* **28**, 609–617.

Erskine, W., and Khan, T. N. 1978. Inheritance of cowpea yields under different soil conditions in Papua New Guinea. *Experimental Agriculture* **14**, 23–28.

Esuruoso, O. F. 1975. Seed-borne fungi of cowpea (*Vigna unguiculata*) in western Nigeria. *Nigerian Journal of Plant Protection* **1**, 87–90.

Evans, A. M. 1976. Species hybridization in the genus *Vigna*. In IITA (International Institute of Tropical Agriculture), *Proceedings of a Collaborators' Meeting on Grain Legume Improvement*, 9–13 June 1975, Ibadan, Nigeria, IITA, 31–34.

Evans, L. T. 1974. The physiological basis of crop yield. In Evans, L. T., ed. *Crop Physiology: Some Case Histories*. London, UK, Cambridge University Press, 327–355.

Ezedinma, F. O. C. 1961. The nutrient requirements of the cowpea (*Vigna sinensis* Endl.) in southern Nigeria. Ibadan, Nigeria, University of Ibadan. (M.Sc. thesis.)

Ezedinma, F. O. C. 1964a. Effect of preparatory cultivation on the general performance and yield of cowpeas. *Nigerian Agricultural Journal* **1**, 21–25.

Ezedinma, F. O. C. 1964b. Effects of inoculation with local isolates of cowpea *Rhizobium* and application of nitrate–nitrogen on the development of cowpeas. *Tropical Agriculture (Trinidad)* **41**, 243–249.

Ezedinma, F. O. C. 1966. The distribution of dry weight changes among shoot components of semi-upright cowpeas [*Vigna unguiculata* (L.) Walp.] during vegetative development. *Horticultural Research* **6**, 91–99.

Ezedinma, F. O. C. 1974. Effect of close spacing on cowpeas (*Vigna unguiculata*) in southern Nigeria. *Experimental Agriculture* **10** (4), 289–298.

Ezueh, M. I. 1977. Grain legume improvement at the National Cereals Research Institute, Moor Plantation. *Tropical Grain Legume Bulletin* **8**, 12.

Ezueh, M. I. 1980. Soil systemic insecticides for cowpea pest control. *African Journal of Plant Protection* **2**, 107–113.

Ezueh, M. I., and Taylor, T. A. 1981. Field resistance in cowpea, *Vigna unguiculata* to the cowpea moth, *Cydia ptychora*. *Annals of Applied Biology* **99**, 307–312.

Ezueh, M. I., and Taylor, T. A. 1983. Effects of time of intercropping with maize on cowpea susceptibility to three major pests. *Tropical Agriculture (Trinidad)* **61**, 82–86.

Ezumah, H. C. Population and planting pattern effects of intercropped maize and cowpea. Ibadan, Nigeria, IITA (unpublished).

Fadayomi, D. 1979a. Effect of crop spacing on weed competition and seed yield in cowpea (*Vigna unguiculata* L. Walp.) cv. Ife Brown. *Ife Journal of Agriculture* **1** (1), 45–59.

Fadayomi, D. 1979b. Weed competition and consideration of cost effectiveness of different weed control alternatives in cowpea. Paper prepared for the 9th annual conference of the Weed Science Society, Nigeria.

FAO (Food and Agriculture Organization of the United Nations). 1969. *Food composition tables for use in Africa*. Rome, Italy, FAO.

FAO. 1973. *Production yearbook*. Rome, Italy, FAO, 27, 116–117.

FAO. 1976. *Development of a program promoting the use of organic materials as fertilizers*. Rome, Italy, FAO, 27.

FAO. 1982. Fertilizer use in multiple cropping systems. Report on Expert Consultation held in New Delhi, 3–6 February 1982, Rome, Italy, *FAO Fertilizer and Plant Nutrition Bulletin* **5**.

FAO/WHO (World Health Organization). 1973. *Energy and protein requirements*. Geneva, Switzerland, WHO, Technical report 522.

Faris, D. G. 1964. The chromosome of *Vigna sinensis* (L.) Savi. *Canadian Journal of Genetics and Cytology* **6**, 255–258.

Faris, D. G. 1965. The origin and evolution of the cultivated forms of *Vigna sinensis*. *Canadian Journal of Genetics and Cytology* **7**, 433–452.

Faroda, A. S. 1973. Note on the effect of different levels of nitrogen, phosphorus and potash on fodder production of cowpea (*Vigna sinensis*). *Indian Journal of Agricultural Research* **7** (3–4), 217–218.

Farquhar, G. D., O'Leary, M. H., and Berry, J. A. 1982. On the relationship between carbon isotope discrimination and the intercellular carbon dioxide concentration in leaves. *Australian Journal of Plant Physiology* **9**, 121–137.

Farquhar, G. D., and Richards, R. A. 1984. Isotopic composition of plant carbon correlates with water-use-efficiency of wheat genotypes. *Australian Journal of Plant Physiology,* **11**, 539–552.

Fassuliotis, G. 1974. Host range of the Colombian lance nematode *Hoplolaimus columbus*. *Plant Disease Reporter* **58**, 1000–1002.

Fauquet, C., and Thouvenel, J. C. 1980. *Viral diseases of crop plants in Ivory Coast*. Paris, ORSTOM (Office de la recherche scientifique et technique outre-mer).

Fawole, I., and Afolabi, N. O. 1983. Genetic control of a branching peduncle mutant of cowpea, *Vigna unguiculata* (L.) Walp. *Journal of Agricultural Science Cambridge* **100**, 473–475.

Feder, W. A., and Feldmesser, J. 1956. Root abnormalities caused by burrowing nematode infections. *Phytopathology* **46**, 11.

Fegla, G. T., Shawkat, A. L. B., and Mohammed, S. Y. 1981. Certain viruses affecting cowpea and their effect on growth and root nodulation of cowpea plants. *Mesopotamia Journal of Agriculture* **16** (1), 137–152.

Fennell, J. L. 1948. New cowpeas resistant to mildew. *Journal of Heredity* **39**, 274–279.

Ferreira, B. S. C. 1980. Ocorrência, no Brasil, de *Trissolcus basalis*, parasita de ovos de *Nezara viridula*. *Pesquisa Agropecuaria Brasileira* **15** (1), 127–128.

Fery, R. L. 1980. Genetics of *Vigna*. In Janick, J., ed., *Horticultural Reviews*, Westport, CT, USA, AVI Publishing, 311–394.

Fery, R. L. 1981. Cowpea production in the United States. *HortScience* **16**, 473–474.

Fery, R. L., and Cuthbert, Jr, F. P. 1975. Inheritance of pod resistance to cowpea curculio infestation in southern peas. *Journal of Heredity* **66**, 43–44.

Fery, R. L., and Cuthbert, Jr, F. P. 1978. Inheritance and selection of nonpreference resistance to cowpea curculio in the southernpea [*Vigna unguiculata* (L.) Walp.]. *Journal of the American Society of Horticultural Science* **103**, 370–372.

Fery, R. L., and Cuthbert, Jr, F. P. 1979. Measurement of pod-wall resistance to the cowpea curculio in the southern pea [*Vigna unguiculata* (L.) Walp.]. *HortScience* **14**, 29–30.

Fery, R. L., and Dukes, P. D. 1977a. An assessment of two genes for Cercospora leaf spot resistance in the southern pea [*Vigna unguiculata* (L.) Walp.]. *HortScience* **12**, 454–456.

Fery, R. L., and Dukes, P. D. 1977b. Cercospora leaf spot of southern pea: studies of yield loss and genetics of resistance. *HortScience* **12** (2), 234.

Fery, R. L., and Dukes, P. D. 1979. Genetics of root knot resistance in the southernpea [*Vigna unguiculata* (L.) Walp.]. *HortScience* **14**, 406 (abstract).

Fery, R. L., and Dukes, P. D. 1980. Inheritance of root-knot resistance in the cowpea [*Vigna unguiculata* (L.) Walp.]. *Journal of the American Society of Horticultural Science* **105**, 671–674.

Fery, R. L., and Dukes, P. D. 1982. Inheritance and assessment of a second root-knot resistance factor in southernpea [*Vigna unguiculata* (L.) Walp.]. *HortScience* **17**, 152 (abstract).

Fery, R. L., and Dukes, P. D. 1984, 'Carolina Cream' southernpea. *HortScience* **19**, 456–457.

Fery, R. L., and Schalk, J. A. 1984. Southern green stink bug: identification of resistance in cowpea. *HortScience* **19**, 211.

Fery, R. L., Dukes, P. D., and Cuthbert, Jr, F. P. 1976. The inheritance of Cercospora leaf spot resistance in southernpea [*Vigna unguiculata* (L.) Walp.]. *Journal of the American Society for Horticultural Science* **101**, 148–149.

Fery, R. L., Dukes, P. D., and Cuthbert, Jr, F. P. 1977. Yield loss of southern pea (*Vigna unguiculata*) caused by Cercospora leaf spot. *Plant Disease Reporter* **61**, 741–743.

Fery, R. L., Dukes, P. D., and Ogle, W. L. 1982. 'Colossus-80' southernpea. *HortScience* **17**, 263–264.

Fischer, H. U., and Lockhart, B. E. 1976a. A strain of cowpea aphid-borne mosaic virus isolated from cowpeas in Morocco. *Phytopathologische Zeitschrift* **85**, 43–48.

Fischer, H. U., and Lockhart, B. E. 1976b. A strain of cucumber mosaic virus isolated from cowpea in Morocco. *Phytopathologische Zeitschrift* **85**, 132–138.

Fischer, R. A., and Maurer, R. 1978. Drought resistance in spring wheat cultivars. 1: grain yield responses. *Australian Journal of Agricultural Research* **29**, 897–912.

Flight, C. 1970. Excavations at Kintampo. *West African Archaeological Newsletter* **12**, 71–73.

Flight, C. 1976. The Kintampo culture and its place in the economic prehistory of West Africa. In Harlan, J. R., de Wet, J. M. J., and Stemler, A. B. L., eds, *Origins of African Plant Domestication*. The Hague, Mouton Publishers, 212–221.

Fortuner, R. 1976. *Pratylenchus zeae*. St. Albans, UK, CIH (Commonwealth Institute of Helminthology), CIH Descriptions of plant-parasitic nematodes 6 (77).

Fox, R. L., and Kang, B. T. 1977. Some major fertility problems of tropical soils. In Vincent, J. M., Whitney, A. S., and Bose, J., eds, *Expoiting the Legume–Rhizobium Symbiosis in Tropical Agriculture*, Honolulu, University of Hawaii, College of Tropical Agriculture, Miscellaneous publication 145, 183–210.

Frahm-Leliveld, J. A. 1965. Cytological data on some wild tropical *Vigna* species and cultivars from cowpea and asparagus bean. *Euphytica* **14**, 251–270.

Francki, R. I. B., Mossop, D. W., and Hatta, T. 1979. Cucumber mosaic virus. Kew, UK, CMI (Commonwealth Mycological Institute), CMI/AAB Descriptions of plant viruses 213.

Franckowiak, J. D., and Barker, L. N. 1974. Inheritance of testa colour in cowpea, *Vigna unguiculata* (L.) Walp. *Agronomy Abstracts* 52–53.

Franckowiak, J. D., Ojomo, O. A., and Barker, L. N. 1973. Ife Brown a new cowpea cultivar for western Nigeria. Paper prepared for the 9th annual meeting of the Agricultural Society of Nigeria, Ilorin.

Frankel, O. H., and Bennett, E. 1970. Genetic resources. In Frankel, O. H., and Bennett, E., eds, *Genetic Resources in Plants—their Exploration and Conservation*, Oxford, Blackwell, IBP Handbook, **11**, 7–17.

Frederick, L. R. 1978. Effectiveness of rhizobia–legume association. In Andrew, C. S., and Kamprath, E. J., eds, *Mineral Nutrition of Legumes in Tropical and Subtropical Soils*, Perth, Australia, Commonwealth Scientific and Industrial Research Organization.

Freyman, S., and Venkateswarlu, J. 1977. Intercropping on rainfed red soils of the Deccan plateau, India. *Canadian Journal of Plant Science* **57** (3), 697–706.

Fromme, F. D. 1924. The rust of cowpeas. *Phytopathology* **14**, 67–79.

Fulton, J. P., and Allen, D. J. 1982. Identification of resistance to cowpea severe mosaic virus. *Tropical Agriculture (Trinidad)* **59**, 66–68.

Fulton, J. P., and Scott, H. A. 1979. A serogrouping concept for legume comoviruses. *Phytopathology* **69**, 305–306.

Fulton, J. P., Gamez, R., and Scott, H. A. 1975. Cowpea chlorotic mottle and bean yellow stipple viruses. *Phytopathology* **65**, 741–742.

Furstenberg, J. P. 1972. Bevolkingsdinamika van nematodes onder verbouing van sekere gewasse. University of Potchefstroom. (Ph.D. thesis.)

Gabr, M. 1981. Malnutrition during pregnancy and lactation. *World Review of Nutrition and Dietetics* **36**, 90–99.

Gabriel, C. J., Black, L. L., and Derrick, K. S. 1981. Observations on some viruses infecting cowpeas in Louisiana. *Phytopathology* **71** (2), 217.

Gadre, K. M., and Umrani, N. K. 1972. Water availability periods for crop planting in Sholapur district. *Research Journal of Mahatma Phule Agricultural University* **3** (2), 73–81.

Gamez, R. 1976. Los virus del frijol en Centro América. 4: Algunas propiedades y transmisión por insectos crisomélidos del virus del moteado amarillo del frijol. *Turrialba* **26**, 160–161.

Ganacharya, N. M., and Mali, V. R. 1981. Comparative studies on two isolates of tobacco ringspot virus from cowpea. *Indian Phytopathology* **34** (1), 112 (abstract).

Gangwar, B., and Karla, G. S. 1979. Studies on mixed cropping of legumes with maize under rainfed conditions. *Madras Agricultural Journal*, **66** (7), 425–429.

Gardner, M. W. 1925. *Cladosporium* spot on cowpea. *Phytopathology* **15**, 453–462.

Gardner, M. W., and Kendrick, J. B. 1925. Bacterial spot of cowpea and lima bean. *Journal of Agricultural Research* **31**, 841–863.

Gatehouse, A. M. R., and Boulter, D. 1983. Assessment of the antimetabolic effects of trypsin inhibitors from cowpea (*Vigna unguiculata*) and other legumes on the development of the bruchid beetle *Callosobruchus maculatus*. *Journal of the Science of Food and Agriculture* **34**, 345–350.

Gay, J. D. 1969. Effect of seed maturation on the infectivity of cowpea chlorotic mottle virus. *Phytopathology* **58**, 1609–1615.

Gay, J. D. 1970. Fungicidal activity of potassium azide as a seed treatment. *Plant Disease Reporter* **54**, 604–605.

Gay, J. D. 1971. An apparent new race of cowpea rust on *Vigna*. *Plant Disease Reporter* **55**, 384–386.

Gay, J. D., Johnson, A. W., and Chalfant, R. B. 1973. Effects of a trap crop on the interaction and distribution of cowpea virus by soil and insect vectors. *Plant Disease Reporter* **57**, 684–688.

Gay, J. D., and Windstead, E. E. 1970. Seedborne viruses and fungi from southernpea seed grown in eight states. *Plant Disease Reporter* **54**, 243–245.

Geervani, P., and Theophilus, F. 1980. Effect of home processing on the protein quality of selected legumes. *Journal of Food Science* **45**, 707–710.

Geervani, P., and Theophilus, F. 1981. Studies on digestibility of selected legume carbohydrates and their impact on the pH of the gastrointestinal tract in rats. *Journal of the Science of Food and Agriculture* **32**, 71–78.

George, D. C. 1922. Notes on plant diseases in Arizona. In Arizona Commission on Agriculture and Horticulture, 14th Annual Report, Phoenix, AZ, Department of Agriculture.

George, L. V. 1948. New pea for nematode stand. *Southern Seedsman (Texas)* **11**, 13.

Germani, G., and Dhery, M. 1973. Observations and experimentation on the part played by nematodes in two diseases of peanut in the Upper Volta region: chlorosis and clump. *Oleagineux* **27**, 235–242.

Ghosh, P., Mitra, G. C., and Sharma, A. K. 1979. Embryoculture as a method of securing viable mutants in *Vigna sinensis* var. Black and *Vigna radiata* (B-105) after EMS treatment. *Proceedings of the Indian National Science Academy B* **45**, 605–612.

Gibbs, A. J. 1977. Tobamovirus group. Kew, UK, CMI (Commonwealth Mycological Institute), CMI/AAB Descriptions of plant viruses 184.

Gibson, A. H. 1980. Host determinants in nodulation and nitrogen fixation. In Summerfield, R. J., and Bunting, A. H., eds, *Advances in Legume Science*. London, UK, Her Majesty's Stationery Office, 69–75.

Gill, P. S., Singh, K., and Tripathi, H. P. 1977. Effect of seed rates and row spacings on the forage yield of cowpea [*Vigna unguiculata* (L.) Walp.] varieties. *Haryana Agriculture University Journal of Research* **7**, 1–6.

Gilmer, R. M., Whitney, W. K., and Williams, R. J. 1974. Epidemiology and control of cowpea mosaic in western Nigeria. In IITA, *Proceedings of the 1st grain legume improvement workshop*, IITA, Ibadan, Nigeria, 29 October–2 November 1983.

Giri, G., and De, R. 1981. Short season fodder legume effects on the grain yield and nitrogen economy of barley under dryland conditions. *Journal of Agricultural Science Cambridge* **96** (2), 457–461.

Gitaitis, R. D. 1983. Two resistant responses in cowpea induced by different strains of *Xanthomonas campestris* pv. *vignicola*. *Plant Disease* **67**, 1025–1028.

Gitaitis, R. D., and Nilakhe, S. S. 1982. Detection of *Xanthomonas campestris* pv. *vignicola* in Southern pea seed. *Plant Disease* **66**, 20–22.

Givord, L. 1981. Southern bean mosaic virus located from cowpea (*Vigna unguiculata*) in the Ivory Coast. *Plant Disease* **65** (9), 755–756.

Glaze, N. C. 1970. Evaluation of herbicides on southern peas. *Proceedings of the Southern Weed Science Society* **23**, 192–194.

Godfrey-Sam-Aggrey, W. 1973. Effects of fertilizers on harvest time and yield of cowpeas (*Vigna unguiculata*) in Sierra Leone. *Experimental Agriculture* **9** (4), 315–320.

Godfrey-Sam-Aggrey, W. 1975. Nitrogen fertilizer management for cowpea (*Vigna unguiculata*) production in Sierra Leone. *Zeitschrift Fuer Acker- und Pflanzenbau* **141**, 169–177.

Godse, D. B., Wani, S. P., Patil, R. B., and Bagyaraj, D. J. 1978. Response of cowpea [*Vigna unguiculata* (L.) Walp.] to *Rhizobium*–VA mycorrhiza dual inoculation. *Current Science* **47**, 784–785.

Goldsworthy, P. R. 1982. Objectives and achievements in the improvement of grain legumes. *Proceedings of the Nutrition Society* **41**, 27–39.

Goldsworthy, P. R., and Redden, R. 1982. Cowpea improvement at IITA 1970–1980. *Tropical Grain Legume Bulletin* 24.

Gonzalez, C., Moreno, R., and Gamez, R. 1975. Identification, incidence, and distribution of a virus of bean (*Vigna sinensis*) in Costa Rica. *Proceedings of the American Phytopathology Society* **2**, 75.

Gonzalez, R., Cardona, C., and Van Schoonhoven, A. 1982a. Evaluación de los daños causados en frijol por larvas y adultos de los crisomélidos *Diabrotica balteata* L. y *Cerotoma facialis* E. *Turrialba* **32** (4), 433–439.

Gonzalez, R., Cardona, C., and Van Schoonhoven, A. 1982b. Morfología y biología de los crisomélidos *Diabrotica balteata* L. y *Cerotoma facialis* E. como plagas del frijol comun. *Turrialba* **32** (3), 257–264.

Goodey, J. B., Franklin, M. T., and Hooper, D. J. 1965. *T. Goodey's The Nematode Parasites of Plants Catalogued under Their Hosts*, 3rd edn. Farnham Royal, UK, CAB (Commonwealth Agricultural Bureaux).

Graham, P. H. 1982. Plant factors affecting symbiotic nitrogen fixation in legumes. In Graham, P. H., and Harris, S. C., eds, *Biological Nitrogen Fixation Technology for Tropical Agriculture*: Papers Presented at a Workshop Held at CIAT, 9–13 March 1981, Cali, Colombia, CIAT (Centro International de Agricultura Tropical), 27–37.

Graham, T. W. 1951. Nematode root rot of tobacco and other plants. *Bulletin of the South Carolina Agriculture Experiment Station* **390**, 2500.

Graham, T. W., and Holdeman, O. L. 1953. The sting nematode, *Benololaimus gracilis* Steiner, a parasite on cotton and other crops in South Carolina. *Phytopathology* **43**, 434–439.

Grancher, C. V., and Bonhomme, R. 1974. Utilisation de l'energie solarire par une culture de *Vigna sinensis*. 3: evolution du rendement energetique pendant la phase vegetative. *Annales Agronomie* **25**, 811–819.

Greathead, D. J. in press. The natural enemies of *Striga* spp. and the prospects for their utilisation as biological control agents. In ICSU/UNESCO (International Council of Scientific Unions/United Nations Educational, Scientific and Cultural Organization), *Proceedings of the Workshop on the Biology and Control of Striga*, Dakar, Senegal, ICSU.

Grundlach, C. M., and Chambliss, O. L. 1977. Resistance in the southernpea, *Vigna unguiculata* (L.) Walpers to the cowpea curculio, *Chalcodermus aeneus* Boheman: the role of tannin. *HortScience* **12**, 234.

Guimarães, C. M., Watt, E. E., and Araújo, J. P. P. de. 1982. Avaliação de germoplasma do caupi [*Vigna unguiculata* (L.) Walp.] para resistancia a seca. Goiânia, Goiás, Brazil, EMBRAPA/CNPAF, Resumos de RENAC.

Gundy, S. D., Van, and Rackham, R. L. 1961. Studies on the biology and pathogenicity of *Hemicycliphora arenaria*. *Phytopathology* **51**, 393–397.

Gunn, C. R. 1973. Recent nomenclatural changes in *Phaseolus* L. and *Vigna* Savi. *Crop Science* **13**, 496.

Gupta, D. C. 1979. Studies on the pathogenicity and relative susceptibility of some varieties of cowpea [*Vigna unguiculata* (L.) Walp.] against *Meloidogyne javanica*. *Forage Research* **5**, 141–145.

Gupta, D. C., and Yadav, B. S. 1980. Pathogenicity of *Rotylenchulus reniformis* on cowpea. *Nematologia Mediterranea* **8**, 91–93.

Gupta, M. D., and Summanwar, A. S. 1980. The location of two mosaic viruses in cowpea seeds. *Seed Science and Technology* **8** (2), 203–206.

Gupta, P., and Edward, J. C. 1973. Studies on the biology of *Heterodera vigni* (Heteroderidae: Nematodae). 1: life cycle. *Indian Journal of Nematology* **3**, 99–108.

Gupta, P., Singh, K. P., and Edward, J. C. 1975. Studies on the effect of some soil borne fungi on the development of *Heterodera vigni* on cowpea. *Indian Journal of Nematology* **5**, 132–135.

Gupta, S. P., and Lodhi, G. P. 1979. Variability for fodder yield and its components in cowpea. *Indian Journal of Agricultural Science* **46**, 407–410.

Hadley, P., Roberts, E. H., Summerfield, R. J., and Minchin, F. R. 1983. A quantitative model of reproductive development in cowpea [*Vigna unguiculata* (L.) Walp.] in relation to photoperiod and temperature, and implications for screening germplasm. *Annals of Botany* **51**, 531–543.

Haizel, K. A. 1974. The agronomic significance of mixed cropping. 1: maize interplanted with cowpeas. *Ghana Journal of Agricultural Science* **7** (3), 169–178.

Haji, F. N. P. 1984. Unpublished research data.

Hall, A. E. 1981. Adaptation of annual plants to drought in relation to improvements in cultivars. *HortScience* **16**, 37–38.

Hall, A. E., and Dancette, C. 1978. Analysis of fallow-farming systems in semi-arid Africa using a model to simulate the hydrologic budget. *Agronomy Journal* **70**, 816–823.

Hall, A. E., and Grantz, D. A. 1981. Drought resistance of cowpea improved by selecting for early appearance of mature pods. *Crop Science* **21**, 461–464.

Hall, A. E., Dancette, C., and Turk, K. J. 1977. Crop adaptation to semi-arid environments. In University of California, Riverside, Proceedings of the International Symposium of Rainfed Agriculture, Riverside, CA, University of California, 398–418.

Hall, A. E., Foster, K. W., and Waines, J. G. 1979. Crop adaptation to semi-arid environments. *Ecological Studies* **34**, 148–179.

Halteren, P. van. 1971. Insect pests of cowpea *Vigna unguiculata* (L.) Walp. in the Accra plains. *Ghana Journal of Agricultural Science* **4**, 121–123.

Hamblin, J. 1980. Breeding legumes for improved systems of crop protection. In Summerfield, R. J., and Bunting, D. A. H., eds, *Advances in Legume Science*, London, UK, Her Majesty's Stationery Office, 243–250.

Hamid Abbas El Faki, Venkataraman, L. V., and Desikachar, H. S. R. 1984. Effect of processing on the in vitro digestibility of proteins and carbohydrates in some Indian legumes. *Qualitas Plantarum Plant Foods for Human Nutrition* **34**, 127–133.

Hanchinal, R. R., and Goud, J. V. 1978. Inheritance in *Vigna*. *Indian Journal of Genetics and Plant Breeding* **38**, 339–342.

Hanchinal, R. R., Habib, A. F., and Goud, J. V. 1979. Correlation and path analysis in cowpea, *Vigna unguiculata* (L.) Walp. *Mysore Journal of Agricultural Science* **13**, 250–257.

Hansford, C. G. 1932. Annual report of the mycologist. In Department of Agriculture, Annual Report of the Department of Agriculture, Uganda, for the year ended 31 December 1931 (part 2), 59–60.

Haque, I., and Gbla, D. S. 1978. Effect of fractional application of nitrogen on cowpeas in Sierra Leone. *Tropical Grain Legume Bulletin* **11**, 11–12.

Haque, S. Q., and Chenulu, V. V. 1972. Seed transmission of cowpea mosaic virus. *Tropical Agriculturc* **49**, 73–75.

Haque, S. Q., and Persad, G. C. 1975. Some observations on the beetle transmitted cowpea mosaic virus. In Bird, J., and Maramorosch, K., eds, *Tropical Diseases of Legumes*, New York, NY, Academic Press, 119–120.

Hare, W. W. 1953. A new race of *Fusarium* causing wilt of cowpea. *Phytopathology* **43**, 291.

Hare, W. W. 1956. Some characters identified in cowpeas segregating for resistance to *Fusarium* wilt. *Phytopathology* **46**, 14 (Abstract).

Hare, W. W. 1957a. Inheritance of resistance to *Fusarium* wilt in cowpeas. *Proceedings of the Association of Southern Agricultural Workers* **54**, 219 (appeared also as an abstract in *Phytopathology* **47**, 312–313).

Hare, W. W. 1957b. Mississippi Crowder, a new disease-resistant cowpea. *Phytopathology* **47**, 565–566.

Hare, W. W. 1959. Resistance to root-knot nematodes in cowpea. *Phytopathology* **49**, 318 (abstract).

Hare, W. W. 1967. A combination of disease resistance in a new cowpea, Mississippi Silver. *Phytopathology* **57**, 460.

Harlan, J. R., and de Wet, J. M. J. 1971. Toward a rational classification of cultivated plants. *Taxon* **20**, 509–517.

Harlan, J. R., and Stemler, A. 1976. The races of sorghum in Africa. In Harlan, J. R., de Wet, J. M. J., and Stemler, A. B. L., eds, *Origins of African Plant Domestication*, The Hague, Mouton Publishers, 445–484.

Harland, S. C. 1919a. Inheritance of certain characters in the cowpea (*Vigna sinensis*). *Journal of Genetics* **8**, 101–132.

Harland, S. C. 1919b. Notes on inheritance in cowpea. *Agricultural News, Barbados* **18**, 20.

Harland, S. C. 1920. Inheritance of certain characters in the cowpea (*Vigna sinensis*). *Journal of Genetics* **10**, 193–205.

Harland, S. C. 1922. Inheritance of certain characters in the cowpea. 3: the very small-eye pattern of the seed coat. *Journal of Genetics* **12**, 254.

Harper, J. M. 1981. Extrusion of foods, vol. 1. Boca Raton, FL, CRC Press, Inc.

Harrison, A. N., and Gudauskas, R. I. 1968. Effects of some viruses on growth and seed production of two cowpea cultivars. *Plant Disease Reporter* **52**, 509–512.

Hartel, P. G., and Alexander, M. 1983. Growth and survival of cowpea rhizobia in acid, aluminum-rich soils. *Soil Science Society of America Journal* **47**, 502–506.

Hartel, P. G., and Alexander, M. 1984. Temperature and desiccation tolerance of cowpea rhizobia. *Canadian Journal of Microbiology* **30**, 820–823.

Hawthorne, P. L. 1943. Breeding and improvement of edible cowpeas. *Proceedings of the American Society of Horticultural Science* **42**, 562–564.

Heald, C. M. 1978. Effect of the reniform nematode on vegetable yields. *Plant Disease Reporter* **62**, 902–902.

Hepper, F. N. 1958. Papilionaceae. In Hutchinson, J., and Dalziel, J. M., eds, *Flora of Tropical West Africa* 2nd edn, **1**, 505–586.

Herbaugh, L., Upton, N. P., and Eplee, R. E. 1980. *Striga gesnerioides* in the United States of America. *Proceedings of the Southern Weed Science Society* **33**, 187–190.

Heyns, J. 1962. A report on South African nematodes of the families Longidoridae, Belondiridae and Alaimidae (Nematode: Dorylaimoidae) with descriptions of three species. *Nematologica* **8**, 15–20.

Hill, R. E., and Mayo, Z. B. 1974. Trap corn to control corn rootworms. *Journal of Economic Entomology* **67** (6), 748–750.

Ho, T. H. 1967. The bean-fly *Melanagromyza phaseoli* Coq. and experiments on its control. *Malaysia Agricultural Journal* **46**, 149–157.

Hobbs, H. A., and Fulton, J. B. 1979. Beetle transmission of cowpea chlorotic mottle virus. *Phytopathology* **69**, 255–256.

Hoffmaster, D. E. 1944. Bacterial canker of cowpeas. *Phytopathology* **34**, 439–441.

Hofmann, F. W. 1926. Hybrid vigor in cowpeas. *Journal of Heredity* **17**, 209–211.

Hogger, C. H., and Bird, G. W. 1976. Weed and indicator hosts of plant parasitic nematodes in Georgia cotton and soybean fields. *Plant Disease Reporter* **50**, 23–226.

Holliday, P. 1970. *Fusarium oxysporum* f.sp. *tracheiphilum*. Kew, UK, CMI (Commonwealth Mycological Institute), CMI Descriptions of pathogenic fungi and bacteria 220.

Holliday, R. H. 1976. The efficiency of solar energy conversion by the whole crop. In Duckham, Jones, J. G. W., and Roberts, E. H., eds, *Food Production and Consumption*, Amsterdam, North Holland Publishing Co., 127–146.

Hong, T. D., Minchin, F. R., and Summerfield, R. J. 1977. Recovery of nodulated cowpea plants [*Vigna unguiculata* (L.) Walp.] from waterlogging during vegetative growth. *Plant and Soil* **48**, 661–672.

Hoover, M. W. 1966. Factors influencing consumer preferences on southern peas (cowpeas). *Proceedings of the Florida State Horticultural Society* **69**, 213–215.

Hopkins, J. C. F. 1950. A descriptive list of plant diseases in southern Rhodesia and a list of bacteria and fungi. Harare, Zimbabwe, Government Stationery Office. 72 p.

Hossain, A. M., and Chambliss, O. L. 1983. Response to selection for pod-wall strength in southernpea, *Vigna unguiculata* (L.) Walp. *HortScience* **18**, 600 (abstract).

Huxley, P. A., and Summerfield, R. J. 1976. Effects of daylength and day/night temperatures on growth and seed yield of cowpea cv. K. 2809 grown in controlled environments. *Annals of Applied Biology* **83**, 259–271.

IAR (Institute for Agricultural Research). 1982. *Grain Legumes Improvement Programme, cropping scheme notes*, February 1982. Samaru, Zaria, Nigeria, IAR.

IAR. 1983. *Grain Legumes Improvement Programme, cropping scheme notes*, February 1983. Samaru, Zaria, Nigeria, IAR.

IAR. 1984. *Grain Legumes Improvement Programme, cropping scheme notes*, February 1984. Samaru, Zaria, Nigeria, IAR.

IBPGR (International Board for Plant Genetic Resources). 1981. Working group meeting on *Vigna*, 17–19 September 1981: report, New Delhi, India. Rome, Italy, IBPGR.

Ibrahim, I. K. A. 1978. The current status of root knot nematodes in northern Egypt. Paper prepared for the research planning conference on root knot nematodes, *Meloidogyne* spp., Cairo University, Giza, Egypt, 29 January–2 February 1978.

Ichinohe, M. 1961. Studies on the soybean cyst nematode *Heterodera glycines*. Report of the Hokkaido National Agriculture Experiment Station 56.

Ichinohe, M., and Asal, K. 1956. Studies on the resistance of soybean plant to the nematode *Heterodera glycines*. 1: Varieties 'Dalichihienuki' and 'Nanguntake-date'. Hokkaido National Agriculture Experiment Station, Research Bulletin 71, 67–79.

IITA (International Institute of Tropical Agriculture). 1973. Grain legume improvement program annual report. Ibadan, Nigeria, IITA. 78 p.

IITA. 1974. Grain legume improvement program annual report. Ibadan, Nigeria, IITA.

IITA. 1975. Annual report. Ibadan, Nigeria, IITA.

IITA. 1976. Annual report. Ibadan, Nigeria, IITA.

IITA. 1977. Annual report. Ibadan, Nigeria, IITA.

IITA. 1979. Annual report. Ibadan, Nigeria, IITA.

IITA. 1980. Annual report. Ibadan, Nigeria, IITA.

IITA. 1982. Annual report. Ibadan, Nigeria, IITA.

IITA. 1983a. Annual report. Ibadan, Nigeria, IITA.

IITA. 1983b. Research highlights. Ibadan, Nigeria, IITA.

IITA. 1984. Research highlights. Ibadan, Nigeria, IITA.

IITA. in press. Annual report. Ibadan, Nigeria, IITA.

IITA–SAFGRAD (International Institute of Tropical Agriculture–Semi-Arid Food Grains Research and Development Project). 1980. Annual report. Ouagadougou, Burkina Faso, IITA–SAFGRAD, C28–C49.

IITA–SAFGRAD. 1981. Annual report. Ouagadougou, Burkina Faso, IITA–SAFGRAD, C36–358.

IITA–SAFGRAD. 1982. Annual report. Ouagadougou, Burkina Faso, IITA–SAFGRAD, 105–133.

IITA–SAFGRAD. 1983. Annual report. Ouagadougou, Burkina Faso, IITA–SAFGRAD, 138–188.

Iizuka, N., Rajeshwari, R., Reddy, D. V. R., Goto, T., Muniyappa, V., Bharathan, N., and Ghanekar, A. M. 1982. Natural occurrence of a strain of cowpea mild mottle virus on groundnut (*Arachis hypogaea*) in India. *Phytopathologische Zeitschrift* **109** (3), 243–253.

IRRI (International Rice Research Institute). 1978. Annual report. Los Baños, Philippines, IRRI.

IRRI. 1980. Annual report. Los Baños, Philippines, IRRI.

IRRI. 1983. Annual report. Los Baños, Philippines, IRRI.

Isenmilla, A. E., Babalola, O., and Obigbesan, G. O. 1981. Varietal influence of intercropped cowpea on the growth, yield and water relations of maize. *Plant and Soil* **62** (1), 153–156.

Islam, R., and Ayanaba, A. 1981. Effect of seed inoculation and preinfecting cowpea (*Vigna unguiculata*) with *Glomus mosseae* on growth and seed yield of the plants under field conditions. *Plant and Soil* **61**, 341–350.

Iwaki, M. 1979. Virus and mycoplasma diseases of leguminous crops in Indonesia. *Review of Plant Protection Research* **12**, 88–97.

Iwaki, M., Roechan, M., and Tantera, D. M. 1975. *Virus Diseases of Legume Plants in Indonesia. 1: Cowpea aphid-borne mosaic virus.* Bogor, Indonesia, Contr. Central Research Institute of Agriculture **13**, 1–14.

Iwaki, M., Thongmeearkon, P., Prommin, M., Honda, Y., and Hibi, T. 1982. Whitefly transmission and some properties of cowpea mild mottle virus occurring on soybean in Thailand. *Plant Disease* **66**, 365–368.

Jackai, L. E. N. 1981a. Relationship between cowpea crop phenology and field infestation by the legume pod borer, *Maruca testulalis*. *Annals of the Entomology Society of America*, 402–408.

Jackai, L. E. N. 1981b. Use of an oil-soluble dye to determine the oviposition sites of the legume pod-borer *Maruca testulalis* (Geyer) (Lepidoptera: Pyralidae). *Insect Science Applications* **2** (4), 205–207.

Jackai, L. E. N. 1982. A field screening technique for resistance of cowpea (*Vigna unguiculata*) to the pod borer *Maruca testulalis* (Geyer) (Lepidoptera: Pyralidae). *Bulletin of Entomology Research* **72**, 145–146.

Jackai, L. E. N. 1983. Using trap plants in the control of insect pests of tropical legumes. Paper prepared for the international workshop on integrated pest control of tropical crop legumes, Goiânia, Goiás, Brazil, 4–9 April 1983.

Jackai, L. E. N., and Singh, S. R. 1981. Studies on some behavioural aspects of *Maruca testulalis* on selected species of *Crotalaria* and *Vigna unguiculata*. *Tropical Grain Legume Bulletin* **22**, 1–3.

Jackai, L. E. N., and Singh, S. R. 1983. Varietal resistance in the integrated pest management of cowpea pests. *Insect Science Applications* **41**, 199–204.

Jain, H. K., 1981. Evolving a new production technology of pulses. *Pulse Crops Newsletter* **1**, 1002.

Jain, H. K., and Mehra, K. L. 1980. Evolution, adaptation, relationships and uses of the species of *Vigna* cultivated in India. In Summerfield, R. J., and Bunting, A. H., eds, *Advances in Legume Science*, London, UK, Her Majesty's Stationery Office, 459–468.

Jain, S. 1982. *Studies on heterosis and combining ability for forage and seed yield characters in cowpea* (V. unguiculata). Hissar, India, Hissar Agricultural University. (Thesis abstract.)

Janarthanan, R. 1972. Occurrence of the pigeon pea cyst nematode in Tamil Nadu. *Indian Journal of Nematology* **2**, 215.

Janoria, M. P., and Ali, M. A. 1970. Correlation and regression studies in cowpeas. *JNKVV Research Journal* **4**, 15–19.

Jansen, W. P., and Staples, R. 1971. Specificity of transmission of cowpea mosaic virus by species within the subfamily Galerucinae, family Chrysomelidae. *Journal of Economic Entomology* **64** (2), 365–367.

Jarry, M., and Bonet, A. 1982. La bruche du haricot, *Acanthoscelides obtectus* Say (Col., Bruchidae), est-elle un danger pour le cowpea, *Vigna unguiculata* (L.) Walp.? *Agronomie* **2**, 963–968.

Jatasara, D. S., Lodhi, G. P., and Grewal, R. P. S. 1980. Note on the efficiency of a new crossing technique in cowpea. *Indian Journal of Agricultural Science* **50**, 876–877.

Jayasekara, S. J. B. A. 1984. Upland crops varietal testing in Sri Lanka for the Asian cropping systems network. Paper prepared for the varietal improvement monitoring tour in Nepal and Indonesia, 9–20 July 1984.

Jayasekara, S. J. B. A., and Hettiarachchi, K. 1984. Cowpea cultivation in Sri Lanka: production constraints and present status. Paper prepared for the varietal improvement monitoring tour in Nepal and Indonesia, 9–20 July 1984.

Jenkins, W. F., and Hare, W. W. 1957. Spacing of southern peas. *Proceedings of the American Society of Horticultural Science* **69**, 405–407.

Jensen, H. J. 1953. Experimental greenhouse host range studies of two root lesion nematodes, *Pratylenchus vulnus* and *Pratylenchus penetrans*. *Plant Disease Reporter* **37**, 384–387.

Jensen, H. J. 1972. Nematode pests of vegetable and related crops. In Webster, J. M., ed., *Economic Nematology*, London, UK, Academic Press, 377–408.

Jerath, M. L. 1968. Insecticidal control of *Maruca testulalis* on cowpea in Nigeria. *Journal of Economic Entomology* **61**, 413–416.

Jhooty, J. S., Sokhi, S. S., Sohi, H. S., Bains, S. S., Rewal, H. S., and Munshi, G. B. 1980. The reaction of cowpea germplasm to different diseases. *Crop Improvement* **7**, 18–24.

Jindal, J. K., and Patel, P. N. 1980. Variability in xanthomonads of grain legumes. 1: pathogenic behavior and taxonomy. *Phytopathologische Zeitschrift* **99**, 332–356.

Jindla, L. N., and Singh, K. B. 1970. Inheritance of flower colour, leaf shape, and pod length in cowpea (*Vigna sinensis* L.). *Indian Journal of Heredity* **2**, 45–49.

João, W. S. J., Elías, L. G., and Bressani, R. 1980a. Efecto de diferentes tratamientos dietéticos sobre el consumo de dietas â base de tubérculos y leguminosas. *Archivos Latinoamericanos de Nutrición* **30**, 187–199.

João, W. S. J., Elías, L. G., and Bressani, R. 1980b. Efecto del proceso de cocción-extrusión (Brady crop cooker) Sobre el vàlor nutritivo de mezclas elaboradas a base de frijol caupí (*Vigna sinensis*)–maíz y de frijol caupí–yuca. *Archivos Latinoamericanos de Nutrición* **30**, 539–550.

Johnson, A. W., Rosebery, G., and Parker, G. 1976. A novel approach to *Striga* and *Orobanche* control using synthetic germination stimulants. *Weed Research* **16**, 223–227.

Johnson, A. W., Young, J. R., and Mullinex, B. G. 1981. Applying nematicides through an overhead sprinkler irrigation system for control of nematodes. *Journal of Nematology* **13**, 154–159.

Jones, F. G. W., and Moriarty, F. 1956. Further observations on the effects of peas, beans and vetch upon soil population levels of pea root eelworm, *Heterodera goettingiana* Leebscher. *Nematologica* **1**, 268–273.

Jones, S. T. 1965. Radiation-induced mutations in southern peas. *Journal of Heredity* **56**, 273–276.

Juárez, H. A., Burgos, C. F., and Saunders, J. L. 1982. Maize–cowpea mixed crop system response to insect control and maize population variation. *Journal of Economic Entomology* **75**, 216–219.

Kaiser, W. J. 1972. Diseases of food legumes caused by pea leaf roll virus in Iran. *FAO Plant Protection Bulletin* **20**, 127–133.

Kaiser, W. J. 1979. Natural infection of cowpea and mung bean by alfalfa mosaic virus in Iran. *Plant Disease Reporter* **63**, 414–418.

Kaiser, W. J., and Danesh, D. 1971. Biology of four viruses affecting *Cicer arietinum* in Iran. *Phytopathology* **61** (4), 373–375.

Kaiser, W. J., and Mossahebi, H. 1975. Studies with cowpea aphid-borne mosaic virus and its effect on cowpea in Iran. *FAO Plant Protection Bulletin* **27**, 27–30.

Kaiser, W. J., and Ramos, A. H. 1979. Two bacterial diseases of cowpea in East Africa. *Plant Disease Reporter* **63**, 304–308.

Kaiser, W. J., and Schalk, J. M. 1975. Transmission studies with pea leaf roll virus on pulse crops in Iran. FAO *Plant Protection Bulletin* **23**, 169–173.

Kaiser, W. J., and Vakili, N. G. 1978. Insect transmission of pathogenic *Xanthomonas phaseoli*, *X. phaseoli* f. sp. *vignicola* to bean and cowpea in Puerto Rico. *Phytopathology* **68**, 1057–1063.

Kaiser, W. J., Danesh, D., Okhovat, M., and Mossahebi, H. 1968. Diseases of pulse crops (edible legumes) in Iran. *Plant Disease Reporter* **52**, 687–691.

Kaiser, W. J., Wyatt, S. D., and Pesho, G. R. 1982. Natural hosts and vectors of tobacco streak virus in eastern Washington. *Phytopathology* **72** (11), 1508–1512.

Kamara, C. S. 1980a. Mulch-tillage technique in Sierra Leone on cowpea growth and yield. *Tropical Grain Legume Bulletin* **19**, 10–13.

Kamara, C. S., 1980b. Effect of tillage techniques and fertilizer nitrogen on the growth and yield of cowpeas in Sierra Leone. In Rosswall, T., ed., *Nitrogen Cycling in West African Ecosystems*, Stockholm, Sweden, SCOPE/UNEP International Nitrogen Unit, Royal Swedish Academy of Science, 243–248.

Kaminski, E. 1969. Drogi r zwoju i realizacja zadan nadzorn kwarantannowego u Polsce. *Ochrana Rostlin* **13**, 8–12.

Kandaswamy, P., Muthuswamy, P., and Krishnamoorthy, K. K. 1976. Effect of stage of cutting on the crude protein, phosphorous content and yield of cowpea (*Vigna sinensis* L.). *Madras Agricultural Journal* **63** (3), 178–180.

Kang, B. T. 1983. Fertilizer use in multiple cropping systems in Nigeria, Tanzania and Senegal. *FAO Fertilizer and Plant Nutrition Bulletin* **5**, 36–46. (Report of an Expert Consultation held in New Delhi, 3–16 February 1982.)

Kang, B. T., and Fox, R. L. 1975. Influence of soil fertility on the protein and sulfur contents of grain legumes. In Luse, R. A., and Rachie, K. O., eds, *Proceedings of IITA Collaborators' Meeting on Grain Legume Improvement*, 9–13 June 1975, Ibadan, Nigeria, IITA, 105–109.

Kang, B. T., and Nangju, D. 1983. Phosphorus response of cowpea [*Vigna unguiculata* (L.) Walp.]. *Tropical Grain Legume Bulletin* **27**, 11–16.

Kanwar, J. S. 1982. Managing soil resources to meet the challenges to mankind. Presidential address. *Transactions of the 12th International Congress of Soil Sciences*, New Delhi, India, 8–16 February.

Karel, A. K., Lakhani, D. A., and Ndunguru, B. J. 1982. Intercropping of maize and cowpea: effect of plant populations on insect pests and seed yield. In Keswani, C. L., and Ndunguru, B. J., eds, *Intercropping: Proceedings of the Second Symposium on Intercropping in Semi-Arid Areas*, Morogoro, Tanzania, 4–7 August 1980, Ottawa, Canada, IDRC (International Development Research Centre).

Karunaratne, S. M., Gunasena, H. P. M., and Mantriratne, M. A. P. 1978. Growth and yield of cowpea [*Vigna unguiculata* (L.) Walp.] var. MI-35 under coconut. *Journal of the National Agriculture Society of Ceylon (Sri Lanka)*, **15**, 8–19.

Karwasara, S. S., Solanki, K. R., and Jhorar, B. S. 1982. Reaction of cowpea varieties to bacterial blight and cowpea mosaic virus. *Tropical Grain Legume Bulletin* **26**, 15–19.

Kassam, A. H., and Kowal, J. M. 1973. Productivity of crops in the savanna and rain forest zones in Nigeria. *Savanna* **2**, 39–49.

Kassanis, B., and Varma, A. 1975. Sunnhemp mosaic virus. Kew, UK, CMI (Commonwealth Mycological Institute), CMI/AAB Descriptions of plant viruses 153.

Katan, J. 1981. Solar heating (solarization) of soil for control of soilborne pests. *Annual Review of Phytopathology* **19**, 211–236.

Kayumbo, H. Y. 1976. Crop protection in mixed ecosystems. Paper prepared for the symposium on intercropping in semi-arid areas, Morogoro, Tanzania, 10–12 May 1976.

Kendrick, J. B. 1931. Seed transmission of cowpea *Fusarium* wilt. *Phytopathology* **21**, 979–983.

Kendrick, J. B. 1936. A cowpea resistant to *Fusarium* wilt and nematode root knot. *Phytopathology* **26**, 98.

Kennedy, M. B. 1982. Extrusion of cowpeas: effects of extruder feed moisture and barrel temperature on the physical properties of extruded cowpea meal. Athens, GA, University of Georgia (M.Sc. thesis).

Keyser, H. H., Munns, D. N., and Hohenberg, J. S. 1979. Acid tolerance of rhizobia in culture and in symbiosis with cowpea. *Soil Science Society of America Journal* **43**, 719–722.

Khalf-Allah, A. M., Faris, F. S., and Nassar, S. H. 1973. Inheritance and nature of resistance to cucumber mosaic virus in cowpea, *Vigna sinensis*. *Egyptian Journal of Genetics and Cytology* **2**, 274–282.

Khatri, H. L., and Chohan, J. S. 1972. Studies on some factors influencing seed transmission of cowpea mosaic virus in cowpea. *Indian Journal of Mycology and Plant Pathology* **2** (1), 40–44.

Khatri, H. L., and Singh, L. 1974. Studies on mosaic disease of cowpea. *Journal of Research, Punjab Agricultural University* **11**, 289–294.

Kheir, A. M., and Farahat, A. A. 1981. Comparative interaction of *Meloidogyne javanica* and five leguminous hosts. *Phytopathologia Mediterranea* **20**, 141–143.

Kheradnam, M., Bassiri, A., and Niknejad, M. 1975. Heterosis, inbreeding depression, and reciprocal effects for yield and some yield components in a cowpea cross. *Crop Science* **15**, 689–691.

Kheradnam, M., and Niknejad, M. 1971. Combining ability of cowpeas (*Vigna sinensis* L.). *Zeitschrift fuer Pflanzenzuechtung* **66**, 312–316.

Kheradnam, M., and Niknejad, M. 1974. Heritability estimates and correlations of agronomic characters in cowpea (*Vigna sinensis* L.). *Journal of Agricultural Science* **82**, 207–208.

Kitajima, E. W., Noda, H., Lin, M. T., and Costa, C. L. 1979. Um mosaico em feijão-de-asa (*Psophocarpus tetragonolobus*) causado por um isolado do subgrupo severo do virus do mosaico da *Vigna*. *Fitopatologia Brasileira* **4**, 519–524.

Klein, P. 1960. Schaden durch freilebende Nematoden an Spargeljung pflanzen. *Pflanzenschutz Munich* **12**, 128–129.

Knippling, R. F. 1979. The basic principles of insect population suppression and management. USDA (United States Department of Agriculture), *Agricultural Handbook* **512**, 397–401.

Kobayashi, T., Hasegawa, T., and Kegasawa, K. 1972. Major insect pests of leguminous crops in Japan. *Tropical Agriculture Research Series Tokyo* **6**, 109–126.

Koehler, C. S., and Mehta, P. H. 1972. Relationships of insect control attempts by chemicals to components of yield in cowpeas in Uganda. *Journal of Economic Entomology* **65**, 1421–1427.

Kohli, K. S., Singh, C. B., Singh, A., Mehra, K. L., and Magoon, M. L. 1971. Variability of quantitative characters in a world collection of cowpea: interregional comparisons. *Genetica Agraria* **25**, 231–242.

Kolhe, A. K. 1970. Genetic studies in *Vigna* sp. *Poona Agricultural College Magazine* **59**, 126–137.

Kolte, S. J., and Nene, Y. L. 1972. Studies on symptoms and mode of transmission of the leaf crinkle virus of urd bean (*Phaseolus mungo*). *Indian Phytopathology* **25**, 401–404.

Kolte, S. J., and Nene, Y. L. 1975. Host range and properties of urd bean leaf crinkle virus. *Indian Phytopathology* **28**, 430–431.

Koshy, P. K., and Swarup, G. 1972. Susceptibility of plants to pigeon pea cyst nematode, *Heterodera cajani*. *Indian Journal of Nematology* **2**, 1–6.

Kowal, J. M., and Knabe, D. T. 1972. An agroclimatological atlas of the northern states of Nigeria with explanatory notes. Zaria, Nigeria, Ahmadu Bello University.

Kozak, M. 1977. Responses of yields and nutrient contents of wheat, maize and cowpea to potassium fertilization. *Agroken, Talajtan* **26** (3–4), 363–378.

Kremer, R. J., and Peterson, H. L. 1983. Field evaluation of selected rhizobia in an improved legume inoculant. *Agronomy Journal* **75**, 139–143.

Krishnaswamy, N., Nambiar, K. K., and Mariakulandai, A. 1945. Studies in cowpea [*V. unguiculata* (L.) Walp.]. *Madras Agricultural Journal* **33**, 145–160, 193–200.

Krutman, S., Araujo, S. C., and Desouza, M. S. 1975. Melhoramento do feijoeiro de

Macassar *Vigna unguiculata* Walp, hibirdaçao e seleçao. *Ciencia e Cultura,* **27,** 256.

Kuhn, C. W. 1963. Field occurrence and properties of the cowpea strain of southern bean mosaic virus. *Phytopathology* **53,** 732–733.

Kuhn, C. W. 1964. Purification, serology and properties of a new cowpea virus. *Phytopathology* **54,** 843–857.

Kuhn, C. W. 1968. Identification and specific infectivity of a soybean strain of cowpea chlorotic mottle virus. *Phytopathology* **58,** 1441–1442.

Kuhn, C. W., and Dawson, W. O. 1973. Multiplication and pathogenesis of cowpea chlorotic mottle virus and southern bean mosaic virus in single and double infections in cowpea. *Phytopathology* **63,** 1380–1385.

Kuhn, C. W., Brantley, B. B., Demski, J. W., and Pio-Ribeiro, G. 1984. Pinkeye Purple Hull-BVR, White Acre-BVR and Corona cowpeas. *HortScience* **19,** 592.

Kuhn, C. W., Brantley, B. B., and Sowell, G. 1966. Southern pea viruses: identification, symptomatology and sources of resistance. *Georgia Agricultural Experiment Station Research Bulletin* 157.

Kuhn, C. W., Wyatt, S. D., and Brantley, B. B. 1981. Genetic control of symptoms, movement and virus accumulation in cowpea plants infected with cowpea chlorotic mottle virus. *Phytopathology* **71,** 1310–1315.

Kulthe, K. S., and Mali, V. R. 1979. Occurrence of tobacco mosaic virus on cowpea (*Vigna unguiculata*) in India. *Tropical Grain Legume Bulletin* **16,** 8–13.

Kumar, A., and Mishra, S. N. 1981. Note on genetic variability for forage yield components in cowpea. *Indian Journal of Agricultural Science* **51,** 807–809.

Kumar, A., Mishra, S. N., and Verma, J. S. 1982. Studies on genetic diversity in cowpea. *Crop Improvement* **9,** 160–163.

Kumar, A., Mishra, S. N., and Verma, J. S. 1983. Correlation and path analysis in cowpea. *Crop Improvement* **10,** 36–39.

Kumar, B. B., and Pillai, P. B. 1979. Effect of N, P and K on the yield of cowpea variety. *Agricultural Research Journal Kerala* **17** (2), 194–199.

Kumar, P., Prakash, R., and Haque, M. F. 1976. Floral biology of cowpea (*V. sinensis* L.). *Tropical Grain Legume Bulletin* **6,** 9–11.

Kumar, P., Prakash, R., and Haque, M. F. 1978. Line × tester analysis of combining ability in cowpea. *Proceedings of the Bihar Academy of Agricultural Science b26,* 12–19.

Kuo, J., and Pate, J. S. submitted for publication. Unusual network of internal phloem in the pod mesocarp of cowpea [*Vigna unguiculata* (L.) Walp. (Fabaceae)].

Kvičala, B. A., Smrz, J., and Blanco, N. 1970. Some properties of cowpea mosaic virus isolated in Cuba. *Phytopathologische Zeitschrift* **69** (3), 223–235.

Kivičala, B. A., Smrz, J., and Blanco, N. 1973. A beetle-transmitted virus disease of cowpea in Cuba. *FAO Plant Protection Bulletin* **21,** 27–29, 39.

Lackey, J. A. 1977. A synopsis of Phaseoleae (Leguminosae, Papilionoideae). Ames, IA, Iowa State University. (PhD. thesis.)

Lackey, J. A. 1978. Leaflet anatomy of Phaseoleae (Leguminosae, Papilionoideae) and its relation to taxonomy. *Botanical Gazette* **139,** 436–446.

Lackey, J. A. 1981. Phaseoleae DC. In Polhill, R. M., and Raven, P. H., eds, *Advances in Legume Systematics,* Part 1. London, UK, Her Majesty's Stationery Office, 301–310.

Ladeinde, T. A. O., Watt, E., and Onajole, A. A. O. 1980. Segregation pattern of three different sources of male-sterile genes in *Vigna unguiculata. Journal of Heredity* **71,** 431–432.

Ladipo, J. L. 1975. Southern bean mosaic virus (cowpea strain) in Nigeria. *Nigeria Agricultural Journal* **12**, 25–28.

Ladipo, J. L. 1976. A vein-banding strain of cowpea aphid-borne mosaic virus in Nigeria. *Nigerian Journal of Science* **10** (1–2), 77–88.

Ladipo, J. L. 1977. Seed transmission of cowpea aphid-borne mosaic virus in some cowpea cultivars. *Nigerian Journal of Plant Protection* **3**, 3–10.

Ladipo, J. L., and Allen, D. J. 1979a. Identification of resistance to cowpea aphid-borne mosaic virus. *Tropical Agriculture (Trinidad)* **56**, 353–359.

Ladipo, J. L., and Allen, D. J. 1979b. Identification of resistance to southern bean mosaic virus in cowpea. *Tropical Agriculture (Trinidad)* **56**, 33–40.

Lagoke, S. T. O., Choudhary, A. H., and Chandra-Singh, D. J. 1982. Chemical weed control in rainfed cowpea in the guinea savanna zone of Nigeria. *Weed Research* **22** (1), 17–22.

Laing, D. R., Kretchmer, P. J., Zuluaga, S., and Jones, P. G. 1983. Field bean (*Phaseolus vulgaris*). In Smith, W. H., and Yashida, S., eds, *Potential Productivity of Field Crops under Different Environments*, Los Baños, Philippines, IRRI, 227–248.

Lakshmi, P. V., and Goud, J. V. 1977. Variability in cowpea (*V. sinensis* L.). *Mysore Journal of Agricultural Science* **11**, 145–147.

Lal, R. 1974. *International expert consultation on use of improved technology in rainfed areas of tropical Asia*, Hyderabad, India. Rome, Italy, FAO.

Lal, R. 1976. No-tillage effects on soil properties under different crops in western Nigeria. *Soil Science Society of America Journal* **40** (5), 762–768.

Lal, R. 1979. Modification of soil fertility characteristics by management of soil physical properties. In Lal, R., and Greenland, D. J., eds, *Soil Physical Properties and Crop Production in the Tropics*, Chichester, UK, John Wiley and Sons, 397–405.

Lal, R., De, R., and Singh, R. K. 1978a. Legume contribution to the fertilizer economy in legume–cereal rotations. *Indian Journal of Agricultural Science* **48** (7), 419–424.

Lal, R., Maurya, P. R., and Osei-Yeboah, S. 1978b. Effects of no tillage and ploughing on efficiency of water use in maize and cowpea. *Experimental Agriculture* **14**, 113–120.

Lal, S., Singh, M., and Pathak, M. M. 1976. Combining ability of cowpea. *Indian Journal of Genetics and Plant Breeding* **35**, 375–378.

Lamptey, P. N., and Hamilton, R. I. 1974. A new cowpea strain of southern bean mosaic virus from Ghana. *Phytopathology* **64**, 1100–1104.

La Rue, T. A. 1980. Host plant genetics and enhancing symbiotic nitrogen fixation. In Stewart, W. D. P., and Gallon, J. R., eds, *Nitrogen Fixation: Proceedings of the Phytochemical Society of Europe, Sussex, September 1979*, San Francisco, CA, Academic Press, 355–364.

Latham, M. C. 1979. *Human Nutrition in Tropical Africa*. Rome, Italy, FAO.

Lathman, D. H. 1934. Life history of a Cercospora leaf-spot fungus of cowpea. *Mycologia* **26**, 516–527.

Lawani, S. M. 1980. Collaboration and the quality of research productivity. *IITA Research Briefs* **1** (3), 6–8.

Lawhale, A. D. 1982. Note on genetic variability in quantitative characters of cowpea in M3 generation. *Indian Journal of Agricultural Science* **52**, 22–23.

Layzell, D. B., and La Rue, T. A. G. 1982. Modeling C and N transport to developing soybean fruits. *Plant Physiology* **70**, 1290–1298.

Lefebvre, C. L., and Sherwin, H. S. 1945. Observations on the bacterial canker of cowpeas. *Phytopathology* **35**, 487.

Lefebvre, C. L., and Sherwin, H. S. 1950. Inheritance of resistance to bacterial canker (*Xanthomonas vignicola*) in cowpea, *Vigna sinensis*. *Phytopathology* **40**, 17–18 (abstract).

Lefebvre, C. L., and Stevenson, J. A. 1945. The fungus causing zonate leaf spot of cowpea. *Mycologia* **37**, 37–45.

Lefebvre, C. L., and Weimer, J. L. 1939. *Choanephora cucurbitarum* attacking cowpeas. *Phytopathology* **29**, 898–901.

Lehman, P. S., Barker, K. R., and Huisingh, D. 1970. The effect of races of *H. glycines* on nodulation and nitrogen fixation on soybean. *Phytopathology* **60**, 1299–1300.

Leleji, O. I. 1975. Inheritance of three agonomic characters in cowpea [*Vigna sinensis* L. (Savi)]. *Madras Agricultural Journal* **62**, 95–97.

Leleji, O. I. 1979. Cowpea breeding activities in the Institute for Agricultural Research, Ahmadu Bello University, Zaria. Paper prepared for the 1st annual research conference, IITA, Ibadan, Nigeria.

Leleji, O. I. 1983. Cowpea breeding and testing at the Institute for Agricultural Research, Ahmadu Bello University. In IITA, *Proceedings of the 1st Grain Legume Improvement Workshop*, IITA, Ibadan, Nigeria, 29 October–2 November 1983, 56–61.

Le Marchand, G., Maréchal, R., and Baudet, J. C. 1976. Observations sur quelques hybrides dans le genre *Phaseolus*. 3: *Phaseolus lunatus*: nouveaux hybrides et considerations sur les affinités interspecifiques. *Bulletin de Recherches Agronomiques de Gembloux* **11**, 183–200.

Leonard, M. D., and Mills, A. S. 1931. A preliminary report on the lima bean pod borer and other legume pod borers in Puerto Rico. *Journal of Economic Entomology* **24**, 466–473.

Le Pelley, R. H. 1959. *Agricultural Insects of East Africa*. Nairobi, Kenya, East Africa High Commission.

Ligon, L. L. 1958. Characteristics of cowpea varieties (*Vigna sinensis*). *Oklahoma Agriculture Experiment Station Bulletin* B-518.

Lima, J. A. A., and Gonçalves, M. F. B. 1980. Transmisibilidade de "cowpea mosaic virus" pelo manhoso *Chalcodermus bimaculatus*. *Fitopatologia Brasileira* **5**, 414–415.

Lima, J. A. A., and Nelson, M. R. 1977. Etiology and epidemiology of mosaic of cowpea in Ceará, Brazil. *Plant Disease Reporter* **61**, 864–867.

Lima, J. A. A., and Souza, C. A. V. 1980. Comovirus do subgrupo severo de 'cowpea mosaic virus' isolado de *Canavalia ensiformis* no Ceará. *Fitopatologia Brasileira* **5**, 417.

Lima, J. A. A., Oliveira, F. M., Kitajima, E. W., and Lima, M. G. A. 1981. Propiedades biológicas e citológicas de um potyvirus isolado de feijão-de-corda no Ceará. *Fitopatologia Brasileira* **6**, 205–216.

Lima, J. A. A., Purcifull, D. E., Hiebert, E. 1979. Purification, partial characterization, and serology of blackeye cowpea mosaic virus. *Phytopathology* **69**, 1252–1258.

Lima, J. A. A., Silveira, L. F. S., and Santos, M. L. B. 1983. Cultivares de feijão-de-corda com resistência de campo ao mosaico ameralo. *Fitopatologia Brasileira* **8**, 614.

Lima, M. G. A., de, Daoust, R. A., and Soper, R. S. 1984. Patogenicidade de fungos a *Elasmopalpus lignosellus* e outros lepidópteros pragas do caupi (*Vigna unguicu-lata* Wal.) pulverizados diretamente numa torre calibrada. Paper prepared for the 9th congress for Brazilian entomology, Londrina, Brazil, 178.

Lin, M. T. 1979. Purification and serology of legume and corn viruses in Brazil. *Fitopatologia Brasileira* **4**, 203–213.

Lin, M. T., Anjos, J. R. N., Kitajima, E. W., and Rios, G. P. 1979. Um novo potyvirus isolado do caupi e potencialmente importante para a cultura do feijão no Brasil. *Fitopatologia Brasileira* **4**, 120–121.

Lin, M. T., Anjos, J. R. N., and Rios, G. P. 1981a. Serological grouping of cowpea severe mosaic virus isolates from central Brazil. *Phytopathology* **71**, 435–438.

Lin, M. T., Anjos, J. R. N., and Rios, G. P. 1982a. Cowpea severe mosaic virus in five legumes in central Brazil. *Plant Disease* **66**, 67–70.

Lin, M. T., Hill, J. H., Kitajima, E. W., and Costa, C. L. 1984. Two new serotypes of cowpea severe mosaic virus. *Phytopathology* **74**, 581–585.

Lin, M. T., Kitajima, E. W., and Rios, G. P. 1981b. Serological identification of several cowpea viruses in central Brazil. *Fitopatologia Brasileira* **6**, 73–85.

Lin, M. T., Santos, A. A., and Kitajima, E. W. 1981c. Host reactions and transmission of two seed-borne cowpea viruses from central Brazil. *Fitopatologia Brasileira* **6**, 193–203.

Lin, M. T., Santos, A. A., and Muñoz, J. O. 1982b. Ocorrência do vírus do mosaico do pepino em caupi no Estado do Piauí. Goiânia, Goiás, Brazil, EMBRAPA/CNPAF, RENAC, 101.

. Linford, M. G., and Oliveira, J. M. 1940. *Rotylenchulus reniformis* nov. gen. n. sp., a nematode parasite of roots. *Proceedings of the Helminthological Society of Washington* **7**, 35–42.

Linford, M. G., and Yap, F. 1940. Some host plants of the reniform nematode in Hawaii. *Proceedings of the Helminthological Society of Washington* **7**, 42–44.

Linford, M. G., Yap, F., and Oliveira, J. M. 1938. Reduction of soil populations of the root knot nematode during decomposition of organic matter. *Soil Science* **45**, 127–141.

Lister, R. M., and Thresh, J. M. 1955. A mosaic disease of leguminous plants caused by a strain of tobacco mosaic virus. *Nature* **115**, 1047.

Litsinger, J. A. 1982. Evaluation of insect control recommendations in a cropping systems program. In IRRI, *Report of a Workshop on Cropping Systems Research in Asia*, Los Baños, Philippines, IRRI, 441–451.

Litsinger, J. A., Bendong, J. F., and Cruz, C. C.de la. 1983. Verifying and extending integrated pest control technology to small-scale farmers. Paper prepared for the international workshop on integrated pest control for tropical crop legumes, Goiânia, Goiás, Brazil, 4–9 April 1983.

Littleton, E. J., Dennett, M. D., Elston, J., and Monteith, J. L. 1979a. The growth and development of cowpeas (*Vigna unguiculata*) under tropical field conditions. 1: leaf area. *Journal of Agricultural Science Cambridge* **93**, 291–307.

Littleton, E. J., Dennett, M. D., Elston, J., and Monteith, J. L. 1981. The growth and development of cowpeas (*Vigna unguiculata*) under tropical field conditions. 3: photosynthesis of leaves and pods. *Journal of Agricultural Science Cambridge* **97**, 529–550.

Littleton, E. J., Dennett, M. D., Monteith, J. L., and Elston, J. 1979b. The growth and development of cowpeas (*Vigna unguiculata*) under tropical field conditions. 2: accumulation and partitioning of dry weight. *Journal of Agricultural Science Cambridge* **93**, 309–320.

Louis, I. H., and Sundaram, M. K. 1975. Induction of white eye mutant in cowpea [*Vigna sinensis* L. (Savi)]. *Madras Agricultural Journal* **62**, 95–97.

Lovisolo, O., and Conti, M. 1966. Identification of an aphid-transmitted cowpea mosaic virus. *Netherlands Journal of Plant Pathology* **72**, 265–269.

Luck, R. F., van den Bosch, R., and Garcia, R. 1977. Chemical insect control—a troubled pest management strategy. *BioScience* **27** (9), 606–611.

Lukoki, L. 1980. Deux legumineuses a graines alimentaires, *Vigna radiata* (L.) Wilczek et *Vigna mungo* (L.) Hepper. Possibilités d'hybridations. Modifications des équilibres en acides aminés par le transfert de peptides libres. Gembloux, Belgium, Faculté des sciences agronomiques de Gembloux. (Thèse de doctorat.)

Luse, R. L., Kang, B. T., Fox, R. L., and Nangju, D. 1975. Protein quality in grain legumes grown in the lowland humid tropics, with special reference to West Africa. In IPI (International Potash Institute), *Fertilizer Use and Protein Production*: 11th Colloquium, Bornholm, Denmark, IPI.

Lush, W. M., and Evans, L. T. 1980. Photoperiodic regulation of flowering in cowpeas [*Vigna unguiculata* (L.) Walp.] *Annals of Botany* **46**, 719–725.

Lush, W. M., and Evans, L. T. 1981. Domestication and improvement of cowpea. *Euphytica* **30**, 579–587.

Lush, W. M., and Wien, H. C. 1980. The importance of seed size in early growth of wild and domesticated cowpeas. *Journal of Agricultural Science Cambridge* **94**, 177–182.

Lush, W. M., Evans, L. T., and Wien, H. C. 1980. Environmental adaptation of wild and domesticated cowpeas [*Vigna unguiculata* (L.) Walp.]. *Field Crops Research* **3**, 173–187.

Luttrell, E. S., and Weimer, J. L. 1952. *Macrophomina* stem canker and ashy stem blight of cowpea. *Plant Disease Reporter* **36**, 194–195.

Mackie, W. W. 1934. Breeding for resistance in blackeye cowpeas to *Fusarium* wilt, charcoal rot, and nematode root knot. *Phytopathology* **24**, 1135.

Mackie, W. W. 1939. Breeding for resistance in blackeye cowpeas to cowpea wilt, charcoal rot and root nematodes. *Phytopathology* **29**, 826.

Mackie, W. W. 1946. Blackeye beans in California. *California Experiment Station Bulletin* 696.

Mackie, W. W., and Smith, F. L. 1935. Evidence of field hybridization in beans. *Journal of the American Society of Agronomy* **27**, 903–909.

Madamba, C. P. 1981. Distribution and identification of *Meloidogyne* spp. in the Philippines and five other Asian countries. *Philippine Agriculturalist* **64**, 21–39.

Mafuka, M. M. 1984. Study on the physiology, survival, competitive ability and cross infectivity of rhizobia of the cowpea group. Louvain-le-Neuve, Belgium, Université. (Thèse de doctorat.)

Magalhães, A. A., Choudhury, M. M., Millar, A. A., and Albuquerque, M. M. 1982. Efeito do deficit de água no solo sobre o ataque de *Macrophomina phaseolina* em feijão. *Pesquisa Agropecuaria Brasileira* **17**, 407–411.

Mahadevappa, M., Gopalareddy, T., and Shankar Reddy, B. T. 1976. Breeding cowpea (*V. sinensis*) for rice fallow and parallel multiple cropping in sugarcane. *Madras Agriculture Journal* **63**, 171–175.

Mahler, R. L., and Wollum, A. G. 1981. The influence of soil water potential and soil texture on the survival of *Rhizobium japonicum* and *Rhizobium leguminosarum* isolates in the soil. *Soil Society of America Journal* **45**, 761–766.

Maitra, S. C., and Jana, M. K. 1968. X-ray experiment in cowpea (*Vigna catjang*). *Harvester 10th Anniversary* 55–57.

Mak, C., and Yap, T. C. 1977. Heterosis and combining ability of seed protein, yield and yield components in long bean. *Crop Science* **17**, 339–341.

Mak, C., and Yap, T. C. 1980. Inheritance of seed protein content and other agronomic characters in long bean (*Vigna sesquipedalis* Fruw.). *Theoretical and Applied Genetics* **56**, 233–239.

Malek, R. B., and Jenkins, W. R. 1964. Aspects of the host parasite relationships of nematodes and hairy vetch. *New Jersey Agricultural Experiment Station Bulletin* 813.

Mali, V. R., and Kulthe, K. S. 1980a. A seedborne potyvirus causing mosaic of cowpea in India. *Plant Disease* **64**, 925–927.

Mali, V. R., and Kulthe, K. S. 1980b. Comparative studies on three seedborne virus isolates from cowpea. *Indian Phytopathology* **33** (3), 415–418.

Mali, V. R., Khalikar, P. V., and Gaushal, D. H. 1983. Seed transmission of poty- and cucumo-virus in cowpea in India. *Indian Phytopathology* **36** (2), 343.

Mali, V. R., Patil, F. S., and Gaushal, D. H. 1981. Immunity and resistance to bean yellow mosaic, cowpea aphid-borne mosaic and tobacco ringspot viruses in cowpea. *Indian Phytopathology* **34** (4), 521–522.

Mann, A. 1914. Coloration of the seed coat of cowpeas. *Journal of Agricultural Research* **2**, 33–57.

Marcano, C. L., and Linares, S. P. J. 1956. Orinoco, Auruca, and Caroni, new cowpea (*Vigna sinensis*) varieties. *Agronomia Tropical (Maracay)* **6** (2), 87–97.

Marcarian, V. 1982. Management of the cowpea—*Rhizobium* symbiosis under stress conditions. In Graham, P. H., and Harris, S. C., eds, *Biological Nitrogen Fixation Technology for Tropical Agriculture*: Papers presented at a Workshop held at CIAT, 9–13 March 1981. Cali, Colombia, CIAT (Centro Internacional de Agricultura Tropical), 279–286.

Marchant, J. A. 1980. Electrostatic spraying—some basic principles. Proceedings of the British Crop Protection Conference, *Weeds* **3**, 987–997.

Maréchal, R. 1970. Données cytologiques sur les aspects de la sous-tribu des Papilionaceae–Phaseoleae–Phaseolinae. *Bulletin du Jardin botanique national de Belgique* **40**, 307–348.

Maréchal, R. 1976. Studies in Phaseolinae. In Luse, R. A., and Rachie, K. O., eds, *Proceedings of IITA Collaborators' Meeting on Grain Legume Improvement*, 9–13 June 1975, Ibadan, Nigeria, IITA, 35–38.

Maréchal, R. 1982. Arguments for a global conception of the genus *Vigna*. *Taxon* **31** (2), 280–283.

Maréchal, R., Mascherpa, J. M., and Stainier, F. 1978a. Etude taxonomique d'un groupe complexe d'espèces des genres *Phaseolus* et *Vigna* (Papilionaceae) sur la base de données morphologiques et polliniques, traitées par l'analyse informatique. Boissiera 28, 1–273.

Maréchal, R., Mascherpa, J. M., and Stainier, F. 1978b. Combinaisons et noms nouveaux dans les genres *Phaseolus, Minkelersia, Macroptilium, Ramirezella* et *Vigna*. *Taxon* **27** (2–3), 199–202.

Martin, J. H., Leonard, W. H., and Stamp, D. L. 1976. *Principles of Field Crop Production*. London, UK, Collier Macmillan.

Matteson, P. C. 1982. The effects of intercropping with cereals and minimal permethrin application on insect pests of cowpea and their control enemies in Nigeria. *Tropical Pest Management* **28** (4), 372–380.

Matthews, G. A. 1981. Improved systems of pesticide application. *Philosophical Transactions of the Royal Society* B259, 163–173.

Matthews, R. E. F. 1979. *Classification and Nomenclature of Viruses*. New York, NY, S. Karger.

Maurya, P. R., and Lal, R. 1979. Effects of bulk density and soil moisture on radicle elongation of some tropical crops. In Lal, R., and Greenland, D. J., eds, *Soil Physical Properties and Crop Production in the Tropics*, Chichester, UK, John Wiley and Sons, 337–347.

McDonald, D. 1970. Survey of cowpea market samples for seed borne fungi. Paper prepared for the Ford Foundation IRAT/IITA seminar on grain legume research in West Africa, University of Ibadan, Ibadan, Nigeria.

McFerson, J., Bliss, F. A., and Rosas, J. C. 1982. Selection for enhanced nitrogen fixation in common beans, *Phaseolus vulgaris*. In Graham, P. H., and Harris, S. C., eds, *Biological Nitrogen Fixation Technology for Tropical Agriculture*: Papers presented at a Workshop Held at CIAT, 9–13 March, 1981. Cali, Colombia, CIAT (Centro Internacional de Agricultura Tropical), 39–44.

McGovran, E. R. 1969. *Insect Pest Management and Control*. Washington, DC, National Academy of Sciences, Publication 1695.

McLaughlin, M. R., Thongmeearkom, P., Milbrath, G. M., and Goodman, R. M. 1977. Isolation and some properties of a yellow sub-group member of cowpea mosaic virus from Illinois. *Phytopathology* 67 (7), 844–847.

McLean, D. M. 1941. Studies on mosaic of cowpea, *Vigna sinensis*. *Phytopathology* 31, 420–429.

McWatters, K. H. 1977. Performance of defatted peanut, soybean, and field pea meals as extenders in ground beef patties. *Journal of Food Science* 42, 1492–1495.

McWatters, K. H. 1978. Cookie baking properties of defatted peanut, soybean, and field pea flours. *Cereal Chemistry* 55, 853–863.

McWatters, K. H. 1980. Replacement of milk protein with protein from cowpea and field pea flours in baking powder biscuits. *Cereal Chemistry* 57, 223–226.

McWatters, K. H. 1982a. Peanut and cowpea meals as a replacement for wheat flour in cake-type doughnuts. *Peanut Science* 9, 46–50.

McWatters, K. H. 1982b. Quality characteristics of cake-type doughnuts containing peanut, cowpea, and soybean flours. *Peanut Science* 9, 101–103.

McWatters, K. H. 1983. Compositional, physical, and sensory characteristics of akara processed from cowpea paste and Nigerian cowpea flour. *Cereal Chemistry* 60, 333–336.

McWatters, K. H., and Heaton, E. K. 1979. Quality characteristics of ground beef patties extended with moist-heated and unheated seed meals. *Journal of the American Oil Chemists' Society* 56 (1), 86A–90A.

Medrano, S. C., Avila, L. R., and Villasmil, P. J. J. 1973. Determination of the critical period of weed competition in cowpea. *Revista de la Facultad de Agronomía*, Universidad de Zulia, 2, 7–13.

Mehndiratta, P. D., and Singh, K. B. 1971. Genetic diversity in respect of grain yield and its components in cowpea germplasm from the Punjab. *Indian Journal of Genetics and Plant Breeding* 31, 388–392.

Mehra, K. L. 1963. Consideration on the African origin of *Elucine coracana* (L.) Gaertn. *Current Science* 32, 300–301.

Mehra, K. L., Singh, C. B., and Kohli, K. S. 1969. Divergence and distribution of fodder characters in world collection of cowpea. *Indian Journal of Heredity* 1, 1–6.

Mehra, K. L., Singh, C. B., Kohli, K. S., and Magoon, M. L. 1974. Divergence and distribution of pigmentation on plant parts in a world collection of cowpea *V. sinensis*. *Genetica Iberica* 26–27, 79–97.

Mehta, P. N. 1970. Investigations with the agricultural potential of the cowpea *Vigna unguiculata* (L.) Walp. in Uganda. Kampala, Uganda, University of East Africa. (M.Sc. thesis.)

Meiners, J. P. 1981. Genetics of disease resistance in edible legumes. *Annual Reviews of Phytopathology* **21**, 189–209.

Messina, F. J. in press. Influence of cowpea pod maturity on the oviposition and larval survival of a bruchid beetle *Callosobruchus maculatus*. *Entomological Experiments and Applications*.

Messina, F. J., and Renwick, J. A. A. 1983. Effectiveness of oils in protecting stored cowpeas from the cowpea weevil (Coleoptera: Bruchidae). *Journal of Economic Entomology* **76**, 634–636.

Messina, F. J., and Renwick, J. A. A. submitted for publication. Role of crowding in the dispersal polymorphism of a bruchid beetle. *Annals of the Entomology Society of America*.

Mew, T. W., and Elazegui, F. A. 1982. *Disease problems in upland crops grown before and after wetland rice*. Los Baños, Philippines, IRRI, 99–113.

Mew, T. W., and Rosales, A. M. 1984. Relationship of soil microorganisms to rice sheath blight development in irrigated and dryland rice cultures. In Food and Fertilizer Technology Center (FFTC), *Soil-borne Crop Diseases in Asia*, Taipei, Taiwan. FFTC, Book Series 26, 147–158.

Mew, T. W., and Rosales, A. M. in press. Influences of *Trichoderma* on survival of *Thanatephorus cucumeris* in association with rice in the tropics. *Phytopathology*.

Miller, A. C., Taylor, J. C., Hawthorne, P. L., and Woodward, R. S. 1962. Lagreen, a new wilt-resistant southern pea. *Louisiana Agricultural Experiment Station Circular* 73.

Miller, J. C., Jr., Scott, J. S., Zary, K. W., and O'Hair, S. K. 1982. The influence of available nitrate levels on nitrogen fixation in three cultivars of cowpea. *Agronomy Journal* **74**, 14–18.

Miller, L. I. 1951. A report on the effect of ethylene dibromide soil treatment on root knot control, nodulation and yield of peanuts. *Virginia Journal of Science* **2**, 109–112.

Minchin, F. R., and Pate, J. S. 1975. Effects of water, aeration, and salt regime on nitrogen fixation in a nodulated legume-definition of an optimum root environment. *Journal of Experimental Botany* **26**, 60–69.

Minchin, F. R., and Summerfield, R. J. 1976. Symbiotic nitrogen fixation and vegetative growth of cowpea [*Vigna unguiculata* (L.) Walp.] in waterlogged conditions. *Plant and Soil* **25**, 113–127.

Minchin, F. R., and Summerfield, R. J. 1978. Potential yield improvement in cowpea (*Vigna unguiculata*): the role of nitrogen nutrition. *Annals of Applied Biology* **88**, 468–473.

Minchin, F. R., Summerfield, R. J., Eaglesham, A. R. J., and Stewart, K. A. 1978. Effects of short term waterlogging on growth and yield of cowpea (*Vigna unguiculata*). *Journal of Agricultural Science Cambridge* **90**, 335–336.

Minges, P. A., ed. 1972. *Descriptive list of vegetable varieties introduced since 1936 by public and private breeders in North America*. Washington, DC, American Seed Trade Association, Inc., St. Joseph, MI, American Society of Horticultural Science.

Mishra, S. N., Rastogi, R., and Verma, J. S. 1982. Exposed stigma flower mutant in cowpea. *Tropical Grain Legume Bulletin* **26**, 2–3.

Mishra, S. N., Rastogi, R., and Verma, J. S. 1984. Note on parental selectivity: a factor affecting success in cowpea crossing. *Tropical Grain Legume Bulletin* 29.

Mital, S. P., Dabas, B. S., and Thomas, T. A. 1980. Evaluation of the germplasm of vegetable cowpea for selecting desirable stocks. *Indian Journal of Agricultural Science* **50**, 323–326.

Mohdnoor, R. B. 1980. Effect of plant density on the dry seed yield of cowpeas in Malaysia. *Tropical Grain Legume Bulletin* 17–18, 11–13.

Molina, M. R., Argueta, C. E., and Bressani, R. 1976. Protein–starch extraction and nutritive value of the black-eyed pea (*Vigna sinensis*) and its protein concentrates. *Journal of Food Science* **41**, 928–932.

Monteith, J. L. 1972. Solar radiation and productivity in tropical ecosystems. *Journal of Applied Ecology* **9**, 747–766.

Monteith, J. L. 1975. Times, rates and limits in crop ecology. In IITA, *Physiology Program Formation Workshop*, Ibadan, Nigeria, IITA, 5–10.

Monteith, J. L. 1977. Climate. In Alvim, P., ed., *Ecophysiology of Tropical Crops*, London, UK, Academic Press, 1–27.

Monteith, J. L. 1978. Reassessment of maximum growth rates for C3 and C4 crops. *Experimental Agriculture* **14**, 1–5.

Moody, K. 1973. Weed control in cowpeas. *Proceedings of the Weed Science Society of Nigeria* **3**, 14–22.

Moody, K., and Whitney, W. K. 1974. The effect of weeds on insect damage to developing cowpea and soybean seeds. *Proceedings of the Weed Science Society of Nigeria* **4**, 16–26.

Moore, W. F. 1974. Studies on techniques of hybridization and the inheritance of resistance to *Verticillium* wilt in protepea (cowpea). State College, MI, Mississippi State University. (Ph.D. thesis.)

Moraes, G. J. de, Magalhães, A. A. de, and Oliveira, C. A. V. 1981. Resistência de variedades de *Vigna unguiculata* ao ataque de *Liriomyza sativae* (Diptera, Agromyzidae). *Pesquisa Agropecuaria Brasileira* **16** (2), 219–222.

Moraes, G. J. de, and Oliveira, C. A. V. 1981. Comportamento de variedades de *Vigna unguiculata* Walp. em relação ao ataque de *Empoasca kraemeri* Ross & Moore, 1957. *Anais da Sociedade Entomologica do Brasil* **10** (2), 255–259.

Moraes, G. J. de, and Ramalho, F. de S. 1980. Alguns insetos associados a *Vigna unguiculata* Walp. no Nordeste. Goiânia, Goiás, Brazil, EMBRAPA/CPATSA, *Bol. de Pesquisa* **1**.

Moraes, G. J. de, Oliveira, C. A. V., Albuquerque, M. M., Salviano, L. M. C., and Possídio, P. L. de. 1980. Efeito da época de infestação de *Empoasca kraemeri* Ross & Moore, 1957 (*Cigarrinha verde* do feijoeiro) (Homoptera: Typhlocibidae) na cultura de *Vigna unguiculata* Walp (Feijão macassar). *Anais da Sociedade Entomologica do Brasil* **9** (1), 67–74.

Moraes, G. J. de, Sardana, B., and Oliveira, C. A. V. 1982. Nível de dano econômico de *Empoasca kraemeri* em *Vigna unguiculata*. *Pesquisa Agropecuaria Brasileira* **17** (12), 1701–1705.

Morse, W. J. 1920. Cowpeas: culture and varieties. *United States Department of Agriculture Farmers' Bulletin* 1148.

Mortensen, J. A., and Brittingham, W. H. 1952. The inheritance of pod colour in the southern pea, *Vigna sinensis*. *Proceedings of the American Society of Horticultural Science* **59**, 451–456.

Moshy, A. J., Leigh, T. F., Foster, K. W., and Schriber, F. 1983. Screening selected cowpea, *Vigna unguiculata* (L.) Walp., lines for resistance to *Lygus hesperus* (Heteroptera: Miridae). *Journal of Economic Entomology* **76**, 1370–1373.

Mughogho, S. K., Awai, J., Lowendorf, H. S., and Lathwell, D. J. 1982. The effects of fertilizer nitrogen and *Rhizobium* inoculation on the yield of cowpeas and subsequent crops of maize. In Graham, P. H., and Harris, S. C., eds, *Biological Nitrogen Fixation Technology for Tropical Agriculture*: Papers Presented at a Workshop Held at CIAT, 9–13 March, 1981. Colombia, CIAT (Centro Internacional de Agricultura Tropical), 297–301.

Mukherjee, P. 1968. Pachytene analysis in *Vigna*. Chromosome morphology in *Vigna sinensis* (cultivated). *Science and Culture* **34**, 252–253.

Mukiibi, J. 1969. Control of leaf diseases of cowpeas caused by fungal pathogens. Paper prepared for the 12th meeting of specialist committee on agricultural botany, Kampala, Uganda, September 1969.

Mukiibi, J. 1979. Some effects of mancozeb, DDT and weeding on disease and pest incidence and on yield of cowpeas, *Vigna unguiculata* (L.) Walp. Paper prepared for the symposium on grain legume improvement in East Africa, Nairobi, Kenya, 20–24 August 1979.

Mulongoy, K., Ayanaba, A., Asanuma, S., and Ranga Rao, V. 1982. Cowpea nodulation and response to inoculation in West Africa. Paper prepared for the 1st OAU/STRC Inter-African Conference on Biofertilizers, Cairo, Egypt, 22–26 March 1982.

Mulongoy, K., Ayanaba, A., and Pulver, E. 1981. Exploiting the diversity in the cowpea–rhizobia symbiosis for increased cowpea production. In Emejuaiwe, S. O., Ogunbi, O., and Sanni, S. O., ed., *GIAM VI : Global* Impacts of Applied Microbiology. London, UK. Academic Press.

Muniyappa, V., and Reddy, D. V. R. 1983. The transmission of cowpea mild mottle virus in a nonpersistent manner. *Plant Disease* **67**, 391–393.

Munns, D. N. 1977. Soil acidity and related problems. In Vincent, J. M., Whitney, A. S., and Boss, J., ed. Exploiting the Legume–*Rhizobium* Symbiosis in Tropical Agriculture, Honolulu, HI, University of Hawaii, College of Tropical Agriculture, Miscellaneous publication 145, 211–236.

Munns, D. N., and Mosse, B. 1980. Mineral nutrition of legume crops. In Summerfield, R. J., and Bunting, A. H., eds, *Advances in Legume Science*, London, UK, Her Majesty's Stationery Office, 115–125.

Murdock, G. P. 1959. *Africa: its peoples and their culture history*. New York, NY, McGraw-Hill.

Muruli, R. S., Pathak, D. M., Mukunya, A. R., Karel, S. O., and Ssali, H. 1980. Cowpea research in Kenya. *Tropical Grain Legume Bulletin* **19**, 3–4.

Musselman, L. J., and Parker, C. 1981. Studies on indigo witchweed, the American strain of *Striga gesnerioides* (Scrophulariaceae). *Weed Science* **29**, 594–596.

Musselman, L. J., Parker, C., and Dixon, N. 1981. Notes on autogamy and flower structure in agronomically important species of *Striga* (Scrophulariaceae) and *Orobanche* (Orobanchaceae). *Beitrage Zur Biologie der Pflanzen* **56**, 329–343.

Musselman, L. J., Parker, C., and Dixon, N. 1984. Some taxonomic problems in the genus *Striga* with particular reference to Africa. In Parker, C., Musselman, L. J., Polhill, R. M., and Wilson, A. K., eds, *Proceedings Third International Symposium on Parasitic Weeds*, 7–9 May 1984, Aleppo, Syria, ICARDA (International Center for Research in the Dry Areas), 53–57.

Myikado, M., Nokayama, I., Yoshioka, H., and Nakatani, N. 1979. The Piperaceae armides 1: structure of pipercide, a new insecticidal armide from *Piper nigrum* L. *Agricultural and Biological Chemistry* **43**, 1609–1611.

Nadal, A. M., and Carangal, V. R. 1977. Promising varieties of rainfed legumes grown at zero tillage after paddy rice. *International Rice Research Institute Newsletter* **2** (5), 25–26.

Nair, K. P. 1979. *Mixed Farming. Intensive multiple cropping with coconut in India.* Berlin and Hamburg, Verlag Paul Parey, 44–55.

Nair, K. P. P., Patel, U. K., Singh, R. P., and Kaushik, M. K. 1979. Evaluation of legume intercropping in conservation of fertilizer nitrogen in maize culture. *Journal of Agricultural Science Cambridge* **93**, 189–194.

Nangju, D. 1979a. Effect of density, plant type, and season on growth and yield of cowpea. *Journal of the American Society of Horticultural Science* **104** (4), 466–470.

Nangju, D. 1979b. Effect of tillage methods on growth and yield of cowpea and soybean. In Lal, R., ed., *Soil Tillage and Crop Production*, Ibadan, Nigeria, IITA, Proceedings series **2**, 93–108.

Nangju, D. 1980. Effect of plant density, spatial arrangement, and plant type on weed control in cowpea and soybean. In Akobundu, I. O., ed. *Weeds and their Control in the Humid and Subhumid Tropics*, Ibadan, Nigeria, IITA, Proceedings series **3**, 288–299.

Nangju, D., Flinn, J. C., and Singh, S. R. 1979. Control of cowpea pests by utilization of insect resistant cultivars and minimum insecticide application. *Field Crops Research* **2**, 373–385.

Nangju, D., Little, T. M., and Anjorin-Ohu, A. 1975. Effect of plant density and spatial arrangement on seed yield of cowpea [*Vigna unguiculata* (L.) Walp.]. *Journal of the American Society of Horticultural Science* **100** (5), 467–470.

Nangju, D., Nwanze, K. F., and Singh, S. R. 1979. Planting date, insect damage and yield of cowpea in western Nigeria. *Experimental Agriculture* **15**, 1–10.

Nangju, D., Rachie, K. O., Watt, E. E., and Akinpelu, M. A. 1976. *Performance of the 1974 first and second international cowpea uniform cultivar trials.* Ibadan, Nigeria, IITA.

Nangju, D., Watt, E. E., Rachie, K. O., and Akinpelu, M. A. 1977. *Results of the 1975 international cowpea uniform cultivar trials.* Ibadan, Nigeria, IITA.

Nariani, T. K., and Kandaswamy, T. K. 1961. Studies on a mosaic disease of cowpea. *Indian Phytopathology* **14**, 777–782.

Nariani, T. K., Viswanath, S. M., Chenulu, V. V., and Moharir, A. V. 1980. Purification and serology of cowpea (*Vigna sinensis* Savi) mosaic virus. *Current Science* **49** (17), 680.

Narsinghani, V. G., and Kumar, S. 1976. Mutation studies in cowpea. *Indian Journal of Agricultural Science* **46**, 61–64.

Nassar, M. A. S. 1978. Studies of reciprocal interspecific hybridizations between *Phaseolus coccineus* L. and *Phaseolus vulgaris* L. UK, University of Southampton. (Ph.D. thesis.)

Ndunguru, B. J., and Summerfield, R. J. 1975a. Comparative laboratory studies of cowpea (*Vigna unguiculata*) and soybean (*Glycine max*) under tropical temperature conditions. 1: germination and hypocotyl elongation. *East Africa Agriculture and Forestry Journal* **41**, 58–64.

Ndunguru, B. J., and Summerfield, R. J. 1975b. Comparative laboratory studies of cowpea (*Vigna unguiculata*) and soybean (*Glycine max*) under tropical temperature conditions. 2: contribution of cotyledons to early seedling growth. *East Africa Agriculture and Forestry Journal* **41**, 65–71.

Neumann, P. M., and Nooden, L. D. 1984. Pathway and regulation of phosphate translocation to the pods of soybean explants. *Physiologia Plantarum* **60**, 166–170.

Neves, B. P. das. 1982. Determinação de resistência varietal ao "manhoso" (*Chalcodermus* sp.) em caupi (*Vigna unguiculata*). Goiânia, Goiás, Brazil EMBRAPA/

CNPAF, 1st Reunião Nacional de Pesquisa de Caupi (RENAC), 64–65.

Neves, B. P. das. 1983. Relatório anual individual. Goiânia, Goiás, Brazil, EMBRAPA/CNPAF.

Neves, B. P. das., and Watt, E. E. 1981. Damage from the cowpea curculio (*Chalcodermus* sp.) in southern peas and common beans. *Annual Report of the Bean Improvement Cooperative* **24**, 13–14.

Neves, B. P. das., Araújo, J. P. de, and Watt, E. E. 1982. Estudo da resistência varietal e dos danos causados pela *Maruca testulalis* em caupi. Goiânia, Goiás, Brazil EMBRAPA/CNPAF, 1st Reunião Nacional de Pesquisa de Caupi (RENAC), 66.

Neves, M. C. P. 1978. Carbon and nitrogen nutrition of cowpea (*Vigna unguiculata*). Reading, UK, University of Reading. (Ph.D. thesis.)

Neves, M. C. P., Minchin, F. R., and Summerfield, R. J. 1981. Carbon metabolism, nitrogen assimilation and seed yield of cowpea (*Vigna unguiculata*) plants dependent on nitrate-nitrogen or on one of two strains of *Rhizobium*. *Tropical Agriculture (Trinidad)*, **58**, 115–132.

Newson, L. D., and Herzog, D. C. 1977. Trap crops for control of soybean pests. *La. Agr.* **20**, 14–15.

Ng, N. Q. 1982. Genetic resources programme of IITA. *Plant Genetic Resources Newsletter* **49**, 26–31.

Ngundo, B. W. 1972. Effects of *Meloidogyne* spp. on production of beans in Kenya. Reading, UK, European Society of Nematologists, 3–8 September (abstract).

Ngundo, B. W., and Taylor, D. P. 1974. Effect of *Meloidogyne* spp. on bean yields in Kenya. *Plant Disease Reporter* **58**, 1020–1023.

Nielsen, C. L., and Hall, A. E. in press. Responses of cowpea [*Vigna unguiculata* (L.) Walp.] in the field to high night air temperatures during flowering. 1: thermal regimes of production regions and field experimental systems. *Field Crops Research*.

Nielsen, C. L., and Hall, A. E. 1985. Responses of cowpea [*Vigna unguiculata* (L.) Walp.] in the field to high night air temperatures during flowering. 2: plant responses. *Field Crops Research*, **10**, 181–196.

Nigh, E. L., Jr. 1966. *Rhizobium* nodule formation on alfalfa as influenced by *Meloidogyne javanica*. *Nematologica* **12**, 96.

Nilakhe, S. S., and Chalfant, R. B. 1982a. Cowpea cultivars screened for resistance to insect pests. *Journal of Economic Entemology* **75**, 223–227.

Nilakhe, S. S., and Chalfant, R. B. 1982b. Lepidopterous pests of cowpeas: seasonal occurrence, distribution pattern, and comparison of ground cloth and sweep-net sampling for larvae. *Journal of Economic Entomology* **75**, 9725.

Nilakhe, S. S., Chalfant, R. B., and Day, A. 1982. Soil insect pests in cowpea fields with emphasis on distribution of wireworms. *Journal of the Georgia Entomological Society* **17**, 145–149.

Nilakhe, S. S., Chalfant, R. B., and Singh, S. V. 1981a. Damage to southern peas by different stages of the southern green stink bug. *Journal of the Georgia Entomological Society* **16**, 409–414.

Nilakhe, S. S., Chalfant, R. B., and Singh, S. V. 1981b. Evaluation of southern green stink bug damage to cowpeas. *Journal of Economic Entomology* **74**, 589–592.

Njoku, E. 1959. An analysis of plant growth in some West African species. 1: growth in full daylight. *Journal of the West African Science Association* **5**, 37–56.

Nnadi, L. A. 1978. Nitrogen economy in selected farming systems of the savanna region. Paper prepared for the United Nations Environmental Protection Workshop on N recycling in West African Ecosystem, Ibadan, Nigeria, IITA, 11–15 December 1978.

Nnanyelugo, D. O. 1982. Nutritional practices, food intake measurements and their relationship to socio-economic grouping, location and their apparent nutritional inadequacy in children. *Appetite* 3, 229–241.

Nogueira, O. L. 1981. Cultura do feijão caupi no estado do Amazonas. Manaus, Amazonas, EMBRAPA/UEPAE, Circ. Técnica 4.

Nooden, L. D., and Murray, B. J. 1982. Tranmission of the monocarpic senescence signal via the xylem in soybean. *Plant Physiology* 69, 754–756.

Norton, J. D. 1961. Inheritance of growth habit and flowering response of the southern pea, *Vigna sinensis* Endl. Baton Rouge, LA, Louisiana State University. (Ph.D. thesis.)

Ntare, B. R. 1982. Evaluation of early generation selection procedures in cowpea *Vigna unguiculata* (L.) Walp. Ibadan, Nigeria, University of Ibadan. (Ph.D. thesis.)

Nutman, P. S. 1976. Alternative sources of nitrogen for crops. *Journal of the Royal Agriculture Society of England* 137, 86–94.

Obilana, A. T. 1983. *Striga* studies and control in Nigeria. In ICRISAT (International Crops Research Institute for the Semi-Arid Tropics), Proceedings Second International *Striga* Workshop, IDRC/ICRISAT, Ouagadougou, Upper Volta, 5–8 October 1981, Patancheru, India, ICRISAT, 87–98.

Ochieng, R. S. 1977. Studies on the bionomics of two major pests of cowpea [*Vigna unguiculata* (L.) Walp.]: *Ootheca mutabilis* Sahlb. (Coleop.: Chrysomelidae) and *Anoplocnemis curvipes* Fabricius (Hemipt.: Coreidae). Ibadan, Nigeria, University of Ibadan. (Ph.D. thesis.)

Ogborn, J. E. A. 1979. The potential use of germinators for *Striga* control by African peasant farmers. In Musselman, L. H., Worsham, A. D., and Eplee, R. E., eds, *Proceedings 2nd International Symposium on Parasitic Weeds*, Raleigh, NC, 16–19 July 1979, Raleigh, NC, North Carolina State University, 29–36.

Ogborn, J. E. A. in press. Research priorities in agronomy. In ICSU/UNESCO (International Council of Scientific Unions/United Nations Educational, Scientific and Cultural Organization), Proceedings of the Workshop on the Biology and Control of *Striga*, Dakar, Senegal, ICSU).

Ogunfowora, A. O. 1976. Research on *Meloidogyne* spp. at the Institute of Agricultural Research and Training, University of Ife, Moor Plantation, Ibadan. Paper prepared for the research planning conference on root-knot nematodes, Ibadan, IITA, 7–11 June 1976, 9–13.

O'Hair, S. K., and Miller, J. C., Jr. 1982. Effects of virus infection on nitrogen fixation in cowpea. *Journal of the American Society of Horticultural Science* 107, 516–519.

Ojehomon, O. O. 1968a. The development of the inflorescence and extra-floral nectaries of *Vigna unguiculata* (L.) Walp. *Journal of the West African Science Association* 13, 92–110.

Ojehomon, O. O. 1968b. Flowering, fruit production and abscission in cowpea, *Vigna unguiculata* (L.) Walp. *Journal of the West African Science Association* 13, 227–234.

Ojomo, O. A. 1971. Inheritance of flowering date in cowpeas [*Vigna unguiculata* (L.) Walp.] *Tropical Agriculture (Trinidad)* 48, 277–282.

Ojomo, O. A. 1972. Inheritance of seed coat thickness in cowpeas. *Journal of Heredity* 63, 147–149.

Ojomo, O. A. 1973. Breeding and improvement of cowpeas in Western State of

Nigeria. In IITA, *Proceedings of the 1st Grain Legume Improvement Workshop*, IITA, Ibadan, Nigeria, 29 October–2 November 1973. Ibadan, Nigeria, IITA, 21–25.

Ojomo, O. A. 1974. Yield potential of cowpeas, *Vigna unguiculata* (L.) Walp.: results of mass and bulk pedigree selection methods in Western Nigeria. *Nigerian Agricultural Journal* **11**, 150–156.

Ojomo, O. A. 1975. The current outlook on cowpea improvement at University of Ife. In Luse, R. A., and Rachie, K. O., eds, *Proceedings of IITA Collaborators' Meeting on Grain Legume Improvement*, 9–13 June 1975, Ibadan, Nigeria, IITA.

Ojomo, O. A. 1977. Morphology and genetics of two gene markers, 'swollen stem base' and 'hastate leaf' in cowpea, *Vigna unguiculata* (L.) Walp. *Journal of Agricultural Science* **88**, 227–231.

Ojomo, O. A., Raji, J. A., and Omueti, O. 1977. Heterosii and combining ability for some plant and seed characters in cowpea (*Vigna unguiculata* L. Walp.) from a diallel cross. *Nigerian Journal of Genetics* **1**, 84–89.

Okigbo, B. N., and Greenland, D. J. 1976. *Intercropping Systems in Tropical Africa*. Madison, WI, American Society of Agronomy.

Okpala, R. O. 1981. Host range and survival of *Colletotrichum capsici* (Syd.) Butler & Bisby and its interaction with cowpea lines in northern Nigeria. Zaria, Nigeria, Ahmadu Bello University. (M.Sc. thesis.)

Oladiran, A. O., and Okusanya, B. O. 1980. Effect of fungicides on pathogens associated with basal stem rots of cowpea in Nigeria. *Tropical Pest Management* **26** (4), 403–409.

Olive, L. S., Brin, D. C., and Lefebvre, C. L. 1945. A leaf spot of cowpea and soybean caused by an undescribed species of *Helminthosporium*. *Phytopathology* **35**, 822–931.

Oliveira, F. J. de, Santos, J. H. R. dos, Alves, J. F., Paiva, J. B., and Assunção, M. V. 1984. Perdas de peso em sementes de cultivares de caupi, atacadas pelo caruncho. *Pesquisa Agropecuaria Brasileira* **19** (1), 47–52.

Olowe, T. 1978. Importance of root-knot nematodes on cowpea, *Vigna unguiculata* (L.) Walp. in Nigeria. Paper prepared for the 3rd research planning conference on root-knot nematodes, Ibadan, Nigeria, IITA, 16–20 November 1981, 85–109.

Olunuga, B. A., and Akobundu, I. O. 1980. Weed problems and control practices in field and vegetable crops in Nigeria. In Akobundu, I. O., ed., *Weeds and their Control in the Humid and Subhumid Tropics*, Ibadan, Nigeria, IITA, Proceedings series 3, 138–146.

Olusanya, E. O. 1977. *Manual on Food Consumption Surveys in Developing Countries*. Ibadan, Nigeria, University of Ibadan.

Omueti, J. O., and Oyenuga, V. A. 1970. Effect of phosphorus fertilizer on the protein and essential components of the ash of groundnut and cowpeas. *West African Biology and Applied Chemistry Journal* **13** (1), 14–19.

Onayemi, O., Pond, W. G., and Krook, L. 1976. Effects of processing on the nutritive value of cowpeas (*Vigna sinensis*) for the growing rat. *Nutrition Reports International* **13**, 299–305.

Onesirosan, P. T. 1975. Seed-borne and weed-borne inoculum in web blight of cowpea. *Plant Disease Reporter* **59** (4), 338–339.

Onesirosan, P. T. 1977. Comparison of *Rhizoctonia solani* isolates from web blight and basal canker of cowpea and from soil. *Plant and Soil* **46**, 135–143.

Onesirosan, P. T., and Barker, L. N. 1971. Stem anthracnose of cowpea in Nigeria. *Plant Disease Reporter*, **55**, 820.

Onesirosan, P. T., and Sagay, S. O. 1975. Survival of two pathogens of cowpea over the dry season. *Plant Disease Reporter* **59**, 820–822.

Onuorah, P. E. 1973. *Pythium* seed decay and stem rot of cowpea [*Vigna sinensis* (Linn.) Savi] in southern Nigeria. *Plant and Soil* **39**, 187–191.

Onwueme, I. C. 1974. Retardation of hypocotyl growth by prior heat stress in seedlings of cowpea, melon, and tomato. *Journal of Agricultural Science Cambridge* **83**, 409–413.

Oostenbrink, M. 1955. Ein inoculatieproef met het erwtencystenaaltie *Heterodera goettingiana* Libscher. *Tijdschrift Plantenziekten* **61**, 65–68.

Orton, W. A. 1902. The wilt disease of the cowpea and its control. *United States Department of Agriculture, Bureau of Plant Industries Bulletin* 17.

Orton, W. A. 1913. The development of disease resistant varieties of plants. *Compter renderments et rapport*: 4ᵉ *conference genetique*, Paris, 18–23 September 1911. Nasson et Cie, 247–261.

Osa-Afiana, L. O., and Alexander, M. 1982. Differences among cowpea rhizobia in tolerance to high temperature and desiccation in soil. *Applied and Environmental Microbiology* **43**, 435–439.

Osiname, O. A. 1978. The fertilizer (NPK) requirement of Ife-Brown cowpea [*Vigna unguiculata* (L.) Walp.]. *Tropical Grain Legume Bulletin* **11**, 13–15.

Oyekan, P. O. 1977. Reaction of some cowpea varieties to *Fusarium oxysporum* f. sp. *tracheiphilum* in Nigeria. *Tropical Grain Legume Bulletin* **8**, 47–49.

Oyekan, P. O. 1979. Chemical control of web blight and leaf spot of cowpea in Nigeria. *Plant Disease Reporter* **63**, 574–577.

Ozenda, P., and Capdepon, M. 1972. Recherches sur les phanerogames parasites. 2: une Scrophulariaceae holoparasite meconnue, *Striga gesnerioides*. *Phytomorphology* **22**, 314–324.

Pakistani Botanical Society. 1982. Abstracts of papers presented to the 1st all Pakistan conference of plant scientists. *Pakistan Journal of Botany* **14**, 1–50.

Palti, J., and Rotem, J. 1983. Cultural practices for the control of crop diseases. In CMI (Commonwealth Mycological Institute), *Plant Pathologist's Pocketbook*, 2nd ed. Kew, UK, CMI, 183–195.

Pandey, N. D. 1962. Studies on the morphology, bionomics and control of some Indian Agromyzidae. *Agra University Journal of Research and Science* **11**, 39–43.

Pandey, R. K., Herrera, W. A. T., and Pendleton, J. W. 1984. Drought response of grain legumes under irrigation gradient: yield and yield components. *Agronomy Journal* **76**, 549–553.

Pandey, R. K., Herrera, W. T., Villegas, A. N., and Pendleton, J. W. 1983. *Adaptation of food legumes to water stress: use of line source sprinkler system*. Los Baños, Philippines, IRRI, 16 April.

Pandey, R. P., Nair, P. K. R., and Tewari, J. P. 1981. Correlation of morphophysiological and sink parameters in cowpea. *Indian Journal of Agricultural Science* **51**, 221–224.

Pandita, M. L., Vashistha, R. N., Bhutani, R., and Batra, B. R. 1982. Genetic variability studies in cowpea (*Vigna sinensis* L. Savi) under dry farming conditions. *Hissar Agricultural University Journal of Research* **12**, 241–245.

Pant, K. C., Chandel, K. P. S., and Joshi, B. S. 1982. Analysis of diversity in Indian cowpea genetic resources. *SABRAO Journal* **14**, 102–111.

Parh, I. A. 1983. Species of *Empoasca* associated with cowpea, *Vigna unguiculata* (L.) Walp. in Ibadan, and three ecological zones in south-western Nigeria (Homoptera: Cicadellidae). *Revue de Zoologie Africaine* **97**, 1.

Parker, C. 1983a. Factors influencing *Striga* seed germination and host parasite specificity. In ICRISAT (International Crops Research Institute for the Semi-Arid Tropics), *Proceedings Second International* Striga *Workshop*, IDRC/ICRISAT, Ouagadougou, Upper Volta, 5–8 October 1981, Patancheru, India, ICRISAT, 31–35.

Parker, C. 1983b. *Striga*—analysis of past research and summary of the problem. In ICRISAT (International Crops Research Institute for the Semi-Arid Tropics), *Proceedings Second International* Striga *Workshop*, IDRC/ICRISAT, Ouagadougou, Upper Volta, 5–8 October 1981, Patancheru, India, ICRISAT, 9–16.

Parker, C. in press. The physiology of *Striga* species: present state of knowledge and priorities for future research. In ICSU/UNESCO (International Council of Scientific Unions/United Nations Educational, Scientific and Cultural Organization), *Proceedings of the Workshop on the Biology and Control of* Striga, Dakar, Senegal, ICSU.

Parker, C., and Reid, D. C. 1979. Host specificity in *Striga* species—some preliminary observations. In Musselman, L. J., Worsham, A. D., and Eplee, R. E., eds, *Proceedings 2nd International Symposium on Parasitic Weeds*, Raleigh, NC, 16–19 July 1979, Raleigh, NC, North Carolina State University, 79–90.

Parmar, P. S. 1981. Mixed cropping of sugarcane and pulses in Banaskantha. *Pulse Crops Newsletter* **1**, 77–78.

Partridge, J. E., and Keen, N. T. 1976. Association of the phytoalexin kievitone with single-gene resistance of cowpeas to *Phytophthora vignae. Phytopathology* **66**, 426–429.

Passioura, J. B. 1982. The role of root system characteristics in the drought resistance of crop plants. In IRRI, *Drought Resistance in Crops with Emphasis on Rice*, Los Baños, Philippines, IRRI, 71–82.

Pate, J. S. 1984. The carbon and nitrogen nutrition of fruit and seed: case studies of selected grain legumes. In Murray, D. R., ed., *Seed Physiology*, New York, USA, Academic Press.

Pate, J. S. in press (a). The carbon and nitrogen nutrition of fruit and seed—case studies of selected grain legumes. In Murray, D. R., ed., *Seed Physiology*, Vol. 1, New York, NY, Academic Press.

Pate, J. S. in press (b). Physiology of pea—a comparison with other legumes in terms of economy of carbon and nitrogen in plant functioning. Paper prepared for the 40th Easter School in Agricultural Science on improvement of the pea crop, University of Nottingham, 2–6 April 1984.

Pate, J. S., and Kuo, J. 1981. Anatomical studies of legume pods – a possible tool in taxonomic research. In Polhill, R. M., and Raven, P. H., ed., *Advances in Legume Systematics,* Kew, UK, Royal Botanic Gardens, 903–912.

Pate, J. S., and Minchin, F. R. 1980. Comparative studies of carbon and nitrogen nutrition of selected grain legumes. In Summerfield, R. J., and Bunting, A. H., eds., *Advances in Legume Science*, London, UK, Her Majesty's Stationery Office, 105–114.

Pate, J. S., Peoples, M. B., and Atkins, C. A. 1983. Post anthesis economy of carbon in a cultivar of cowpea. *Journal of Experimental Botany* **34**, 544–562.

Pate, J. S., Peoples, M. B., and Atkins, C. A. 1984. Spontaneous phloem bleeding from cryopunctured fruits of a ureide-producing legume. *Plant Physiology* **74**, 599–605.

Pate, J. S., Peoples, M. B., Van Bel, A. J. E., Kuo, J., and Atkins, C. A. in press. Diurnal water balance of the cowpea fruit. *Plant Physiology.*

Pate, J. S., Sharkey, P. J., and Atkins, C. A. 1977. Nutrition of a developing legume fruit. Functional economy in terms of carbon, nitrogen, water. *Plant Physiology* **59**, 506–510.

Patel, P. N. 1982a. Genetics of host reactions to three races of the bacterial pustule pathogen in cowpea. *Euphytica* **31**, 805–814.

Patel, P. N. 1982b. Genetics of cowpea reactions to two strains of cowpea mosaic virus from Tanzania. *Phytopathology* **72**, 460–462.

Patel, P. N., and Jindal, J. K. 1982. Distinguishing reactions to the bacterial blight and pustule organisms of cowpea in pods of Phaseolus vulgaris. *Zeitschrift für Pflanzenkrankheiten und Pflanzenschutz* **89**, 406.

Patel, P. N., and Kuwite, C. 1982. Prevalence of cowpea aphid-borne mosaic virus and two strains of cowpea mosaic virus in Tanzania. *Indian Phytopathology* **35** (3), 467–472.

Patel, P. N., Walker, J. C., Hagedorn, D. J., Garcia, Carlos Deleon, and Teliz-Ortiz, M. 1964. Bacterial brown spot of bean in central Wisconsin. *Plant Disease Reporter* **48**, 335–337.

Pathak, R. S. 1979. Cowpea improvement in Kenya with general reference to insect pest resistance. Paper prepared for the research conference on host plant resistance, IITA, Ibadan, Nigeria, 15–19 October 1979.

Patil, F. S., and Mali, V. R. 1982. Natural occurrence of tobacco ringspot virus (TRSV) on cowpea in India. *Indian Phytopathology* **35** (3), 399–403.

Pawar, N. B., Shirsat, A. M., and Ghur Gule, J. N. 1977. Effect of seed inoculation with *Rhizobium* on grain yield and other characters of cowpea (*Vigna unguiculata*). *Tropical Grain Legume Bulletin* **7**, 3–5.

Peacock, F. E. 1956. The reniform nematode in the Gold Coast. *Nematologica* **1**, 307–310.

Pegg, K. G., and Alcorn, J. L. 1967. *Ascochyta* disease of French beans. *Queensland Agricultural Journal* **93**, 321–323.

Peoples, M. B., Pate, J. S., and Atkins, C. A. 1983. Mobilization of nitrogen in fruiting plants of a cultivar of cowpea. *Journal of Experimental Botany* **34**, 563–578.

Peoples, M. B., Pate, J. S., and Atkins, C. A. in press (a). The effect of nitrogen source on transport and metabolism of nitrogen in fruiting plants of cowpea [*Vigna unguiculata* (L.) Walp.].

Peoples, M. B., Pate, J. S., Atkins, C. A., and Murray, D. R. in press (b). Economy of water, carbon, and nitrogen in the developing cowpea fruit. *Plant Physiology*.

Perez, J. E., and Cortes-Monllor, A. 1971. Further studies on a mosaic virus of cowpea from Puerto Rico. *Journal of the Agriculture University of Puerto Rico* **55** (2), 184–191.

Perfect, T. J., Cook, A. G., and Critchley, B. R. 1978. Effects of intercropping on pest incidence. Ibadan, Nigeria, IITA. (mimeo.)

Perrin, R. M. 1977. Studies on host-plant selection by the cowpea seed moth, *cydia ptychora*. *Annals of Applied Biology* **87**, 175–182.

Perrin, R. M. 1978. Varietal differences in the susceptibility of cowpea to larvae of the seed moth, *cydia ptychora* (Meyrick) Lepidoptera: Tortricidae. *Bulletin of Entomological Research* **68**, 47–56.

Phadnis, B. A., and Thombre, P. G. 1971. Colchicine induced mutants in cowpea (*Vigna sinensis*). *Indian Journal of Horticulture* **28**, 303.

Phatak, H. C. 1974. Seed-borne plant viruses: identification and diagnosis in seed health testing. *Seed Science and Technology* **2**, 3–155.

Phatak, H. C., Diaz-Ruiz, J. R., and Hull, R. 1976. Cowpea ringspot virus: a seed transmitted cucumo virus. *Phytopathologische Zeitschrift* **87**, 132–142.

Pieterse, A. H., and Pesch, C. J. 1983. The witchweeds (*Striga* spp.)—a review. *Abstracts on Tropical Agriculture* **9** (8), 9–37.

Pio-Ribeiro, G., and Kuhn, C. W. 1980. Cowpea stunt: heterozygous and differential reactions of cowpea cultivars. *Phytopathology* **70**, 244–249.

Pio-Ribeiro, G., and Paguio, O. R. 1980. Informaçoes bibliográficas sobre as denominaçoes 'cowpea mosaic virus' e 'cowpea severe mosaic virus'. *Fitopatologia Brasileira* **5** (3), 373–376.

Pio-Ribeiro, G., Kuhn, C. W., and Brantley, B. B. 1980. Cowpea stunt: inheritance pattern of the necrotic synergistic reaction. *Phytopathology* **70**, 250–252.

Pio-Ribeiro, G., Wyatt, S. D., and Kuhn, C. W. 1978. Cowpea stunt: a disease caused by a synergistic interaction of two viruses. *Phytopathology* **68**, 1260–1265.

Piper, C. V. 1912. Agricultural varieties of the cowpea and immediately related species. United States Department of Agriculture, Bureau of Plant Industries Bulletin 229.

Piper, C. V. 1913. The wild prototype of the cowpea. United States Department of Agriculture, Bureau of Plant Industries, Circular 124.

Pizzamiglio, M. A. 1979. Aspéctos da biologia de *Empoasca kraemeri* (Ross & Moore, 1957) (Hom. Cicadellidae) em *Phaseolus vulgaris* e ocorrência de parasitismo em ovos. *Anais da Sociedade Entomologica do Brasil.* **8** (2), 369–372.

Platt, B. S. 1975. Tables of representative values of foods commonly used in tropical countries. London, UK, Her Majesty's Stationery Office, Medical Research Council, Special report 302.

Poku, J. A., and Akobundu, I. O. 1983. Chemical weed control in cowpea in the subhumid and guinea savanna zones of Nigeria. Paper prepared for the 2nd biennial conference of the West African Weed Science Society, Abidjan, Ivory Coast.

Poku, J. A., and Akobundu, I. O. 1984. Effect of tillage and herbicide on cowpea weed control and yield in the subhumid tropics. Weed Science Society of America 267 (abstract).

Ponte, J. J. 1966. Uma nova enfermedade do feijão de corda, *Vigna sinensis Endl. Boletim da Sociedade Cearense de Agronomía* **7**, 35–38.

Ponte, J. J. 1974. Variedades de feijão macassar resistentes ao carvao (*Entyloma vignae*). *Fitopatologia* **9**, 65.

Ponte, J. J. 1976. Doencas do feijão macassar (*Vigna sinensis* L. Savi). Curso Fitopatologia Aplicada a Cultura Tropicais (mimeo).

Ponte, J. J., Santos, A. A., and Chagas, J. M. F. 1974. Incidência de carvao do feijão macassar no estados do Piauí e Rio Grande do Norte. *Fitopatologia* **9**, 66.

Ponte, J. J., and Santos, C. D. G. dos. 1982. Comportamento de novos hibridos de feijão macassar *Vigna unguiculata* Walp. em relação ao parasitismo de nematoides das galhas, *Meloidogyne* spp. Paper prepared for the 6th annual meeting on nematology in Brazil, 8–12.

Porter, W. M. 1973. Documentation and maintenance of the germplasm collection of *Vigna unguiculata*. In IITA, *Proceedings of the 1st Grain Legume Improvement Workshop*, IITA, Ibadan, Nigeria, 29 October–2 November 1973, Ibadan, Nigeria, IITA, 17.

Porter, W. M., Rachie, K. O., Rawal, K. M., Wien, H. C., Williams, R. J., and Luse, R. A. 1974. *Cowpea germplasm catalogue no. 1*. Ibadan, Nigeria, IITA.

Prabhu, A. S., Albuquerque, F. C., and Lima, E. F. 1979. Avaliação da resistência a *Entyloma vignae* em feijão caupi. *Fitopatologia Brasileira* **4**, 375–378.

Prakash, C. S. 1980. Evaluation of genotypes and inheritance studies on resistance to bacterial blight (Xanthomonas vignicola Burkh.) in cowpea (*Vigna unguiculata* L. Walp.). *Mysore Journal of Agricultural Science* **14**, 462.

Prakash, J., and Joshi, R. D. 1980. Some aspects of seed transmission in cowpea. *Seed Science and Technology* **8**, 393–399.

Prasad, D., Singh, R. P., Srivastava, K. P., Agnihotri, N. P., and Sethi, C. L. 1982. Evaluation of mephosfolan for nematode and insect control on okra, cowpea, and tomato. *Indian Journal of Nematology* **12**, 277–284.

Prasad, M. M. K. D., and Patel, K. M. 1978. Genetic variability studies in cowpea (*Vigna sinensis* L. savi) through diallel crosses. *Andhra Agricultural Journal* **25**, 214–219.

Prasad, S. K., Dasgupta, D. R., and Mukhopadhyaya, M. C. 1964. Nematodes associated with commercial crops in northern India and host range of *Meloidogyne javanica* (Treub 1885) Chitwood 1949. *Indian Journal of Entomology* **26**, 438–446.

Prasad, S. K., Mathur, V. K., and Chawla, M. L. 1965. Plant parasitic nematodes associated with sugarcane at Douala and Mavana farm. *Labdev. Journal of Science and Technology* **3**, 211.

Pratt, R. C. 1983. Gene transfer between Tepary and common beans. *Desert Plants* **5** (1), 57–63.

Premsekar, S., and Raman, V. S. 1972. A genetic analysis of the progenies of the hybrid *Vigna sinensis* (L.) Savi and *V. sesquipedalis* (L.) Fruw. *Madras Agricultural Journal* **159**, 449–456.

Price, M. L., Hagerman, A. E., and Butler, L. G. 1980. Tannin content of cowpeas, chick-peas, pigeon peas, and mung beans. *Journal of Agriculture and Food Chemistry* **28**, 459–461.

Pulver, E. L. 1977. IITA Grain Legume Improvement Program. Ibadan, Nigeria, IITA, 49–60 (mimeo).

Purseglove, J. W. 1968. *Tropical Crops. Bicotyledons*. London, UK, Longmans.

Purseglove, J. W. 1976. The origins and migration of crops in tropical Africa. In Harlan, J. R., de Wet, J. M. J., and Stemler, A. B. L., eds, *Origins of African Plant Domestication*, The Hague, Mouton Publishers, 291–309.

Purss, G. S. 1957. Stem rot: a disease of cowpeas caused by an undescribed species of *Phytophthora*. *Queensland Journal of Agricultural Science* **14**, 125–154.

Purss, G. S. 1958. Studies on varietal resistance to stem rot (*Phytophthora vignae* Purss) in the cowpea. *Queenland Journal of Agricultural Science* **15**, 1–14.

Quinderá, M. A. W., and Barreto, P. D. 1983. Susceptibilidade do caupi ao *Callosobruchus maculatus*: estudos preliminares. Fortaleza, CE, EPACE, Comun. Técnico 13.

Rabakoarihanta, A., Mok, D. N. S., and Mok, M. C. 1979. Fertilization and early embryo development in reciprocal interspecific crosses of *Phaseolus*. *Theoretical and Applied Genetics* **57**, 59–64.

Rachie, K. O. 1973a. Highlights of grain legume improvement at IITA, 1970–1973. In IITA, *Proceedings of the 1st Grain Legume Improvement Workshop*, IITA, Ibadan, Nigeria, 29 October–2 November 1973, Ibadan, Nigeria, IITA, 1–14.

Rachie, K. O. 1973b. Improvement of food legumes in tropical Africa. In UN (United Nations), *Nutritional Improvement of Food Legumes by Breeding*, New York, NY, UN.

Rachie, K. O. 1977. The nutritional role of grain legumes in the lowland humid tropics. In Ayanaba, A., and Dart, P. J., eds, *Biological Nitrogen Fixation in Farming Systems of the Tropics*, Chichester, UK, John Wiley and Sons, 45–60.

Rachie, K. O. 1978. Productivity of potentials of edible legumes in the lowland tropics. In Jung, G. A., ed., *Crop Tolerance to Suboptimal Land Conditions*. Madison, WI, American Society of Agronomy, Special publication 32, 71–96.

Rachie, K. O., and Rawal, K. M. 1976. Integrated approaches to improving cowpeas,

Vigna unguiculata (L.) Walp. Ibadan, Nigeria, IITA, Technical Bulletin 5.

Rachie, K. O., and Roberts, L. M. 1974. Grain legumes of the lowland tropics. *Advances in Agronomy* **26**, 44–61.

Rachie, K. O., and Silvestre, P. 1977. Grain legumes. In Leakey, C. L. A., and Wills, J. B., eds, *Food Crops of the Lowland Tropics*, London, UK, Oxford University Press, 41–74.

Rachie, K. O., Rawal, K. M., Franckowiak, J. D. 1975a. Rapid hand crossing of cowpeas. Ibadan, Nigeria, IITA, Technical bulletin 2.

Rachie, K. O., Rawal, K., Franckowiak, J. D., and Akinpelu, M. A. 1975b. Two outcrossing mechanisms in cowpeas, *Vigna unguiculata* (L.) Walp. *Euphytica* **24**, 159–163.

Rachie, K. O., Rawal, K., Williams, R. J., Singh, S. R., Nangju, D., Wien, H. C., and Luse, R. A. 1975c. VITA 1 cowpea. *Tropical Grain Legume Bulletin* **1**, 16–17.

Rachie, K. O., Singh, S. R., Williams, R. J., Watt, E., Nangju, D., Wien, H. C., and Luse, R. A. 1976. VITA 4 cowpea. *Tropical Grain Legume Bulletin* **2**.

Radhakrishnan, T., and Jebraj, S. 1982. Genetic variability in cowpea. *Madras Agricultural Journal* **69**, 216–219.

Raheja, A. K. 1976a. Assessment of losses caused by insect pests to cowpea in northern Nigeria. *PANS* **22**, 229–233.

Raheja, A. K. 1976b. ULV spraying for cowpea in northern Nigeria. *PANS* **22**, 327–332.

Raheja, A. K. 1978. Yield losses from pests and the economics of chemical pest control on cowpea in northern Nigeria. In Singh, S. R., van Emden, H. F., and Taylor, T. A., eds, *Pests of Grain Legumes: Ecology and Control*. London, UK, Academic Press, 259–265.

Raheja, A. K., and Hayes, H. M. 1975. Sole crop cowpea production by farmers using improved practices. *Tropical Grain Legume Bulletin* **1** (1), 6.

Raheja, A. K., and Leleji, O. I. 1974. An aphid-borne virus disease of irrigated cowpea in northern Nigeria. *Plant Disease Reporter* **58**, 1080–1084.

Rai, T. B., and Utkhede, P. 1973. Grain legume breeding and testing program in Tanzania. In IITA, *Proceedings of the 1st Grain Legume Improvement Workshop*, IITA, Ibadan, Nigeria, 29 October–2 November 1973, Ibadan, Nigeria, IITA, 79.

Raj, S., and Patel, P. N. 1978. Studies in resistance in crops to bacterial diseases in India. 10: inheritance of multiple resistances in cowpea. *Indian Phytopathology* **31**, 294–299.

Rajan, S. S. 1977. Annual progress report on the food legume improvement program. Yezin, Burma, Agriculture Research Institute, UNDP/FAO Project 72-003, ARI publication 13, 53.

Rajendra, B. R., Mujeeb, K. A., and Bates, L. S. 1979. Genetic analysis of seed-coat types in interspecific *Vigna* hybrids via SEM. *Journal of Heredity* **70**, 245–249.

Rajukanno, K., Vasu devan, P., Saivaraj, K., and Krishnamoorthy, K. K. 1977. Insecticide residues in green gram, black gram and cowpea. *Pesticides* **11** (10), 25–26.

Ramaiah, K. V., Parker, C., Vasudeva Rao, M. J., and Musselman, L. J. 1983. *Striga* identification and control handbook. Patancheru, India, ICRISAT (International Crops Research Institute for the Semi-Arid Tropics), Information bulletin 15.

Ramaiah, M., and Narayanaswamy, P. 1974. A new graft-transmissible disease of cowpea (*Vigna sinensis* Savi). *Current Science* **43**, 417–418.

Ramaiah, M., Narayanaswamy, P., and Subramanian, C. L. 1976. Leaf blight of cowpea (*Vigna sinensis*) caused by *Sclerotium rolfsii* Sacc. *Current Science* **45** (9), 352.

Ramakrishna, Y. S., and Singh, R. P. 1978. Influence of climate on crop production in drylands. *Proceedings of the Indian National Science Academy* **B44** (6), 431–436.

Ramalho, F. S., and Ramos, J. R. 1979. Distribuição de ovos de *Empoasca kraemeri* Ross & Moore, 1957 na planta de feijão. *Anais da Sociedade Entomologica do Brasil* **8** (1), 85–91.

Raman, K. V., Singh, S. R., and van Emden, H. F. 1978. Yield losses in cowpeas following leafhopper damage. *Journal of Economic Entomology* **71**, 836–838.

Rao, K. N., and Raychaudhuri, S. P. 1979. Inhibition of cowpea mosaic viral infection by surfactants. *Indian Journal of Mycology and Plant Pathology* **9**, 209–213.

Rao, M. U., and Prasad, M. 1978. Multiple cropping on uplands with jute as main crop. *Indian Journal of Agronomy* **23** (4), 345–350.

Raski, D. J. 1952. On the host range of the sugar beet nematode in California. *Plant Disease Reporter* **36**, 5–7.

Rawal, K. M. 1975a. Cowpea breeding at IITA: an overview. In Luse, R. A., and Rachie, K. O., eds, *Proceedings of IITA Collaborators' Meeting on Grain Legume Improvement*, 9–13 June 1975, Ibadan, Nigeria, IITA, 6.

Rawal, K. M. 1975b. Natural hybridization among wild, weedy and cultivated *Vigna unguiculata* (L.) Walp. *Euphytica* **24**, 699–707.

Rawal, K. M., Porter, W. M., Franckowiak, J. D., Fawole, I., and Rachie, K. O. 1976. Unifoliate leaf: a mutant in cowpeas. *Journal of Heredity* **67**, 193–194.

Rawal, R. D., and Sohi, H. S. 1983. Studies in the control of leaf spot of cowpea caused by *Septoria vignicola* Rao. *Pesticides* **17** (8), 31–34.

Razak, A. R., and Evans, A. A. F. 1976. An intracellular tube associated with feeding by *Rotylenchulus reniformis* on cowpea root. *Nematologica* **22**, 182–189.

Rebois, R. V., Epps, J. M., and Hartwig, E. E. 1970. Correlation of resistance in soybeans to *Heterodera glycines* and *Rotylenchulus reniformis*. *Phytopathology* **60**, 695–700.

Redden, R. J. 1981. Vegetable cowpea breeding at IITA. *Tropical Grain Legume Bulletin* **23**, 6–10.

Reddy, N. R., Pierson, M. D., Sathe, S. K., and Salunkhe, D. K. 1984. Chemical, nutritional and physiological aspects of dry bean carbohydrates—a review. *Food Chemistry* **13**, 25–68.

Reddy, P. P., and Singh, D. B. 1983. Chemical control of *Meloidogyne incognita* on selected crops. *Nematologica Mediterranea* **11**, 197–198.

Reeder, B. D., Norton, J. D., and Chambliss, O. L. 1972. Inheritance of bean yellow mosaic virus resistance in southern pea, *Vigna sinensis* (Torner). *Journal of the American Society of Horticultural Science* **97**, 235–237.

Reid, D. C., and Parker, C. 1979. Germination requirements of *Striga* spp. In Musselman, L. J., Worsham, A. D., and Eplee, R. E., eds, *Proceedings 2nd International Symposium on Parasitic Weeds*, Raleigh, NC, 16–19 July 1979, Raleigh, NC, North Carolina State University, 202–210.

Rejesus, R. S. 1978. Pests of grain legumes and their control in the Philippines. In Singh, S. R., van Emden, H. F., and Taylor, T. A., eds, *Pests of Grain Legumes: Ecology and Control*. London, UK, Academic Press, 2/7–52.

Remison, S. U. 1978. Neighbour effects between maize and cowpea at various levels of N and P. *Experimental Agriculture* **14**, 205–212.

Rhodes, E. R. 1978. A note of the nitrogen nutrition of a local cowpea cultivar grown in Sierra Leone. *Tropical Grain Legume Bulletin* **11**, 16–18.

Rhodes, E. R., Evenhuis, B., and Taylor, W. E. 1979. A note on cowpea (*Vigna unguiculata*) nutrient composition and its relationship to yield. *Tropical Agriculture (Trinidad)* **56**, 241–243.

Riggs, R. D., and Hamblen, M. L. 1962. *Soybean cyst nematode host studies in the family Leguminosae.* Report of the Arkansas Agricultural Experiment Station, University of Arkansas, Series 110, 1–18.

Rios, G. P. 1983a. Reação de cultivares de caupi [*Vigna unguiculata* (L.) Walp.] a *Sphaceloma* sp. *Fitopatologia Brasileira* **8**, 251–258.

Rios, G. P. 1983b. Relatório anual individual. Goiânia, Goiás, Brazil, EMBRAPA/ CNPAF. 38 p.

Rios, G. P., and Neves, B. P. das. 1982a. Influência do sistema de cultivo no incidência do mosaico severo e populacion de vectores na cultivo do caupi [*Vigna unguiculata* (L.) Walp.]. Goiânia, Goiás, Brazil, EMBRAPA/CNPAF, Resumos de RENAC.

Rios, G. P., and Neves, B. P. das. 1982b. Resistência de linhagens e cultivares de caupi [*Vigna unguiculata* (L.) Walp.] ao virus do mosaico severo (VMSC). *Fitopatologia Brasileira* **7** (2), 175–184.

Rios, G. P., and Neves, B. P. das. in press. Doenças e pragas do caupi. Goiânia, Goiás, Brazil, EMBRAPA/CNPAF.

Rios, G. P., and Watt, E. E. 1980. Identificacion de fuentes de resistência a las principales enfermedades de caupi [*Vigna unguiculata* (L.) Walp.]. *Fitopatologia* **15**, 24.

Rios, G. P., Fernandes, P. M., and Neves, B. P. das. 1982. Desenvolvimento da mancha de *Ascochyta* em caupi [*V. unguiculata* (L.) Walp.]. Goiânia, Goiás, Brazil, EMBRAPA, Reunião Nacional de Pesquisa de Caupi, 112.

Rios, G. P., Watt, E. E., Araújo, J. P. P. de, and Neves, B. P. das. 1980. Identification of sources of resistance to the principal diseases of southern pea [*Vigna unguiculata* (L.) Walp.] in Brazil. *Annual Report of the Bean Improvement Cooperative* **5**, 23.

Rios, G. P., Watt, E. E., Araújo, J. P. P. de., and Neves, B. P. das. 1982. Cultivar CNC 0434—imune ao mosaico severo do caupi. Goiânia, Goiás, Brazil, EMBRAPA/CNPAF, Resumos de RENAC.

Risch, S. 1976. Effect of variety of cowpea (*Vigna unguiculata* L.) on feeding preference of three chrysomelid beetles, *Cerotoma ruficornis, Diabrotica balteata* and *Diabrotica adelpha. Turrialba* **26** (4), 327–330.

Roberts, S. B., Paul, A. A., Cole, T. J., and Whitehead, R. G. 1982. Seasonal changes in activity, birth weight and lactational performance in rural Gambian women. *Transactions of the Royal Society of Tropical Medicine and Hygiene* **76**, 668–678.

Robertson, B., Hall, A. E., and Foster, K. W. submitted for publication. A field technique for screening for genotypic differences in root growth. *Crop Science.*

Robertson, D. G. 1963. Cowpea virus research in Nigeria. In IAR, *Proceedings of the first Nigerian Grain Legume Conference*, Samaru, Zaria, Nigeria, IAR, Ahmadu Bello University, 75–76.

Robertson, D. G. 1965. The local lesion reaction for recognizing cowpea varieties immune from and resistant to cowpea yellow mosaic virus. *Phytopathology* **55**, 923–925.

Robertson, D. G. 1966. Seed-borne viruses of cowpea in Nigeria. Oxford, UK, University of Oxford. (B.Sc. thesis.)

Robinson, K. R. 1970. *The Iron Age of the Southern Lake Area of Malawi.* Malawi, Department of Antiquities of Malawi, Publication 8.

Robinson, P. E. 1961. Root knot nematodes and legumes. *Nature* **189**, 506–507.

Robinson, R. W., Munger, H. M., Whitaker, T. W., and Bohn, G. W. 1976. Genes of the Cucurbitaceae. *HortScience* **11**, 554–568.

Roesingh, C. 1980. Untersuchungen uber die Resistenz de Kuherbse, *Vigna unguicu-*

lata (L.) Walp., gegen *Megalurothrips sjostedtii* (Thysanoptera, Thripidae). Hochschulverlag, Freiburg.

Rogers, K. M., Norton, J. D., and Chambliss, O. L. 1973. Inheritance of resistance to cowpea chlorotic mottle virus in southern pea, *Vigna sinensis. Journal of the American Society of Horticultural Science* **98**, 62–63.

Rohde, W. A., Johnson, A. W., Dowler, C. C., and Glaze, N. C. 1980. Influence of climate and cropping patterns on the efficiency of ethoprop, methyl bromide, and DD MENCS for control of root-knot nematodes. *Journal of Nematology* **12**, 33–39.

Roman, J. 1964. Immunity of sugarcane to the reniform nematode. *Journal of Agriculture of the University of Puerto Rico* **48**, 162–163.

Root, R. B. 1973. Organization of a plant–arthropod association in simple and diverse habitats: the fauna of collards (*Brassica olearaces*). *Ecological Monographs* **43**, 95–124.

Rosario, R. R. D., Lozano, Y., and Noel, M. G. 1981. The chemical and biochemical composition of legume seeds. 2: cowpea. *Philippine Agriculturist* **64**, 49–57.

Ross, J. P. 1959. Nitrogen fertilization on the response of soybeans infected with *Heterodera glycines. Plant Disease Reporter* **43**, 1284–1286.

Rossel, H. W. 1977. Preliminary investigations on the identity and ecology of legume virus diseases in northern Nigeria. *Tropical Grain Legume Bulletin* **8**, 41–46.

Rossel, H. W., and Thottappilly, G. 1985. Virus diseases of important food crops in tropical Africa. Ibadan, Nigeria, IITA. Publication Series (in press).

Rossel, H. W., Thottappilly, G., and Van Lent, J. W. M. in press. Identification and control of the most important virus diseases in IITA's mandated crops. Ibadan, Nigeria, IITA, Publication series.

Rotimi, A. O. 1972. Effects of inoculation with commercial peat-base cowpea *Rhizobium* strain on the development of cowpea varieties. *Nigerian Agricultural Journal* **7**, 174–179.

Roy, R. S., and Richharia, R. H. 1948. Breeding and inheritance studies on cowpea, *V. sinensis. Journal of the American Society of Agronomy* **40**, 479–489.

Rubaihayo, P. R., Radley, R. W., Khan, T. N., Mukiibi, J., Leakey, C. L. A., and Ashley, J. M. 1973. The Makerere program. In UN (United Nations), *Nutritional Improvement of Food Legumes by Breeding*, New York, NY, UN.

Ruhendi, and Litsinger, J. A. 1982. Effect of rice stubble and tillage methods on the preflowering insect pests of grain legumes. In IRRI, *Cropping Systems Research in Asia*. Los Baños, Philippines, IRRI, 85–98.

Rushdi, M. H., Sellam, M. A., Abd-Elrazik, A., Allam, A. D., and Salem, A. 1980. Histological changes induced by *Meloidogyne javanica* and *Fusarium* on roots of selected leguminous plants. *Egyptian Journal of Phytopathology* **12**, 43–47.

Ruthenberg, H. 1980. *Farming Systems in the Tropics*, 3rd edn, Oxford, UK, Clarendon Press.

Rymal, K. S., and Chambliss, O. L. 1976. Cowpea curculio feeding stimulants from southernpea pods. *Journal of the American Society of Horticultural Science* **101**, 722–724.

Rymal, K. S., Chambliss, O. L., and McGuire, J. A. 1981. The role of volatile principles in nonpreference resistance to cowpea curculio in southernpeas, *Vigna unguiculata* (L.) Walp. *HortScience* **16**, 670–672.

Sabet, K. A. 1959. Studies on bacterial diseases of sudan crops. 3: on the occurrence, host range and taxonomy of bacteria causing leaf blight diseases of certain leguminous crops. *Annals of Applied Biology* **47**, 318–331.

Sachchidananda, J., Singh, S., Nam, P., and Verma, V. S. 1973. Bean common mosaic virus on cowpea in India. *Zeitschrift für Planzenkrankheiten und Pflanzenschutz* **80**, 88–92.

Saldanha, C. J. 1963. The genus *Striga* Lour. in western India. *Bulletin of the Botanical Survey of India* **5**, 67–70.

Salem, F. M. 1978. Zur pathogenitat von *Meloidogyne javanica* (Treub) Chitwood (Nematoda Heteroderidae) bei der Kuherbse (*Vigna unguiculata*) in drei verschiedenen Bodentypen. *Anzeiger fuer Schaedlingskunde Pflanzenschutz Umweltschutz* **51**, 60–61.

Sales, F. M., Gonçalves, M. F. B., Martins, O. F. G., and Mendes, C. 1979. Insetos e outros artrópodes de importância agrícola em perímetros irrigados e de sequeiro do Estado do Piauí. *Fitossanidade* **3** (1–2), 129.

Sambandan, R., Rajagopalan, C. K., and Devakumar, J. P. 1965. A note on the quality for pods in vegetative cowpea. *Madras Agricultural Journal* **52** (1), 35–36.

Sanderlin, R. S. 1973. Survival of bean pod mottle and cowpea mosaic viruses in beetles following intrahemocoelic injections. *Phytopathology* **63**, 259.

Santos, A. A. dos. 1982. Doenças do caupi [*Vigna unguiculata* (L.) Walp.] no Estado do Piauí. Goiânia, Goiás, Brazil, EMBRAPA/CNPAF, Reuniâo Nacional de Pesquisa de Caupi, 112.

Santos, A. A. dos, and Freire Filho, F. R. 1984a. Meios de transmissão do vírus do mosqueado amarelo do feijão macassar. *Fitopatologia Brasileira* **9**, 4–6.

Santos, A. A. dos, and Freire Filho, F. R. 1984b. Resultado preliminar sobre resistência de campo de cultivares de feijão macassar ao vírus do mosqueado amarelo. *Fitopatologia Brasileira* **9**, 4–8.

Santos, A. A. dos, Lin, M. T., and Kitajima, E. W. 1980. Serodiagnose de viroses em caupi no Estado do Piauí. *Fitopatologia Brasileira* **5**, 457–458.

Santos, A. A. dos, Lin, M. T., and Kitajima, E. W. 1981. Properties of cowpea rugose mosaic virus. *Phytopathology* **71**, 890.

Santos, A. A. dos, Lin, M. T., and Kitajima, E. W. in press. Caracterização de dois potyvirus isolados de caupi [*Vigna unguiculata* (L.) Walp.] no Estado do Piauí. *Fitopatologia Brasileira* **9**.   '

Santos, A. A. dos, Silva, P. H. S., and Mesquita, R. C. M. 1982. Insetos associados à cultura do caupi (*Vigna unguiculata*) no Estado do Piauí. Goiânia, Goiás, Brazil, EMBRAPA/CNPAF, 1st Reunião Nacional de Pesquisa de Caupi, (RENAC), 60–61.

Santos, J. H. R. dos. 1971. Aspéctos da biologia do *Callosobruchus maculatus* (Fabr. 1792) sobre sementes de *Vigna sinensis* Endl. Piracicaba, São Paulo, Universidade de São Paulo. 87 p. (M.Sc. thesis.)

Santos, J. H. R. dos, and Bastos, J. A. M. 1977. Nível de controle econômico do manhoso, *Chalcodermus bimaculatus* F.1: primeira aproximaçao. Relatório de Pesquisa 1976. Fortaleza, Ceará, Brasil, Univ. Fed. Ceará, Centro de Ciencias Agrarias, SUDENE 59–69.

Santos, J. H. R. dos, Alves, J. F., and Oliveira, F. J. de. 1978. Perda de peso em sementes de *Vigna sinensis* (L.) Savi decorrente do ataque de *Callosobruchus maculatus* (F., 1775) primeira aproximação. *Ciencia Agronomica* **8** (1–2), 51–56.

Santos, J. H. R. dos, Oliveira, F. J. de, Almeida, J. M. de, and Silva, P. C. da. 1977a. Influência do ataque de pulgão, sobre a produção do feijão-de-corda, *Vigna sinensis* (L.) Savi. Relatorio de Pesquisa 1976. Fortaleza, Ceará Brazil, Univ. Fed. Ceará, Centro de Ciencias Agrarias, 80–88.

Santos, J. H. R. dos, and Vieira, F. V. 1971. Ataque do *Callosobruchus maculatus* F. à *Vigna sinensis* Endl. 1: influência sobre o poder germinativo de sementes da cv. 'Serido'. *Ciencia Agronomica* **1** (2), 71–73.

Santos, J. H. R. dos, Teofilo, E. V., Almeida, J. M. de, and Oliveira, F. J. de. 1977b. Percentagens de sementes de cultivares de *Vigna sinensis* (L.) Savi, atacadas pelo *Chalcodermus bimaculatus* F. Relatório de Pesquisa 1976. Fortaleza, Ceará, Brazil, Univ. Fed. Ceará, Centro de Ciencias Agarias, SUDENE, 70–74.

Santos, J. H. R. dos, Vieira, F. V., and Pereira, L. 1977c. Importância relativa dos insetos e ácaros hospedados nas plantas do feijão-de-corda, nos perímetros irrigados do DNOCS, especialmente no Ceará. 1: primeira lista. Fortaleza, Ceará Brasil, Univ. Fed. Ceará, Centro de Ciencias Agrarias, Convênio de Fitossanidade, DNOCS/UFC.

Sarup, P., Jotwani, M. G., and Pradhan, S. 1961. Relative toxicity of some important insecticides to the bean aphid, *Aphis craccivora* Koch. *Indian Journal of Entomology* **22**, 105–108.

Sasaki, T. 1922. Inheritance of eye and flower color in the cowpea (in Japanese). *Journal of the Science of Agriculture Society* **232**, 19–38.

Sasidhar, V. K., and Sadanandan, N. 1976. Tapioca after cowpea gives higher yield. *Indian Farming* **26** (6), 23.

Sasser, J. M. 1979. Economic importance of *Meloidogyne* in tropical countries. In Lamberti, F., and Taylor, C. E., eds, *Root-knot nematodes* Meloidogyne *Species Systematics: Biology and Control*, New York, NY, Academic Press, 359–374.

Satpathy, J. M., Dash, P. K., and Misra, B. 1975. Insecticidal residues in some vegetable crops grown in treated soils. *Indian Journal of Agricultural Science* **44**, 113–115.

Satyanarayan, A., Ramchander, P. R., and Rajendran, R. 1982. Reciprocal differences for yield component characters in the F2 population of cowpea. *Indian Journal of Agricultural Science* **52**, 639–642.

Sauer, C. O. 1952. *Agricultural Origins and Dispersals*. Cambridge, MA, Massachusetts Institute of Technology, MIT Press.

Saunders, A. R. 1952. Complementary lethal genes in the cowpea. *South African Journal of Science* **48**, 195–197.

Saunders, A. R. 1959. Inheritance in the cowpea (*Vigna sinensis* Endb.). 1: colour of the seed coat. *South African Journal of Agricultural Science* **2**, 285–307.

Saunders, A. R. 1960a. Inheritance in the cowpea (*Vigna sinensis* Endb.). 2: seed coat colour pattern; flower, plant and pod colour. *South African Journal of Agricultural Science* **3**, 141–162.

Saunders, A. R. 1960b. Inheritance in the cowpea (*Vigna sinensis* Endb.). 3: mutations and linkages. *South African Journal of Agricultural Science* **3**, 327–348.

Saunders, A. R. 1960c. Inheritance in cowpea (*Vigna sinensis* Endb.). 4: lethal combinations. *South African Journal of Agricultural Science* **3**, 497–515.

Saxena, H. P. 1971. Insect pests or pulse crops. In Indian Agricultural Research Institute (IARI), *New Vistas in Pulse Production*. Delhi, India, IARI, 87–101.

Saxena, H. P. 1978. Pests of grain legumes and their control in India. In Singh, S. R., van Emden, H. F., and Taylor, T. A., eds, *Pests of Grain Legumes: Ecology and Control*. London, UK, Academic Press, 15–23.

Saxena, M. C., and Yadav, D. S. 1979. Parallel cropping with short duration pigeon pea under the humid subtropical conditions at Pantnagar. *Indian Journal of Agricultural Science* **49** (2), 95–99.

Saxena, H. P., Kumar, S., and Prasad, S. K. 1971. Efficacy of some systemic granular insecticides against Galerucid beetle, Kharif pulses. *Indian Journal of Entomology* **33**, 470–471.

Schalk, J. C., and Fery, R. L. 1982. Southern green stink bug and leaffooted bug: effect on cowpea production. *Journal of Economic Entomology* **75**, 72–75.

Schneider, R. W. 1973. Epidemiology, yield loss prediction and control of cercospora leaf spot of cowpea (*Vigna unguiculata*). Champagne–Urbana, IL, University of Illinois. (Ph.D. thesis.)

Schneider, R. W., Sinclair, J. B., and Williams, R. J. 1973. Cowpea (southern) (*Vigna unguiculata*). *Fungicide and Nematicides Tests: Results for 1973*, **29**, 65–66.

Schneider, R. W., Williams, R. J., and Sinclair, J. B. 1976. Cercospora leaf spot of cowpea: models for estimating yield loss. *Phytopathology* **66**, 384–388.

Schubert, K. B., and Ryle, G. J. A. 1980. The energy requirements for nitrogen fixation in nodulated legumes. In Summerfield, R. J., and Bunting, A. H., eds, *Advances in Legume Science*, London, UK, Her Majesty's Stationery Office, 85–96.

Sekar, R., and Sulochana, C. B. 1983. Blackeye cowpea mosaic virus: occurrence in India. *Indian Journal of Plant Pathology* **1** (1), 38–44.

Selassie-Gebre, K., Marchoux, G., and Quiot, J. B. 1976. Methode d'identification du virus du fletrissement de la fève (BBMW) isole de *Vigna sinensis* (L.) Savi. *Annals of Phytopathology* **8**, 331–342.

Sellschop, J. P. F. 1962. Cowpeas, *Vigna unguiculata* (L.) Walp. *Field Crop Abstracts* **15**, 259–266.

Sen, N. K., and Bhowal, J. G. 1960a. Cytotaxonomic studies on *Vigna*. *Cytologia* **25**, 195–207.

Sen, N. K., and Bhowal, J. G. 1960b. Colchichine-induced tetraploids of six varieties of *Vigna sinensis*. *Indian Journal of Agricultural Science* **30**, 149–161.

Sen, N. K., and Bhowal, J. G. 1961. Genetics of *V. sinensis* (L.) Savi. *Genetica* **32**, 247–266.

Sen, N. K., and Bhowal, J. G. 1962. A male sterile mutant cowpea. *Journal of Heredity* **53**, 44–46.

Sen, N. K., and Hari, M. N. 1956. Comparative study of diploid and tetraploid cowpea. *Proceedings of the 43rd Indian Science Congress* **3**, 258 (abstract).

Sene, O. 1967. Determinisme genetique de la precocité chez *Vigna unguiculata* (L.) Walp. *Agronomie Tropical (Paris)* **22**, 309–318.

Sene, O., and N'Diaye, S. M. 1973. Improvement of cowpea (*Vigna unguiculata*) in CNRA, Bambey, Senegal 1970–1973. In IITA, *Proceedings of the 1st Grain Legume Improvement Workshop*, IITA, Ibadan, Nigeria, 29 October–2 November, Ibadan, Nigeria, IITA, 72.

Shackel, K. A., and Hall, A. E. 1979. Reversible leaflet movements in relation to drought adaptation in cowpeas *Vigna unguiculata* (L.) Walp. *Australian Journal of Plant Physiology* **6**, 265–276.

Shackel, K. A., and Hall, A. E. 1983. Comparison of water relations and osmotic adjustment in sorghum and cowpea under field conditions. *Australian Journal of Plant Physiology* **10**, 423–435.

Shackel, K. A., Foster, K. W., and Hall, A. E. 1982. Genotypic differences in leaf osmotic potential among grain sorghum cultivars grown under irrigation and drought. *Crop Science* **22**, 1121–1125.

Shankar, G., Nene, Y. L., and Srivastava, S. K. 1973. A mosaic disease of cowpea (*Vigna sinensis* Savi). *Indian Journal of Microbiology* **13**, 209–211.

Shankar, L., Singh, M., and Pathak, M. M. 1975. Combining ability in cowpea. *Indian Journal of Genetics and Plant Breeding* **35**, 375–378.

Sharma, B. 1969a. Chemically induced mutations in cowpea (*Vigna sinensis* L. Savi). *Current Science* **38**, 520–521.

Sharma, B. 1969b. Induction of a wide-range mutator in *Vigna sinensis*. Paper prepared for a symposium on Radiations and radio mimetic substances in mutation breeding, Bombay, 26–29 September 1969. Bombay, India, Bhabka Atomic Research Centre, 13–22.

Sharma, B., Tikkoo, S. K., and Kant, K. 1971. On the use of *Vigna* mutator in the improvement of cowpea. Paper prepared for the international symposium on use of isotopes and radiation in agriculture and animal husbandry research, New Delhi, India, December 1971, 26–32.

Sharma, N. K., and Sethi, C. L. 1975. Effects of initial inoculum levels of *Meloidogyne incognita* and *Heterodera cajani* on cowpea and on their population development. *Indian Journal of Nematology* **5**, 148–154.

Sharma, R. D. 1981. Suscetibilidade de cultivares de caupi [*Vigna unguiculata* (L.) Walp.] ao nematoide *Meloidogyne javanica* (Treub 1885) Chitwood 1949. Paper prepared for the 5th Brazilian reunion on nematology, 9–13.

Sharma, S. C., and Singh, H. G. 1974. Effect of methods of intercropping maize and cowpea on the quality of forage. *Madras Agricultural Journal* **61** (8), 392–397.

Sharma, S. R., and Varma, A. 1975a. Three sap transmissible viruses from cowpea in India. *Indian Phytopathology* **28** (2), 192–198.

Sharma, S. R., and Varma, A. 1975b. Natural incidence of cowpea viruses and their effect on yield of cowpea. *Indian Phytopathology* **28** (3), 330–334.

Sharma, S. R., and Varma, A. 1976. Cowpea yellow fleck—a whitefly transmitted disease of cowpea. *Indian Phytopathology* **29** (4), 421–423.

Sharma, S. R., and Varma, A. 1982a. Aphid transmission of two cucumo-viruses from plants also infected with tobamovirus. *Zentralblatt fuer Mikrobiologie* **137**, 415–419.

Sharma, S. R., and Varma, A. 1982b. Effect of systemic insecticides on cowpea banding mosaic virus and its transmission by *Aphis craccivora* Koch. *Zentralblatt fuer Mikrobiologie* **137**, 519–523.

Shaw, T. 1976. Early crops in Africa: a review of evidence. In Harlan, J. R., de Wet, J. M. J., and Stemler, A. B. L., eds, *Origins of African Plant Domestication*, The Hague, Mouton Publishers, 107–153.

Shekhawat, G. S., and Patel, P. N. 1977. Seed transmission and spread of bacterial blight of cowpea and leaf spot of green gram in summer and monsoon seasons. *Plant Disease Reporter* **61** (5), 390–392.

Shekhawat, G. S., Patel, P. N., and Raj, S. 1977. Histopathology of cowpea plants infected with *Xanthomonas vignicola*. *Zeitschrift fuer Pflanzenkrankheiten und Pflanzenschutz* **84** (9), 547–558.

Sheoraj, P., and Patel, P. N. 1977. Studies on resistance in crops to bacterial disease in India. 9: sources of multiple disease resistance in cowpea. *Indian Phytopathology* **30**, 207–212.

Sheoraj, P., and Patel, P. N. 1978. Studies on resistance in crops to bacterial diseases in India. 10: inheritance of multiple disease resistance in cowpea. *Indian Phytopathology* **31**, 294–299.

Shepherd, R. J. 1964. Properties of a mosaic virus of cowpea and its relationship to the bean pod mottle virus. *Phytopathology* **54**, 466–473.

Shepherd, R. J. 1971. Southern bean mosaic virus. Kew, UK, CMI (Commonwealth Mycological Institute), CMI/AAB Descriptions of plant viruses 57.

Shepherd, R. J., and Fulton, R. W. 1962. Identity of a seed-borne virus of cowpea. *Phytopathology* **52**, 489–493.

Sher, S. A. 1964. Revision of Hoplolaiminae (Nematoda). 3: Scutellonema Andrassy 1958. *Nematologica* **9**, 421–423.

Sherman, F., and Todd, J. N. 1938. Cowpea curculio. *South Carolina Agricultural Experiment Station Annual Report* **51**, 71–75.

Sherwin, H. S., and Lefebvre, C. L. 1951. Reaction of cowpea varieties to bacterial canker. *Plant Disease Reporter* **35**, 303–317.

Sherwood, F. W., Weldon, V., and Peterson, W. J. 1954. Effect of cooking and of methionine supplementàtion on the growth-promoting property of cowpea (*Vigna sinensis*) protein. *Journal of Nutrition* **52**, 199–208.

Shoyinka, S. A. 1973. Status of virus diseases of cowpea in Nigeria. In IITA, *Proceedings of the 1st Grain Legume Improvement Workshop*, IITA, Ibadan, Nigeria, 29 October–2 November 1973, Ibadan, Nigeria, IITA, 270–273.

Shoyinka, S. A. 1975. Attempted control of virus incidence in cowpeas by the use of barrier crops. Nigerian Society of Plant Protection, NSPP abstracts, Occasional publication 1, 27.

Shoyinka, S. A., Bozarth, R. F., Rees, J., and Okusanya, B. O. 1979. Field occurrence and identification of southern bean mosaic virus (cowpea strain) in Nigeria. *Turrialba* **29** (2), 111–116.

Shoyinka, S. A., Bozarth, R. F., Rees, J., and Rossel, H. W. 1978. Cowpea mottle virus: a seed borne virus with distinctive properties infecting cowpea in Nigeria. *Phytopathology* **68** (5), 693–699.

Siddiqi, M. R. 1972. *Rotylenchulus reniformis*. St. Albans, UK, CIH (Commonwealth Institute of Helminthology), CIH Descriptions of plant parasitic nematodes, 1 (5).

Siddiqi, I. A., Sher, S. A., and French, A. M. 1973. Distribution of plant parasitic nematodes in California. Sacramento, CA, Department of Food and Agriculture.

Silva, A. de B., and Magalhães, B. P. 1980. Insetos nocivos a cultura do feijão caupi (*Vigna unguiculata*) no Estado do Pará. Belém, Pará, Brasil, EMBRAPA/CPATU, *Bol. de Pesquisa* 3.

Silva, J. J. C. da. 1983. Inimigos naturais de *Elasmopalpus lingosellus*. Dourados, Mato Crosso do Sul, Brasil, EMBRAPA/UEPAE, 5–10 (unpublished).

Sinclair, J. B., and Walker, J. C. 1955. Inheritance of resistance to cucumber mosaic in cowpea. *Phytopathology* **45**, 563–564.

Sinclair, M. J., and Eaglesham, A. R. J. 1984. Intrinsic antibiotic resistance in relation to colony morphology in three populations of West African cowpea rhizobia. *Soil Biology and Biochemistry* **16**, 247–251.

Sindhu, J. S., Pathak, M. M., Gangwar, L. C., Singh, K. B., and Singh, V. P. 1978. A fleshy and long podded mutant in cowpea (*Vigna unguiculata* L. Walp.). *Science and Culture* **44**, 418–419.

Singh, A., Ahlawat, I. P. S., and Saraf, C. S. 1975. Studies on weed control in Kharif cowpea. *Indian Journal of Weed Science* **7** (2), 85–88.

Singh, B., Khanna, A. N., and Vaidyn, S. M. 1964. Crossability studies in the genus Phaseolus. *Journal of Postgraduate School* **2**, 47–50.

Singh, B. B., and Mligo, J. L. 1981a. 'TK-1' and 'TK-5': new varieties of cowpeas for Tanzania. *Tropical Grain Legume Bulletin* **22**, 14–16.

Singh, B. B., and Mligo, J. L. 1981b. Yield stability of newly developed breeding lines in Tanzania. *Crop Research Bulletin* **1**, 1–4.

Singh, B. B., and Singh, S. R. 1983. Objectives and achievements of IITA research program on cowpeas and soybeans. In Homes, J. C., and Rahir, W. M., eds, *More Food from Better Technology*, Rome, Italy, FAO, 778–790.

Singh, B. B., Mligo, J. K., Marenge, E. T., and Ibrahim, J. A. 1982a. Sources of resistance to *Phytophthora* stem rot in cowpea. *Tropical Grain Legume Bulletin* **25**, 17–20.

Singh, B. B., Singh, R. R., and Jackai, L. E. N. 1982b. Cowpea breeding for diseases and insect resistance. In FAO (Food and Agriculture Organization of the United Nations), *Breeding for Durable Disease and Pest Resistance*, Rome, Italy, FAO, Plant Production and Protection Paper 55, 139–152.

Singh, B. M., Saharam, G. S., Scood, A. K., and Shvam, K. R. 1978. *Ascochyta* leaf spot of mash and cowpea. *Indian Phytopathology* **31** (3), 388–389.

Singh, C. B., Mehra, K. L., and Kohli, K. S. 1971. Evaluation of a world collection of *V. sinensis* (L.) Savi for pod characters. *SABRAO Newsletter* **3**, 11–16.

Singh, C. B., Mehra, K. L., Kohli, K. S., Singh, A., and Magoon, M. L. 1977. Correlation and factor analysis of fodder yield components in cowpea. *Acta Agronomica Acad. Scientiarum Hungaricae* **26**, 378–388.

Singh, D., and Patel, P. N. 1977. Studies on resistance in crops to bacterial diseases in India. 7: inheritance of resistance to bacterial blight in cowpea. *Indian Phytopathology* **30**, 99–102.

Singh, D. B., and Reddy, P. P. 1982. Cultivars of cowpea resistant to root knot nematodes in India. *Tropical Pest Management* **28**, 185–186.

Singh, H., and Dhooria, M. S. 1971. Bionomics of the pea pod borer *Etiella zinckenella* (Trietschke). *Indian Journal of Entomology* **33**, 123–130.

Singh, H. B., Mital, S. P., Dabas, B. S., and Thomas, T. A. 1974. Breeding for photoinsensitivity and earliness in grain cowpea. *Indian Journal of Genetics and Plant Breeding* **34A**, 771–776.

Singh, I., and Chohan, J. S. 1974. Seedborne fungi on cowpea (*Vigna sinensis*). *Indian Phytopathology* **27**, 239–240.

Singh, J. P., and Musymi, A. B. K. 1979. Control of rusts and downy mildews by a new systemic fungicide, Bayleton. *Pesticides* **13** (4), 51–53.

Singh, K. B., and Jain, R. P. 1972. Heterosis and combining ability in cowpea *Indian Journal of Genetics* **32**, 62–66.

Singh, K. B., and Jindla, L. N. 1971. Inheritance of bud and pod color, pod attachment, and growth habit in cowpeas. *Crop Science* **11**, 928–929.

Singh, K. B., and Mehndiratta, P. D. 1969. Genetic variability and correlation studies in cowpea. *Indian Journal of Genetics and Plant Breeding* **29**, 104–109.

Singh, K. B., and Mehndiratta, P. D. 1970. Path analysis and selection indices in cowpea. *Indian Journal of Genetics and Plant Breeding* **30**, 471–475.

Singh, K. C., and Singh, R. P. 1977. Intercropping of annual grain legumes with sunflower. *Indian Journal of Agricultural Science* **47** (11), 563–567.

Singh, R., and Singh, R. 1974. Natural infection of sickle senna (*Cassia tora* L.) and cowpea (*Vigna sinensis* Savi) plants by some new strains of southern bean mosaic virus. *Partugalia Acta Biologica* **13**, 87–89.

Singh, R. C., Faroda, A. S., Faroda, R. S., and Dahiya, D. R. 1981. Choose your pulse for summer. *Pulse Crops Newsletter* **1**, 79–80.

Singh, S., and Lamba, P. S. 1971. Agronomic studies on cowpea FS-68. 1: effect of soil moisture regimes, seed rates and levels of phosphorus on growth characters and yield. *Haryana Agricultural University Journal of Research* **1** (3), 1–7.

Singh, S. R. 1976. Coordinated minimum insecticide trial: yield performance of insect resistant cowpea cultivars from IITA compared with Nigerian cultivars. *Tropical Grain Legume Bulletin* **5**, 4.

Singh, S. R. 1977a. Cowpea cultivars resistant to insect pests in world germplasm collection. *Tropical Grain Legume Bulletin* **9**, 1–7.

Singh, S. R. 1977b. *Grain legume entomology: training booklet*. Ibadan, Nigeria, IITA.

Singh, S. R. 1980. Biology of cowpea pests and potential for host plant resistance. In

Harris, M. K., ed., *Biology and Breeding for Resistance to Arthropods and Pathogens in Agricultural Plants*. College Station, TX, Texas A&M University Bulletin MP1451.

Singh, S. R., and Allen, D. J. 1979. *Cowpea pests and diseases*. Ibadan, Nigeria, IITA, Manual series 2.

Singh, S. R., and Allen, D. J. 1980. Pests, diseases, resistance and protection in cowpeas. In Summerfield, R. J., and Bunting, A. H., eds, *Advances in Legume Science*, London, UK, Her Majesty's Stationery Office, 419–443.

Singh, S. R., and Taylor, T. A. 1978. Pests of grain legumes and their control in Nigeria. In Singh, S. R., van Emden, H. F., Taylor, T. A., eds, *Pests of Grain Legumes: ecology and control*. London, UK, Academic Press.

Singh, S. R., and van Emden, H. F. 1979. Insect pests of grain legumes. *Annual Reviews of Entomology* **24**, 255–278.

Singh, S. R., Luse, R. A., Leuschner, K., and Nangju, D. 1979. Groundnut oil for control of cowpea weevil. *Journal of Stored Products Research* **14**, 77–80.

Singh, S. R., Singh, B. B., Jackai, L. E. N., and Ntare, B. R. 1983. *Cowpea research at IITA*. Ibadan, Nigeria, IITA, Information series 14, 1–20.

Singh, S. R., van Emden, H. F., and Taylor, T. A., eds, 1978a. *Pests of Grain Legumes: ecology and control*. London, UK, Academic Press.

Singh, S. R., van Emden, H. F., and Taylor, T. A. 1978b. The potential for the development of integrated pest management systems in cowpea. In Singh, S. R., van Emden, H. F., and Taylor, T. A., eds, *Pests of Grain Legumes: ecology and control*, London, UK, Academic Press, 329–336.

Singh, S. R., Williams, R. J., Rachie, K. O., Rawal, K., Nangju, D., Wien, H. C., and Luse, R. A. 1975. VITA 3 cowpea. *Tropical Grain Legume Bulletin* **1**, 18–19.

Singh, S. R., Williams, R. J., Rachie, K. O., Watt, E., Nangju, D., Wien, H. C., and Luse, R. A. 1976. VITA 5 cowpea. *Tropical Grain Legume Bulletin* **5**, 41–42.

Sinha, O. K., and Khare, M. N. 1977a. A site of infection and further development of *Macrophomina phaseolina* and *Fusarium equiseti* in naturally infected cowpea seeds. *Seed Science and Technology* **5** (4), 721–725.

Sinha, O. K., and Khare, M. N. 1977b. Chemical control of *Macrophomina phaseolina* and *Fusarium equiseti* associated with cowpea seeds. *Indian Phytopathology* **30**, 337–340.

Sinha, S. K. 1972. Crop selection for water use efficiency. *Indian Farming* **22** (2), 62–63.

Sinha, S. K. 1973. Yield of grain legumes: problems and prospects. *Indian Journal of Genetics and Plant Breeding* **34A**, 988–994.

Sinha, S. K. 1977. Food legumes: distribution, adaptability and biology of yield. FAO Plant Production and Protection Paper 3.

Sinner, J., Jibrin, A., and Ashraf, M. 1983. Dry season cowpeas at Bida: an initial assessment of exploratory trials. Ibadan, Nigeria, IITA, Farming Systems Program, Discussion paper.

Slack, S. A., and Fulton, J. P. 1971. Some plant virus–beetle vector relations. *Virology* **43**, 728–729.

Slade, R. H. 1977. *Crop areas and yields for 1976/77 in the Funtua, Gombe and Gusau Agricultural development projects*. Kaduna, Agricultural Projects Monitoring, Evaluation and Planning Unit, Evaluation technical note 3.

Smartt J. 1978. The evolution of pulse crops. *Economic Botany* **32**, 185–198.

Smartt, J. 1980. Evolution and evolutionary problems in food legumes. *Economic Botany* **34** (3), 219–235.

Smartt, J. 1984. Gene pools in grain legumes. *Economic Botany* **38** (1), 24–35.

Smith, C. E. 1924. Transmission of cowpea mosaic by bean leaf beetle. *Science* **60**, 268.

Smithson, J. B., Redden, R., and Rawal, K. M. 1980. Methods of crop improvement and genetic resources in *Vigna unguiculata*. In Summerfield, R. J., and Bunting, A. H., eds, *Advances in Legume Science*, London, UK, Her Majesty's Stationery Office, 445–457.

Smithson, J. B., Watt, E. E., and Aggarwal, V. D. 1977. Recent progress in cowpea breeding at IITA. *Tropical Grain Legume Bulletin* **8**, 5–6.

Snowdon, J. D. 1927. Report of the acting mycologist for the period Nov. 10, 1925 to Sept. 30, 1926. In Department of Agriculture, Annual Report Uganda Department of Agriculture for the Year Ended 31st December 1926. Kampala, Uganda, Department of Agriculture, 30–32.

Snyder, W. C. 1942. A seed-borne mosaic of asparagus bean, *Vigna sesquipedalis*. *Phytopathology* **32**, 518–523.

Sohoo, M. S., Arora, N. D., Lodhi, G. P., and Chandra, S. 1971. Genotypic and phenotypic variability in cowpeas (*V. sinensis*) under different environmental conditions. *Journal of Research, Punjab Agricultural University* **8**, 159–164.

Southwood, T. R. E., and Way, M. J. 1970. Ecological background to pest management. In Rabb, R. L., and Guthrie, F. R., eds, *Concepts of Pest Management*, Raleigh, NC, North Carolina State University, 6–28.

Sowell, G. Jr., Kuhn, C. W., and Brantley, B. B. 1965. Resistance of southern pea, *Vigna sinensis*, to cowpea chlorotic mottle virus. *Proceedings of the American Society of Horticultural Science* **86**, 487–490.

Spencer, J. A., and Walters, H. J. 1969. Variation in certain isolates of *Corynespora cassiicola*. *Phytopathology* **59**, 38–60.

Spillman, W. J. 1911. Inheritance of the 'eye' in *Vigna*. *American Naturalist* **45**, 513–523.

Spillman, W. J. 1912. The present status of the genetics problem. *Science* **35**, 757–767.

Spillman, W. J. 1913. Color correlation in cowpea. *Science* **38**, 302.

Spillman, W. J., and Sando, W. J. 1930. Mandelian factors in the cowpea (*Vigna* species). *Michigan Academy of Science, Arts and Letters Papers* **11**, 249–283.

Sprau, F. 1960. Einige Bemerkungen ueber Ditylenchus destructor Thorne den Erreger der Alchenkratze an Kartoffeln, und sein verstaerktes Auftreten in den Jahren 1959 und 1960. *Pflanzenschutz Munich* **12**, 151–153.

Sprent, J. I. 1976. Water deficits and nitrogen-fixing nodules. In Kozlowski, T. T., ed., *Water Deficits and Plant Growth*, volume 4, London, UK, and New York, USA, Academic Press.

Sprent, J. I. 1980. Root nodule anatomy, type of export product and evolutionary origin in some Leguminosae. *Plant, Cell and Environment* **3**, 35–43.

Sprent, J. I., and Thomas, R. J. 1984. Nitrogen nutrition of seedling grain legumes: some taxonomic, morphological and physiological constraints. *Plant, Cell and Environment*, **7**, 637–645.

Srinivasa Rao, P. 1976. Nature of carbohydrates in pulses. *Journal of Agriculture and Food Chemistry* **24**, 958–961.

Sri Ram, C., Singh, C. B., Mehra, K. L., and Magoon, M. L. 1976. Tolerance to leaf hopper (*Empoasca kerri* Pruthi), flea beetle (*Pagria signata* Motasch) and semi-looper (*Plusia nigrisiana* Wlk.) infection in a world collection of cowpea; interregional comparisons. *Genetica Iberica* **28**, 116–130.

Srivastava, B. K. 1964. Pests of pulse crops. In Pant, N. C., *Entomology in India*,

1938–1963. New Delhi, India, Entomology Society of India, 83–91.

Standifer, M. S. 1959. The pathologic histology of bean roots injured by sting nematodes. *Plant Disease Reporter* **43**, 983–986.

Stanton, W. R., Doughty, J., Orraca-Tetteh, R., and Steele, W. M. 1966. *Grain Legumes in Africa*. Rome, Italy, FAO (Food and Agriculture Organization of the United Nations).

Staples, R. R. 1958. Southern Rhodesia, Department of Research and Specialist Services, report for the year ended 30 September 1957. In Ministry of Agriculture, Rhodesia and Nyasaland, *Report for 1956–57*, Harare, Zimbabwe, Ministry of Agriculture, 7–86.

Steele, W. M. 1972. Cowpeas in Africa. Reading, UK, University of Reading. (Ph.D. thesis.)

Steele, W. M. 1976. Cowpeas *Vigna unguiculata*. In Simmonds, N. W., ed., *Evolution of Crop Plants*, London, UK, Longmans, 183–185.

Steele, W. M., and Mehra, K. L. 1980. Structure, evolution and adaptation to farming system and environment in *Vigna*. In Summerfield, R. J., and Bunting, A. H., eds, *Advances in Legume Science*, London, UK, Her Majesty's Stationery Office, 393–404.

Steele, W. M., Allen, D. J., and Summerfield, R. J. in press, Cowpea (*Vigna unguiculata* L. Walp.). In Summerfield, R. J., and Roberts, E. H., eds, *Grain Legume Crops*, London, UK, Collins.

Stewart, K. A. 1976. On the yield of cowpea (*Vigna unguiculata* L. Walp.). Reading, UK, University of Reading (Ph.D. thesis).

Stewart, K. A., Summerfield, R. J., Minchin, F. R., and Ndunguru, B. J. 1980. Effects of contrasting aerial environments on yield potential in cowpea [*Vigna unguiculata* (L.) Walp.]. *Tropical Agriculture (Trinidad)* **57**, 43–52.

Stoop, W. A., and van Staveren, J. P. 1982. Effect of cowpeas in cereal rotations on subsequent crop yields under semiarid conditions in Upper Volta. In Graham, P. H., and Harris, S. C., eds, *Biological Nitrogen Fixation Technology for Tropical Agriculture*: Papers Presented at a Workshop Held at CIAT, March 9–13, 1981. Cali, Colombia, CIAT (Centro Internacional de Agricultura Tropical), 653–657.

Stowers, M. D., Eaglesham, A. R. J., Sinclair, M. J., Goldman, B. J., and Ayanaba, A. in preparation. Diversity of indigenous cowpea rhizobia from West Africa. 1: some physiological and biochemical characteristics.

Strength, D. R., and Chambliss, O. L. 1979. Protein content, protein quality, and yield of selected cultivars of southern peas. *Highlights of Agricultural Research* **24**, 7.

Strider, D. L., and Toler, R. W. 1963. Efficiency of screening southern peas in the seedling stage for *Cladosporium* spot disease. *Plant Disease Reporter* **47**, 493–496.

Strobel, J. W. 1961. *Verticillium* wilt of okra and southern pea in southern Florida. *Proceedings of the Florida Horticultural Society* **74**, 171–175.

Su, H. C. F. 1976. Toxicity of a chemical component of lemon oil to cowpea weevils. *Journal of the Georgia Entomology Society* **11**, 297–301.

Su, H. C. F. 1977. Insecticidal properties of black pepper to rice weevils and cowpea weevils. *Journal of Economic Entomology* **70**, 18–21.

Su, H. C. F. 1984. Comparative toxicity of three peppercorn extracts to four species of stored products insects under laboratory conditions. *Journal of the Georgia Entomology Society* **19**, 190–199.

Su, H. C. F., and Horvat, R. 1981. Isolation, identification and insecticidal properties of *Piper nigrum* amides. *Agriculture and Food Chemistry* **29**, 115–118.

Su, H. C. F., Speirs, R. D., and Mahany, P. G. 1972a. Citrus oils as protectants of black-eyed peas against cowpea weevils: laboratory evaluation. *Journal of Economic Entomology* **65**, 1433–1436.

Su, H. C. F., Speirs, R. D., and Mahany, P. G. 1972b. Toxicity of citrus oils to several stored-product insects: laboratory evaluation. *Journal of Economic Entomology* **65**, 1438–1441.

Subasinghe, S. M. C., and Fellowes, R. W. 1978. Recent trends in grain legumes pest research in Sri Lanka. In Singh, S. R., van Emden, H. F., and Taylor, T. A., *Pests of Grain Legumes: ecology and control.* London, UK, Academic Press, 37–42.

Subero, L. J. 1980. *Corynespora cassiicola*, a pathogen of beans (*Vigna sinensis* L.) in Venezuela. *Fitopatologia* **15**, 17–18.

Summerfield, R. J. 1975. Effects of daylengths and day–night temperatures on cowpeas and soyabeans. In IITA, *Proceedings of a Physiology Program Formulation Workshop*, Ibadan, Nigeria, IITA, 11–14.

Summerfield, R. J. 1980. The contribution of physiology to breeding for increased yields in grain legume crops. In Hurd, R. G., Biscoe, P. V., and Dennis, C., eds, *Opportunities for Increasing Crop Yields*, London, UK, Pitman, 51–69.

Summerfield, R. J., and Roberts, E. H. 1985. *Vigna unguiculata*. In Halevy, A. H., ed., *A Handbook of Flowering*, Boca Raton, FL, CRC Press.

Summerfield, R. J., and Wien, H. C. 1980. Effects of photoperiod and air temperature on growth and yield of economic legumes. In Summerfield, R. J., and Bunting, A. H., eds, *Advances in Legume Science*, London, UK, Her Majesty's Stationery Office, 17–36.

Summerfield, R. J., Dart, P. J., Huxley, P. A., Eaglesham, A. R. J., Minchin, F. R., and Day, J. M. 1977a. Nitrogen nutrition of cowpea (*Vigna unguiculata*). 1: effects of applied nitrogen and symbiotic nitrogen fixation on growth and seed yield. *Experimental Agriculture* **13**, 129–142.

Summerfield, R. J., Huxley, P. A., and Steele, W. 1974. Cowpea (*Vigna unguiculata* L. Walp.). *Field Crop Abstracts* **27** (7), 301–312.

Summerfield, R. J., Huxley, P. A., Dart, P. J., and Hughes, A. P. 1976. Some effects of environmental stress on seed yield of cowpea [*Vigna unguiculata* (L.) Walp.] cv. Prima. *Plant and Soil* **44**, 527–546.

Summerfield, R. J., Minchin, F. R., and Roberts, E. H. 1978. Realisation of yield potential in soyabean [*Glycine max* (L.) Merr.] and cowpea [*Vigna unguiculata* (L.) Walp.] In BCPC (British Crop Protection Council), *Opportunities for Chemical Plant Growth Regulation*, London, UK, BCPC, Monograph 21, 125–134.

Summerfield, R. J., Minchin, F. R., Roberts, E. H., Hadley, P. 1983. Cowpeas. In Smith, W. H., and Yoshida, S., eds, *Potential Productivity of Field Crops under Different Environments*, Los Baños, Philippines, IRRI, 249–280.

Summerfield, R. J., Minchin, F. R., Roberts, E. H., and Wien, H. C. 1977b. Photomorphogenetic effects on cowpea: a comparison of controlled environments and field-grown plants. *Plant Science Letters* **8**, 355–361.

Sundaram, A., Ramakrishna, A., Marappan, P. V., and Balakrishnan, C. 1980. Factor analysis in cowpea. *Indian Journal of Agricultural Science* **50**, 216–218.

Sutton, B. C., and Waterston, J. M. 1966. *Ascochyta phaseolorum*. Kew, UK, CMI (Commonwealth Mycological Institute), CMI/AAB Descriptions of pathogenic fungi and bacteria 81.

Swaans, H., and van Kammen, A. 1973. Reconsideration of the distinction between the severe and yellow strains of cowpea mosaic virus. *Netherlands Journal of Plant Pathology* **79**, 257–265.

Swaine, G. 1969. Studies on the biology and control of pests of seed beans (*Phaseolus vulgaris*) in northern Tanzania. *Bulletin of Entomological Research* **59**, 323–338.

Swami, B. N., and Pal, P. B. 1974. Effect of nitrogen and phosphorus levels on the dry matter and % K in cowpea (*Vigna sinensis*) in different soils of Rajasthan. *Agriculture and Agro-Industries Journal* **7** (8), 28–29.

Swanson, T. A., and van Gundy, S. D. in press. Factors influencing the differentiation of cowpea wilt-races. *Plant Disease*.

Ta'Ama, M. 1983. Yield performance of thrip resistant cultivars under no insecticide application. *Tropical Grain Legume Bulletin* **27**, 26–28.

Taha, A. H. Y., and Kassab, A. S. 1979. Effects of soil type on infectivity of separate and concomitant *Meloidogyne javanica* and *Rotylenchulus reniformis* on *Vigna sinensis*. Ain Shamu University, *Faculty of Agriculture Research Bulletin* **1072**, 1–11.

Taha, A. H. Y., and Kassab, A. S. 1980. Interrelations between *Meloidogyne javanica*, *Rotylenchulus reniformis* and *Rhizobium* sp. on *Vigna sinensis*. *Journal of Nematology* **12**, 57–62.

Tahvanainen, J. O., and Root, R. B. 1972. The influence of vegetational diversity on the population ecology of a specialised herbivore, *Phyllotreta cruciferae* (Coleoptera: Chrysomelidae). *Oecologia (Berlin)* **50**, 321–346.

Taiwo, M. A., and Gonsalves, D. 1982. Serological grouping of isolates of blackeye cowpea mosaic and cowpea aphid-borne mosaic virus. *Phytopathology* **72**, 583–589.

Taiwo, M. A., Gonsalves, D., Provvidenti, R., and Thurston, H. D. 1982. Partial characterization and grouping of isolates of blackeye cowpea mosaic and cowpea aphid-borne mosaic viruses. *Phytopathology* **72**, 590–596.

Taiwo, M. A., Provvidenti, R., and Gonsalves, D. 1981. Inheritance of resistance to blackeye cowpea mosaic virus in *Vigna unguiculata*. *Journal of Heredity* **72**, 433–434.

Talbert, R. E., Wallinder, C. J., and Saunders, P. A. 1979. *Field evaluation of herbicides in vegetable crops*. Arkansas Agricultural Experiment Station, Mimeograph series 281.

Talens, L. T. 1977. Cowpea little leaf disease in the Philippines: possible viral etiology as detected by immunodiffusion technique. *Philippines Phytopathology* **13**, 43–49.

Tanaka, A. 1977. Photosynthesis and respiration in relation to productivity of crops. In Mitsui, A., ed., *Biological Solar Energy Conversion*, New York, NY, Academic Press, 213–229.

Tanaka, Y., Ephrussi, B., Hadarn, E., Hagberg, A., Kemp, T., Love, A., Nachtsheim, H., Pontecowo, G., and Rhoades, M. M. 1957. Report of the International Committee on Genetic Symbols and Nomenclature. *International Union of Biological Science Colloques* **B30**, 1–6.

Taylor, T. A. 1967. The bionomics of *Maruca testulalis* Gey., a major pest of cowpeas in Nigeria. *Journal of the West Africa Science Association* **12**, 112–129.

Taylor, T. A. 1969. On the population dynamics and flight activity of *Taeniothrips sjostedti* on cowpea. *Bulletin of the Entomological Society of Nigeria* **2**, 60–71.

Taylor, T. A. 1977. Mixed cropping as an input in the management of crop pests in tropical Africa. *African Environment* **2** (4), 111–126.

Taylor, T. A. 1978. *Maruca testulalis*: an important pest of tropical grain legumes. In Singh, S. R., van Emden, H. F., and Taylor, T. A., eds, *Pests of Grain Legumes: ecology and control*, New York, NY, Academic Press, 193–200.

Taylor, W. E., and Vickery, B. 1974. Insecticidal prepulte of limonene, a constituent of citrus oil. *Ghana Journal of Agricultural Science* **7**, 61–62.

Tehon, L. R. 1933. Notes on the parasitic fungi of Illinois. *Mycologia* **25**, 237–257.

Tehon, L. R. 1937. Notes on the parasitic fungi of Illinois 6. *Mycologia* **29**, 434–446.

Tehon, L. R., and Stout, G. L. 1928. An ascomycetous leaf spot of cowpeas. *Phytopathology* **18**, 701–704.

Teliz, D., Lownsbery, B. F., and Grogan, R. G. 1966. Transmission of tomato ringspot virus by *Xiphinema americanum. Phytopathology* **56**, 151.

Templeton, G. E. 1982. Biological herbicides: discovery, development, deployment. *Weed Science* **30** (4), 430–433.

Teri, J. M. 1984. Cowpea wilt incited by *Fusarium oxysporum* f. sp. *tracheiphilum* in Tanzania. *Tropical Grain Legume Bulletin* **29**, 25–26.

Thomason, I. J. 1958. The effect of the root-knot nematodes *Meloidogyne javanica* on blackeye bean wilt. *Phytopathology* **48**, 398.

Thomason, I. J. 1959. Chisel application of methyl bromide for control of root-knot nematode and *Fusarium* wilt. *Plant Disease Reporter* **43**, 580–583.

Thomason, I. J., and McKinney, H. E. 1960. Reaction of cowpeas *Vigna sinensis* to root knot nematodes *Meloidogyne* spp. *Plant Disease Reporter* **44**, 51–53.

Thomason, I. J., Erwin, D. C., and Garber, M. J. 1959. The relationship of the root-knot nematode *Meloidogyne javanica* to *Fusarium* wilt of cowpea. *Phytopathology* **49**, 602–606.

Thompson, V. S. 1977. Studies on resistance to cowpea mosaic virus in selected varieties of cowpea [*Vigna unguiculata* (L.) Walp.]. St Augustine, Trinidad, University of the West Indies, Faculty of Agriculture, Department of Crop Science. (M.Sc. thesis.)

Thouvenel, J. C., Monsarrat, A., and Faquet, C. 1982. Isolation of cowpea mild mottle virus from diseased soybeans in the Ivory Coast. *Plant Disease* **66** (4), 336–337.

Tiagi, B. 1956. A contribution to the embryology of *Striga orobachoides* Benth and *Striga euphrasioides* Benth. *Bulletin of the Torrey Botanical Club* **83**, 154–170.

Tikka, S. B. S., and Asawa, B. S. 1978. Note on selection indices in cowpea. *Indian Journal of Agricultural Science* **48**, 767–769.

Tikka, S. B. S., Jaimini, S. N., Asawa, B. M., and Mathur, J. R. 1977. Genetic variability, interrelationships, and discriminant function analysis in cowpea [*V. unguiculata* (L.) Walp.]. *Indian Journal of Heredity* **9**, 1–9.

Tikka, S. B. S., Sharma, R. K., and Mathur, J. R. 1976. Genetic analysis of flower initiation in cowpea [*V. unguiculata* (L.) Walp.]. *Zeitschrift fur Pflanzenzuchtung* **77**, 23–29.

Timsina, J. 1984. Effects of indeterminate cowpea [*Vigna unguiculata* (L.) Walp.] varieties on the succeeding crop of dry seeded rice. Los Baños, Philippines, University of the Philippines. (M.Sc. thesis.)

Todd, J. W., and Canerday, T. D. 1968. Resistance of southern peas to the cowpea curculio. *Journal of Economic Entomology* **61**, 1327–1329.

Toler, R. W. 1964. Identity of a mosaic virus of cowpea (*Vigna sinensis*). *Phytopathology* **54**, 910.

Toler, R. W., and Dukes, P. D. 1965. *Choanephora* pod rot of cowpeas (*Vigna sinensis*). *Plant Disease Reporter* **49**, 347–350.

Toler, R. W., Dukes, P. D., and Jenkins, S. F. Jr. 1966. Growth response of *Fusarium oxysporum* f. sp. *tracheiphilum* wilt of *Vigna sinensis* in vitro to various oxygen and carbon-dioxide tensions. *Phytopathology* **56**, 183–186.

Toler, R. W., Thompson, S. S., and Barber, J. B. 1963. Cowpea (southern pea) diseases in Georgia, 1961–62. *Plant Disease Reporter* **47** (8), 746–747.

Tompkins, C. M., and Gardner, M. W. 1935. Relation of temperature to infection of beans and cowpea seedlings by *Rhizoctonia bataticola*. *Hilgardia* **9**, 219–230.

Trehan, K. B., Bagrecha, L. R., and Srivastava, V. K. 1970. Genetic variability and correlations in cowpea [*Vigna sinensis* (L.) Savi] under rainfed conditions. *Indian Journal of Heredity* **2**, 39–43.

Truong, V. D., Loguren, L. B., Abilay, R. M., and Mendoza, M. T. 1982. Screening for cowpea genotypes low in lipoxygenase. *Philippine Agriculturist* **65**, 153–158.

Tsuchizaki, T., Yora, K., and Asuyama, H. 1970. The viruses causing mosaic of cowpeas and adzuki beans, and their transmissibility through seeds. *Annals of the Phytopathological Society (Japan)* **36**, 112–120.

Turk, K. J., and Hall, A. E. 1980a. Drought adaptation of cowpea; 2: influence of drought on plant-water status and relations with seed yield. *Agronomy Journal* **72**, 421–427.

Turk, K. J., and Hall, A. E. 1980b. Drought adaptation of cowpea; 3: influence of drought on plant growth and relations with seed yield. *Agronomy Journal* **72**, 428–433.

Turk, K. J., and Hall, A. E. 1980c. Drought adaptation of cowpea; 4: influence of drought on water use, and relations with growth and seed yield. *Agronomy Journal* **72**, 434–439.

Turk, K. J., Hall, A. E., and Asbell, C. W. 1980. Drought adaptation of cowpea; 1: influence of drought on seed yield. *Agronomy Journal* **72**, 413–420.

Turner, N. C., and Jones, M. M. 1980. Turgor maintenance by osmotic adjustment: a review and evaluation. In Turner, N. C., and Kramer, P. J., eds, *Adaptation of Plants to Water and High Temperature Stress*, New York, NY, Wiley Interscience, 87–103.

Tyagi, I. D., Parihar, B. P. S., Dixit, R. K., and Singh, H. G. 1978. Component analysis for green fodder yield in cowpea. *Indian Journal of Agricultural Science* **48**, 646–649.

University of Ife. 1972. *Seventh annual research report of the Faculty of Agriculture*. Ile Ife, Nigeria, University of Ife.

USDA (United States Department of Agriculture). 1957. Witchweed (*Striga asiatica*)—a new parasitic plant in the United States. Washington, DC, US Government Printing Office, Research Service special publication 10, 142.

Usua, E. J., and Singh, S. R. 1978. Parasites and predators of the cowpea pod borer, *Maruca testulalis* (Lepidoptera: Pyralidae). *Nigerian Journal of Entomology* **3**, 100–102.

Vakili, N. G. 1977. Field screening of cowpea for *Cercospora* leaf spot resistance. *Tropical Agriculture (Trinidad)* **54**, 69–76.

Vakili, N. G., Kaiser, W. J., Perez, J. E., and Monllor, A. C. 1975. Bacterial blight of beans caused by two *Xanthomonas* pathogenic types from Puerto Rico. *Phytopathology* **65**, 401–403.

Valverde, R. A., Moreno, R., and Gamez, R. 1978. Beetle vectors of cowpea mosaic virus in Costa Rica. *Turrialba* **28**, 90–91.

Valverde, R. A., Moreno, R., and Gamez, R. 1982a. Incidence and some ecological aspects of cowpea severe mosaic virus in two cropping systems in Costa Rica. *Turrialba* **32** (1), 29–32.

Valverde, R. A., Moreno, R., and Gamez, R. 1982b. Yield reduction in cowpea (*Vigna unguiculata* L.Walp.) infected with cowpea severe mosaic virus in Costa Rica. *Turrialba* **32** (1), 89–90.

Van den Bosch, R. 1978. *The Pesticide Conspiracy*. New York, NY, Doubleday.

Van Dobben, W. H. 1962. Influence of temperature and light conditions on dry matter distribution rate of development and yield of agriculture crops. *Netherlands Journal of Agricultural Science* **10**, 377–389.

Van Hoof, H. A. 1963. Overbrenging van het cowpea mosaic virus in Suriname. *Surinaamse Landbouw* **11**, 131–137.

Van Kammen, A. 1971. Cowpea mosaic virus. Kew, UK, CMI (Commonwealth

Mycological Institute), CMI/AAB Descriptions of plant viruses 47.

Van Kammen, A., and de Jager, C. P. 1978. Cowpea mosaic virus. Kew, UK, CMI (Commonwealth Mycological Institute), CMI/AAB Descriptions of plant viruses 197.

Van Rensburg, H. J., and Strijdom, B. W. 1980. Survival of fast and slow growing *Rhizobium* spp. under conditions of relatively mild desiccation. *Soil Biology and Biochemistry* **12**, 353–356.

Van Velsen, R. J. 1962. Cowpea mosaic, a *virus* disease of *Vigna sinensis* in New Guinea. *Papua New Guinea Agricultural Journal* **14**, 153–161.

Varkey, P. A., and Jacob, S. 1979. Screening of cowpea varieties for the rice fallows. *Agricultural Research Journal of Kerala* **17** (1), 120–121.

Varma, A., Sharma, S. R., and Moharir, A. V. 1978. Witchesbroom of cowpea—a mycoplasmal disease. *Current Science* **47** (2), 56–57.

Vavilov, N. I. 1951. The origin, variation, immunity and breeding of cultivated plants. *Chronica Botanica* **13** (1949–1950), 1–364.

Veeraswamy, R. J., Palaniswamy, G. A., and Rathnaswamy, R. 1973. Genetic variability in some quantitative characters of *Vigna sinensis* (L.) Savi. *Madras Agricultural Journal* **60**, 1359–1360.

Veerupakshappa, K., Hiramath, S. R., and Shivashankar, G. 1980. Note on correlation in different segregating generations of cowpea. *Indian Journal of Agricultural Science* **50**: 979–981.

Venugopal, R., and Goud, J. V. 1977. Inheritance of pigmentation of cowpea. *Current Science* **46**, 277.

Verdcourt, B. 1970. Studies in the Leguminosae–Papilionoideae for the flora of tropical East Africa. *Kew Bulletin* **24**, 507–569.

Verma, J. S., and Mishra, S. N. 1981. Cowpea—the nutritious green fodder. *Indian Farmers' Digest* **15** (3), 32–34.

Vidano, C., and Conti, M. 1965. Transmission con afidi d'un 'cowpea mosaic virus' isolato da *Vigna sinensis* Endl. in Italia. *Annali Accademía de Science, Torino* **99**, 1041–1050.

Vieira, F. V., Bastos, J. A. M., and Pereira, L. 1975. Influência do *Chalcodermus bimaculatus* F. sobre o poder germinativo do feijão-de-corda, *Vigna sinensis* (L.) Savi. *Fitossanidade* **1** (2), 47–48.

Vincent, J. M. 1970. *A Manual for Practical Study of Root Nodule Bacteria.* Oxford, UK, Blackwell Scientific Publications, Handbook 15.

Vincent, J. M. 1980. Factors controlling the legume–*Rhizobium* symbiosis. In Newlon, W. E., and Orme-Johnson, W. H., eds, *Nitrogen Fixation*, vol. 2. Baltimore, MD, University Park Press, 103–129.

Vismanathan, T. V. 1978. 'Kanakamani': a dual purpose cowpea. *Indian Farming* **28** (3), 10, 13.

Wahua, T. A. T., Babalola, O., and Aken'Oven, M. E. 1981. Intercropping morphologically different types of maize with cowpeas: LER and growth attributes of associated cowpeas. *Experimental Agriculture* **17** (4), 407–413.

Walker, C. A. Jr., and Chambliss, O. L. 1981. Inheritance of resistance to blackeye cowpea mosaic virus in cowpea [*Vigna unguiculata* (L.) Walp.]. *Journal of the American Society of Horticultural Science* **106** (4), 410–412.

Walker, H. L. 1981. Granular formulation of *Alternaria macrospora* for control of spurred anoda (*Anoda cristata*). *Weed Science* **29** (3), 342–343.

Wall, J. R. 1970. Experimental introgression in the genus *Phaseolus*; 1: effects of mating systems on interspecific gene flow. *Evolution* **24**, 356–366.

Wallace, D. H., Ozbun, J. L., and Munger, H. M. 1972. Physiological genetics of crop yield. *Advances in Agronomy* **24**, 97–146.

Walters, H. J., and Barnett, O. W. 1964. Bean leaf beetle transmission of Arkansas cowpea mosaic virus. *Phytopathology* **54**, 911.

Walters, H. J., and Dodd, N. L. 1969. Identification and beetle transmission of an isolate of cowpea chlorotic mottle virus from *Desmodium*. *Phytopathology* **59**, 1055.

Walters, H. J., and Henry, D. G. 1970. Bean leaf beetle as a vector of the cowpea strain of southern bean mosaic virus. *Phytopathology* **60**, 177–178.

Wardojo, S., Hijink, M. J., and Oostenbrink, M. 1963. Damage to white clover by inoculation of *Heterodera trifolii*, *Meloidogyne hapla* and *Pratylenchus penetrans*. Rome, Italy, FAO, 15th International Symposium on Crop Protection, 101.

Warrag, M. O. A., and Hall, A. E. 1983. Reproductive responses of cowpea to heat stress: genotypic differences in tolerance to heat at flowering. *Crop Science* **23**, 1088–1092.

Warrag, M. O. A., and Hall, A. E. 1984a. Reproductive responses of cowpea (*Vigna unguiculata* L. Walp.) to heat stress, 1: responses to soil and day air temperatures. *Field Crops Research* **8**, 3–16.

Warrag, M. O. A., and Hall, A. E. 1984b. Reproductive responses of cowpea (*Vigna unguiculata* L. Walp.) to heat stress, 2: responses to night air temperature. *Field Crops Research* **8**, 17–33.

Warui, C. M. 1979. Review of cowpea yield potential in Coast Province, Kenya. In IITA, First Annual Research Conference, 15–19 October 1979. Section B: Insect Resistance in Cowpea. Ibadan, Nigeria, IITA.

Watkins, G. M. 1943. Further notes on a cowpea bacterial canker in Texas. *Plant Disease Reporter* **27**, 556.

Watt, E. E. 1978. First annual report on IITA/EMBRAPA/IICA cowpea program in Brasil. Goiânia, Goiás, Brazil, EMBRAPA/CNPAF.

Webber, H. J., and Orton, W. A. 1902. A cowpea resistant to root-knot (*Heterodera radicicola*). *USDA, Bureau of Plant Industries Bulletin* **17**, 23–26.

Weimer, J. L. 1949. Red stem canker of cowpea, caused by *Phytophthora cactorum*. *Journal of Agricultural Research* **78**, 65–75.

Weimer, J. L., and Harter, L. L. 1926. Root rot of bean in California caused by *Fusarium martii phaseoli* Burk and *F. aduncisporum* n. sp. *Journal of Agricultural Research* **32**, 311–319.

Wells, D. G., and Deba, R. 1961. Sources of resistance to the cowpea yellow mosaic virus. *Plant Disease Reporter* **45**, 878–881.

Westphal, E. 1974. *Pulses in Ethiopia: their taxonomy and agricultural significance*. Wageningen, Center for Agricultural Publishing.

Wettstein, R. 1893. Scrophulariaceae. In Engler, A., and Prantl, K., eds, *Die Naturlichen Pflanzenfamien* **4** (36), 39–107.

Whitehead, A. G. 1968. Taxonomy of *Meloidogyne* with description of four new species. *Transactions of the Royal Zoological Society of London* **31**, 260–401.

Whitney, W. K., and Gilmer, R. M. 1974. Insect vectors of cowpea viruses in Nigeria. *Annals of Applied Biology* **77**, 17–21.

Wickens, G. M. 1936. Report of the division of plant pathology for the year ending 31st Dec. 1935. *Rhodesian Agricultural Journal* **34**, 689–696.

Wickra-Masinghe, N., and Fernando, H. E. 1962. Investigations on insecticidal seed dressings, soil treatments and foliar sprays for the control of *Melanagromyza phaseoli* (Tryon) in Ceylon. *Bulletin of Entomological Research* **53**, 223–240.

Wien, H. C. 1973. Grain legume physiology at IITA. In IITA, *Proceedings of the 1st Grain Legume Improvement Workshop*, IITA, Ibadan, Nigeria, 29 October–2 November 1973. Ibadan, Nigeria, IITA, 105–114.

Wien, H. C. 1982. Dry matter production, leaf area development and light interception of cowpea lines with broad and narrow leaflet shape. *Crop Science* **22**, 733–737.

Wien, H. C., and Ackah, E. E. 1978. Pod development period in cowpeas: varietal differences as related to seed characters and environmental effects. *Crop Science* **18**, 791–794.

Wien, H. C., and Littleton, E. J. 1975. GLIP physiology subprogram. In Luse, R. A., and Rachie, K. O., eds, *Proceedings of IITA Collaborators' Meeting on Grain Legume Improvement*, 9–13 June 1975, Ibadan, Nigeria, IITA, 127–129.

Wien, H. C., and Nangju, D. 1976. The cowpea as an intercrop under cereals. Paper prepared for a symposium on intercropping in semiarid areas, Morogoro, Tanzania, 10–11 May 1976.

Wien, H. C., and Roesingh, C. 1980. Ethylene evolution by thrips-infested cowpea provides a basis for thrips resistance screening with ethephon sprays. *Nature* **283**, 192–194.

Wien, H. C., and Smithson, J. B. 1979. The evaluation of genotypes for intercropping. In ICRISAT (International Crops Research Institute for the Semi-Arid Tropics), *Proceedings of an International Workshop on Intercropping*, Patancheru, India, ICRISAT, 105–116.

Wien, H. C., and Summerfield, R. J. 1984. Cowpea (*Vigna unguiculata* L. Walp.). In Goldsworthy, P. R., and Fisher, N. M., eds, *The Physiology of Tropical Field Crops*, Chichester, UK, John Wiley and Sons, 353–384.

Wien, H. C., and Tayo, T. O. 1978. The effect of defoliation and removal of reproductive structures on growth and yield of tropical grain legumes. In Singh, S. R., van Emden, H. F., and Taylor, T. A., eds, *Pests of Grain Legumes: ecology and control*, London, UK, Academic Press, 241–252.

Wien, H. C., Lal, R., and Pulver, E. L. 1979a. Effects of transient flooding on growth and yield of some tropical crops. In Lal, R., and Greenland, D. J., eds, *Soil Physical Properties and Crop Production in the Tropics*, New York, NY, Wiley Interscience, 235–245.

Wien, H. C., Littleton, E. J., and Ayanaba, A. 1979b. Drought stress of cowpea and soybean under tropical conditions. In Mussell, H., and Staples, R. eds, *Stress Physiology in Crop Plants*, New York, NY, Wiley Interscience, 294–301.

Wienk, J. F. 1963. Photoperiodic effects in *Vigna unguiculata* (L.) Walp. *Meded. Landbouwhogeschool Wageningen* **63**, 1–82.

Wight, W. F. 1907. *The history of the cowpea and its introduction into America.* USDA, Bureau of Plant Industries, Bulletin 102, 43–59.

Wilczek, R. 1954. Phaseolinae. In *Flore du Congo Belge et du Rwanda–Urundi* **6**, 260–409.

Wild, H. 1948. A suggestion for the control of tobacco witchweed [*Striga gesnerioides* (Willd.) Vatke] by leguminous trap crops. *Rhodesia Agricultural Journal* **45**, 208–215.

Wilken, G. C. 1972. Micro-climate management by traditional farmers. *Geographical Reviews* **62**, 544–560.

Wilkinson, H. 1927. Annual report of the mycologist. In Department of Agriculture, Annual Report of the Department of Agriculture of Kenya for the year ending 31st Dec. 1926. Nairobi, Kenya, Department of Agriculture.

Williams, C. B., and Chambliss, O. L. 1980. Autocrossing in southernpea. *HortScience* **15**, 179.

Williams, C. E. 1974. *A preliminary study of consumer preferences in the choice of cowpeas: Western and Kwara states headquarters and areas of Nigeria.* Ibadan, Nigeria, Department of Agricultural Economics and Extension, University of Ibadan.

Williams, K. J. O. 1972. *Meloidogyne javanica.* St. Albans, UK, CIH (Commonwealth Institute of Helminthology), CIH Descriptions of plant-parasitic nematodes 1 (3).

Williams, K. J. 1973. *Meloidogyne incognita.* St. Albans, UK, CIH (Commonwealth

Institute of Helminthology), CIH Descriptions of plant-parasitic nematodes 2 (18).

Williams, K. J. 1974. *Meloidogyne hapla* nematodes. St. Albans, UK, CIH (Commonwealth Institute of Helminthology), CIH Descriptions of plant-parasitic nematodes 3 (31).

Williams, K. J. 1975. *Meloidogyne arenaria*. St. Albans, UK, CIH (Commonwealth Institute of Helminthology), CIH Descriptions of plant-parasitic nematodes 5 (62).

Wiliams, R. J. 1973a. Cowpea seedling disease control. Paper prepared for the 3rd annual conference of the Nigerian Society for Plant Protection, Benin, February 1973. (Abstract also appeared in NSPP Occasional publication 1, 13, 1975.)

Williams, R. J. 1973b. The anthracnose of cowpea. Paper prepared for the 3rd annual conference of the Nigerian Society for Plant Protection, Benin, February 1973. (Abstract also appeared in NSPP Occasional publication 1, 13.)

Williams, R. J. 1975a. Control of cowpea diseases in the IITA Grain Legume Improvement Program. In Bird, J., and Maramorosch, K., eds, *Tropical Diseases of Legumes*, New York, NY, Academic Press, 139–146.

Williams, R. J. 1975b. Control of cowpea seedling mortality in southern Nigeria. *Plant Disease Reporter* **59**, 245–248.

Williams, R. J. 1975c. Diseases of cowpea (*Vigna unguiculata* L. Walp.) in Nigeria. *PANS* **21** (3), 263–267.

Williams, R. J. 1975d. An international testing program for resistance to cowpea diseases. *Tropical Grain Legume Bulletin* **2**, 5–7.

Williams, R. J. 1975e. Screening cowpeas for resistance to cowpea mosaic virus. Nigerian Society of Plant Protection, NSPP abstracts, Occasional publication 1, 28.

Williams, R. J. 1976. Review of the major diseases of soybean and cowpea with special reference to geographical distribution, means of dissemination and control. *African Journal of Plant Protection* **1** (1), 83–96.

Williams, R. J. 1977. Identification of resistance to cowpea (yellow) mosaic virus. *Tropical Agriculture (Trinidad)* **54**, 61–68.

Williams, R. J., and Ayanaba, A. 1975. Increased incidence of *Pythium* stem rot in cowpea treated with benomyl and related fungicides. *Phytopathology* **65**, 217–218.

Wilson, G. F., and Caveness, F. E. 1980. The effect of rotation crops on the survival of root-knot, root-lesion and spiral nematodes. *Nematropica* **10**, 56–61.

Wilson, R. D. 1936. A bacterial disease of sake beans. *Proceedings of the Royal Society of New South Wales* **69**, 215–223.

Wolf, F. A., and Lehman, S. O. 1920. Notes on little known plant diseases in North Carolina in 1920. *North Carolina Agricultural Experiment Station* **1920**, 55–58.

Woolley, J. N. 1976. Breeding cowpea for resistance to *Maruca testulalis*. Methods and preliminary results. *Tropical Grain Legume Bulletin* **4**, 13–14.

Wraight, S. P., Daoust, R. A., Magalhães, B. P., and Roberts, D. W. 1983. Preliminary laboratory studies of a recently isolated mononematous *Hirsutella* species from *Empoasca kraemeri*. Paper prepared for the 16th annual meeting of the Society for Invertebrate Pathology, Ithaca, NY, Cornell University, 46.

Xi, Z., Xu, S., and Mang, K. 1982. An isolate causing systemic necrosis mosaic on cowpea in Peking suburb. *Acta Phytopathologica Sinica* **12** (4), 38–42.

Yadahalli, Y. N., Rahendra, B. R., and Jayaram, G. 1973. Cowpea and black gram could be profitably grown after paddy under tank achkats. *Current Research* **2** (11), 92.

Yadava, R. B. R., and Patil, B. D. 1984. Screening of cowpea [*V. unguiculata* (L.) Walp.] varieties for drought tolerance. *Zeitschrift fuer Pflanzenzuchtung*, **93**, 259–262.

Yadava, R. B. R., Verma, O., and Sastry, J. A. 1980. A note on the effect of growth regulators on seed yields of cowpea [*V. unguiculata* (L.) Walp.] plants. *Seed Research* **8**, 88–90.

Yarnell, S. H. 1948. The southern tomato exchange program. *Proceedings of the American Society of Horticultural Science* **52**, 375–382.

Yarnell, S. U. 1965. Cytogenetics of the vegetable crops. 4: legumes. *Botanical Review* **31**, 247–330.

Youngblood, J. P., and Chambliss, O. L. 1973. Inheritance of cowpea curculio resistance in southern peas. *HortScience* **8**, 283 (abstract).

Yu, T. F. 1946. A mosaic disease of cowpea (*Vigna sinensis* Endl.). *Annals of Applied Biology* **33**, 450–454.

Zablotowicz, R. M., and Focht, D. D. 1979. Denitrification and anaerobic, nitrate-dependent acetylene reduction in cowpea *Rhizobium*. *Journal of General Microbiology* **111**, 445–448.

Zafar, A. M. 1977. Some preliminary observations on cowpea bean (*Vigna sinensis*) varieties. *Journal of Agricultural Research (Pakistan)* **15** (1), 75–77.

Zamora, A. F., and Fields, M. L. 1979a. Microbiological and toxicological evaluation of fermented cowpeas (*Vigna sinensis*) and chick-peas (*Cicer arietinum*). *Journal of Food Science* **44**, 928–929.

Zamora, A. F., and Fields, M. L. 1979b. Nutritive quality of fermented cowpeas (*Vigna sinensis*) and chick-peas (*Cicer arietinum*). *Journal of Food Science* **44**, 234–236.

Zannou, T. A. 1981. Resistance to root-knot nematodes *Meloidogyne javanica* (Treub. 1885) Chitwood 1949 and *M. incognita* (Kofoid and White 1919) Chitwood 1949 in cowpea, *Vigna unguiculata* (L.) Walp. Ibadan, Nigeria, University of Ibadan. (Thèse de ingenieur agronome.)

Zary, K. W., and Miller, J. C. Jr. 1980. The influence of genotype on diurnal and seasonal patterns of nitrogen fixation in southernpea [*Vigna unguiculata* (L.) Walp.]. *Journal of the American Society of Horticultural Science* **105**, 699–701.

Zary, K. W., Miller, J. C., Jr., Weaver, R. W., and Barnes, L. W. 1978. Intraspecific variability for nitrogen fixation in southernpea [*Vigna unguiculata* (L.) Walp.]. *Journal of the American Society of Horticultural Science* **103**, 806–808.

Zaveri, P. P., Patel, P. K., and Yadavendra, J. P. 1980. Diallel analysis of flowering and maturity in cowpea. *Indian Journal of Agricultural Science* **50**, 807–810.

Zaveri, P. P., Patel, P. K., Yadavendra, J. P., and Shah, R. M. 1983. Heterosis and combining ability in cowpea. *Indian Journal of Agricultural Science* **53**, 793–796.

Zettler, F. W., and Evans, I. R. 1972. Blackeye cowpea mosaic virus in Florida: host range and incidence in certified cowpea seed. *Proceedings of the State Horticultural Society* **85**, 99–101.

Zhukovskii, P. M. 1962. *Cultivated Plants and their Wild Relatives*. Cambridge, UK, Commonwealth Bureau of Plant Breeding (translated by P. S. Hudson).

Index